VEGETABLE CROP PESTS

VEGETABLE CROP PESTS

Edited by

Roderick G. McKinlay

The Scottish Agricultural College,
West Mains Road, Edinburgh EH9 3JG, Scotland

MACMILLAN
PRESS

First published 1992

Published by
MACMILLAN ACADEMIC AND PROFESSIONAL LTD
Houndmills, Basingstoke, Hampshire RG21 2XS
and London
Companies and representatives
throughout the world

Filmset by Wearside Tradespools, Boldon, Tyne and Wear

ISBN 978-1-349-09926-9   ISBN 978-1-349-09924-5 (eBook)
DOI 10.1007/978-1-349-09924-5

A catalogue record for this book is available from the British Library

The mention of a product is not an endorsement,
and the contributors, editor and publishers will not be
held liable for any dosages or treatments mentioned in
the text. If in doubt, please check the source literature
and manufacturer's recommendations.

Any view expressed is the responsibility of the individual contributor.
The publication of an individual opinion is not an endorsement
by either the editor or the publishers.

Contents

Preface

In terms of global crop production, vegetables are important: in 1988, world production of vegetables (including melons, potatoes, pulses and sugar beet) was 60% of world cereal production (*FAO Yearbook*, Volume 42). Vegetable crops are grown in both the tropics and the temperate latitudes, with some crops being grown more in one area than in the other. For instance, the production of pulse crops and other vegetables, such as cabbages, pumpkins, squashes, gourds, cucumbers and gherkins, is more important in Asia, but the production of potatoes, sugar beet and tomatoes is more important in Europe. While there is no clear distinction between vegetables grown in the tropics and vegetables grown in the temperate latitudes, this book addresses primarily the pest problems of temperate vegetables.

The book covers all the major invertebrate pests of economic importance on vegetable crops grown in temperate countries (mainly insects, but also mites, nematodes and millepedes). The entry for each pest includes geographical distribution, description, life cycle, plant damage and control. Control includes any natural, biological, cultural, chemical, integrated and legislative methods. Because of the increasing importance of environmental issues in agriculture, emphasis is given to non-chemical methods of managing pests.

The contents are divided into nine chapters. Chapter 1 is an introductory chapter covering the origin of cultivated vegetables, the invertebrate pests which are found on temperate vegetables and the methods used to control the pests. The next eight chapters are devoted to the pests of particular crops. One chapter (Chapter 7) is devoted to the pests of monocotyledon crops (onions, leeks, sweet corn). The remaining chapters are concerned with the pests of dicotyledon crops: Chapter 2 with pests of chenopodiaceous crops; Chapter 3 with pests of composite crops; Chapter 4 with pests of cruciferous crops; Chapter 5 with pests of cucurbit crops; Chapter 6 with pests of leguminous crops; Chapter 8 with pests of solanaceous crops; and Chapter 9 with pests of umbelliferous crops. Because some pests attack several crops, the possibility existed that consideration of these pests would be duplicated across several chapters. This potential duplication has been limited by considering the pest under one important crop only with cross-referring occurring during consideration of the other affected crops.

I thank all the authors who have contributed to this volume and helped me to compile such an authoritative reference, which should be of value to students of agriculture and horticulture as well as researchers involved in the study of vegetable crops world-wide.

I should also like to thank David Grist of The Macmillan Press Ltd for his help and advice during the compilation of this volume. And, finally, I should like to thank my wife and family for their enduring patience and understanding.

Edinburgh, 1991   R. G. McK.

The Contributors

A. J. Biddle
Processors & Growers Research
  Organisation
Great North Road
Thornhaugh
Peterborough PE8 6HJ
UK

T. H. Coaker
Dept of Zoology
University of Cambridge
Pembroke Street
Cambridge CB2 3DX
UK

D. A. Cooke
Broom's Barn Experimental
  Station
Higham
Bury St Edmunds
Suffolk IP28 6NP
UK

A. M. Dewar
Broom's Barn Experimental
  Station
Higham
Bury St Edmunds
Suffolk IP28 6NP
UK

P. R. Ellis
Horticulture Research
  International
Wellesbourne
Warwick CV35 9EF
UK

B. Emmett
Agricultural Development and
  Advisory Service
MAFF
Government Buildings
Lawnswood
Leeds LS16 5PY
UK

S. Finch
Horticulture Research
  International
Wellesbourne
Warwick CV35 9EF
UK

J. A. Hardman
Horticulture Research
  International
Wellesbourne
Warwick CV35 9EF
UK

S. H. Hutchins
Dow Chemical USA
9002 Purdue Road
Indianapolis
IN 46268-1189
USA

R. G. McKinlay
The Scottish Agricultural College
West Mains Road
Edinburgh EH9 3JG
UK

A. M. Spaull
The Scottish Agricultural College
West Mains Road
Edinburgh EH9 3JG
UK

R. W. Straub
New York State Agricultural
  Experiment Station
Hudson Valley Laboratory
PO Box 727
Highland
NY 12528-0727
USA

A. R. Thompson
Horticulture Research
  International

Wellesbourne
Warwick CV35 9EF
UK

J. A. Wightman
Legumes Improvement Programme
ICRISAT
Patancheru PO
Andhra Pradesh 502 324
India

A. York
Department of Entomology
Purdue University
West Lafayette
IN 47907
USA

CHAPTER 1

Introduction

T. H. Coaker

1.1 CLASSIFICATION, ORIGINS AND PRODUCTION OF VEGETABLES

Vegetables and their cultivars can be grouped according to their similarities of use, appearance, morphology, environmental sensitivity, life cycle and other criteria, each grouping of vegetables reflecting its adaptation, and general cultural and management needs (Peirce, 1987). Some classification systems are more useful than others but they all attempt to reduce a large number of different crops into logical associations.

Classification can be based on botanical relationships, which reflect evolutionary pathways that can be useful to both botanists and growers, since crop management systems may be influenced by botanical similarities. For example, crop rotation as a cultural control practice for pest and disease management may be dependent on botanically unrelated plants (Coaker, 1987), while weeds may act as alternative hosts to organisms that attack related crop plants. Botanical classification of varieties or cultivars also distinguishes their specific characteristics, which may otherwise appear identical. Grouping by the edible part of the plant provides some value in crop management, handling and marketing: for example, similarities among crops from which a particular part is harvested, such as roots, leaves, fruit or seed. Each category provides the grower with knowledge of the cultural techniques and the type and degree of pest or damage control required. Analogous to grouping by the edible part is a grouping based on primary use: for example, as green vegetables or salad crops. Each group contains crops that can be consumed raw or cooked, but this distinction is not absolute, as many crops can be used in different ways. The life cycle of the crop may also be used to classify vegetables. Most are grown as annuals whether they are true annuals or biennials, and a few are grown as perennials in temperate climates. Flowering in biennial species tends to be temperature-dependent and occasionally day-length-dependent, thereby producing seeds in the following year. The timing of flowering and fruit set can, therefore, influence the susceptibility of the plant to attack by particular pest species or its response to the damage they cause.

Vegetables originate from wild plants found from the tropics to the temperate regions and have been selected to perform best under local conditions in which they are grown. However, these modifications have not changed their inherent adaptations to certain temperature requirements: for example, tomatoes are a warm-season crop grown over a wide range of latitudes. Such adaptations to temperature also allow vegetables to be separated broadly into warm- and cool-season crops, which can be further subdivided into those that are hardy and frost tolerant, those that have seed that germinates and grows at low temperatures, and those that have produce that can be stored at cold temperatures.

In the following section on the origins and production of the principal perennial and annual vegetables (Simmonds, 1976; Simpson and Conner-Ogorzaly, 1986), grouping is by botanical families. The principal producers of the vegetable crops discussed have been taken from the *FAO Production Year Book* (1987), which also provides the annual hectarage, yield and production figures of many vegetable crops.

1.1.1 Perennial Vegetables

Perennial vegetables include plants from taxonomically diverse families that are relatively long-lived (replanted every 15–25 years) and are mostly indigenous to temperate zones, such as asparagus, globe artichokes and rhubarb. Several other vegetables are botanically classified as perennials but are grown as annuals—for example, Jerusalem artichokes, horseradish and potatoes. A number of perennial vegetables are also tropical and therefore sensitive to low temperatures.

Asparagus (Asparagus officinalis L.: Liliaceae)

Native to the temperate scrub communities of Europe, western Asia and northern Africa, asparagus has changed little from its wild form, which is still found in certain saline areas. It was once used as a medicinal plant, but was considered a delicacy by the Ancient Greeks, as it is today. The plant has the ability to sprout rapidly from a perennial underground system of rhizomes and is cultivated in areas where the plant dies back during the winter and sends new succulent shoots (spears) from a tiny stem (crown) in the spring. The young expanded spears are cut when they emerge from the ground. If they are allowed to develop, the plant becomes a large woody bush which flowers to produce small round berries. Propagation is from seed or one-year-old crowns. Asparagus is, in general, an expensive vegetable, occurring for a limited period in the spring, when only a proportion of the spears can be harvested without damaging the plant. As well as being sold fresh, the spears can be canned and sometimes frozen. Commercial production is important in Europe, Asia, Australia, New Zealand and North America.

Globe Artichoke (Cynara scolymus L.: Compositae)

A native of the Mediterranean region and the Canary Isles, the globe artichoke was cultivated thousands of years ago as a leaf plant known as cadoon and used for its edible roots and petioles. It was used as a luxury food in second-century Rome. The plant resembles a large, semi-prostrate thistle, the flower stalk terminating in a large globular inflorescence covered in many involucral bracts. The immature flower head together with the thickened receptacle and fleshy bases of the involucral leaves are eaten as a vegetable either raw or cooked. The first record of the modern type of flower-bearing heads came from Italy in AD 1400. It is still used as a luxury vegetable and is grown world-wide, with most of the global production in Italy, France and Spain.

Rhubarb (Rheum rhaponticum L.: Polygonaceae)

Native to the cool regions of central Asia, where it still grows wild, rhubarb is now widely grown in temperate Eurasia and North America, where the summers average less than 24 °C and the winters less than 4 °C. Dried rhizomes of medicinal rhubarb (*R. palmatum* L. and *R. officinale* L.), which have powerful laxative effects, were used by the Chinese in 2700 BC. These species are natives of China and Tibet and were introduced as plants into Britain, also for medicinal purposes, in the sixteenth century. Only from the eighteenth century was the garden rhubarb recognised for its culinary properties, since when the succulent, acidic leaf stalk has been eaten.

1.1.2 Annual Vegetables

Onions and Related Crops (Amaryllidaceae)

Among the most popular vegetables are onions (*Allium* L. spp.) and close relatives garlic, leeks, shallots and chives. With their oniony or garlicky flavours and odours of alkyl sulphides, they are used for food flavouring in most countries. They are grown mainly as temperate crops but are also widely grown in subtropical regions. The main producers of onions are China, India, the USSR, the USA and Japan; and of garlic, Spain, Argentina and Italy. The edible *Allium* spp. have their centre of origin in Central Asia except for leeks and chives, which originated in the Near East and Mediterranean regions. The use of *Allium* spp. as food, medicine or religious objects dates back to 3200 BC. They were common vegetables of the Middle Ages in Europe, being included in writings by herbalists.

Sweet Corn (Zea mays *L.: Gramineae*)

Although the origin of maize (*Zea mays*) is unknown, it was cultivated in Mexico over 7000 years ago. Maize was brought by Columbus to Europe, where it is less important than in other parts of the world. It has become a world staple food, exceeded in importance only by wheat and rice. Sweet corn is unripe maize, which is less nutritious than the mature grain. It is grown in temperate to subtropical regions, the main production areas being South America, parts of the USA, and East and South Africa.

Carrots and Related Crops (Umbelliferae)

Carrots (*Daucus carota* L. var. *sativa* DC.) and parsnips (*Pastinaca sativa* L.) are the two most widely grown umbelliferous, annual, root vegetables. Both are cool-weather crops and are grown in the temperate climates of Europe during the spring, summer and autumn and in the subtropical climates of Asia during the winter. Wild forms of both crops are known in Europe and west Asia.

The first carrot roots consumed were probably purple in colour and were grown in Afghanistan around AD 1000. Yellow types were subsequently isolated and grown in Syria in the ninth or tenth century, and introduced into China in the fourteenth century and into Europe 100 years later. The first orange types appeared in the Netherlands in the seventeenth century, and, because of their improved colour and flavour, rapidly became popular in Europe and were then introduced into North America and other parts of the world.

Parsnips resemble carrots but are pale yellow and sweeter and not as popular. They were first used by the Greeks and Romans for both medicine and food, and by the mid-sixteenth century were cultivated throughout Europe as a staple food by poorer people.

Celery (*Apium graveolens* L. var. *dulce* Pers.) seeds are used as a herb like many other members of the Umbelliferae, but the crop is usually grown as a salad vegetable, when the swollen petioles are consumed. This species occurs wild in Europe and Asia Minor and has a similar history of use to its other relatives. Celeriac (*A. graveolens* L. var. *rapaceum* DC.) has swollen edible roots, and parsley (*Petroselinum crispum* Nym.) and chervil (*Anthriscus cerefolium* Hoffm.) are used for flavouring but are less commonly grown than other members of the Umbelliferae.

Lettuce and Related Crops (Compositae)

Among the major salad plants are lettuce and related crops. They are high-risk vegetables for growers, as their succulent leaves and petioles are susceptible to pest damage, poor production and handling systems, and environmental extremes.

Lettuce (*Lactuca sativa* L.) is native to southern Europe and western Asia and is derived from the wild lettuce (*L. scaricola* L.), a common weed in both the Old and New Worlds. Some forms of lettuce have been cultivated since 4500 BC and were utilised throughout Europe by Christian times. Lettuce has long been a favourite of the home gardener, but more recently it has become an important commercial crop and has been developed in several forms, which include: cold and heat tolerance for winter and summer production; and firm and loose heads, the former being suitable for mechanical planting, harvesting and packaging.

Chicory and endive (*Cichorium* L. spp.) are native to Europe and the eastern Mediterranean, respectively. Chicory has been cultivated at least since Greek and Roman times and endive was used by the ancient Egyptians.

Cucurbits (Cucurbitaceae)

The major cucurbit crops are taxonomically very diverse but are biologically similar and have a similar range of pests. The crops are usually produced in relatively small quantities for local consumption. They are nevertheless important items of diet in many countries, being eaten fresh (musk melon, water melon, cucumber), cooked (squash, pumpkin, marrow) or pickled (gherkin, cucumber).

The cucumber (*Cucumis sativus* L.) is thought to be indigenous to India, the water melon (*Citrullus vulgaris* Schrad.) to Africa and the pumpkins and squashes (*Cucurbita* L. spp.) to Central and South America. The cucumber was introduced into Europe during biblical times and into England in the fourteenth century, but was not cultivated until its reintroduction 250 years later.

China is the main producer in the world of all cucurbits. Other major producers are the USSR and Japan (cucumbers and gherkins), Romania and Egypt (pumpkins, etc.) and the USA and Spain (melons).

Cruciferous Crops (Cruciferae)

The Cruciferae is a diverse group of plants of European and Asia Minor origin. The number of cruciferous crops exceeds that of any other economically important plant group. A number of them are of world-wide importance as food plants, being cultivated from the Arctic to the subtropics and at higher altitudes in the tropics.

Brassica oleracea L. and *B. campestris* L. appear to be the genetic sources of most of the edible cruciferous crops, their development from wild species probably taking place in different parts of the Mediterranean region, where they are distributed along the coasts.

Most of the leafy vegetables belong to *B. oleracea* and are grouped together as brassica or cole crops; they include kale, Brussels sprouts, cabbage, kohlrabi, broccoli and cauliflower. Each type is a different modification of the leaf and shoot system and all contain glucosinolates, which in the presence of the enzyme myrosinase hydrolyse to isothiocyanates, thiocyanates, nitriles and goitrin. These secondary plant chemicals, which contribute to the characteristic odours and flavour of cruciferous plants, also help to identify them as host plants to many insect species (Finch, 1980).

B. oleracea was cultivated by the Greeks as early as 650 BC, as either cabbage or kohlrabi, but, although selection occurred at an early date, each type has its own history of domestication. The modern variety of headed cabbage was probably developed in Germany, where both the red and white forms were grown in AD 1160. The cabbage has long been an important item of the European diet. The cold-tolerance of brassica crops has made them popular in the northern temperate zone but has nevertheless limited their range of successful cultivation. Brussels sprouts (*B. oleracea* var. *gemmifera* Zenker) were apparently selected from a mutant cabbage plant that appeared in a European garden in 1750. Kohlrabi is often given a separate varietal name (*B. oleracea* var. *caulo-rapa* DC.), and although vaguely described by Pliny (AD 23–79), an unequivocal description appeared in the late sixteenth century. Both cauliflower (*B. oleracea* var. *botrytis* L.) and broccoli (*B. oleracea* var. *italica* Plenck) are later selections than cabbage and kale, with cauliflower recorded by the Arabians in the twelfth century and in European texts in the seventeenth century.

Brussels sprouts, cabbages and cauliflowers have been bred to mature at different times of the year and for improved quality to meet marketing and processing requirements. Also, with increased mechanisation, the need for uniformity in heading, curding and sprout production has emphasised the breeding of cultivars which can be harvested in one operation. Processing has also encouraged a market demand for mini-cauliflowers, small cabbage heads and Brussels sprouts, and with improved low-temperature storage facilities and acceptance of frozen vegetables, breeding programmes tend towards the production of cultivars which can be harvested in late summer and early autumn to avoid winter losses.

The major producers of cabbages are the USSR, China, Korea and Japan, and of cauliflowers are China, India, Italy and France.

The two most important root crops in this family are turnips and swedes. Turnips (*B. campestris* L.) are thought to have grown wild in Europe and Asia, being referred to in writings from India about 2000 BC. The swede (*B. napus* L.) may have arisen by spontaneous hybridisation of turnips and kale, possibly in mediaeval gardens. Radish (*Raphanus sativus* L.) is grown world-wide, and inscriptions from Egyptian tombs about 4000 years old show it was valued at that time. In China, Japan and other oriental countries, radishes are among the most important vegetables. The peoples of these countries prefer the very large white- or black-skinned forms to the small red spheres consumed in Europe and North America.

Beet and Related Crops (Chenopodiaceae)

Beta L. is an Old World genus, virtually confined to Europe, which includes beetroot, sugar beet, mangold and fodder beet. It dates from prehistoric times, and chard (*B. vulgaris* L. var. *cicla* L.) has been known since 300 BC. It was initially grown for its root, but later its leaves were used as pot herbs. The Romans used beet, probably *B. maritima* L., a wild beet from the Mediterranean coasts, and it was taken from Italy to northern Europe by the Barbarian invaders. Red beet (*B. vulgaris*) was also used by the Romans in the second and third centuries and was recorded in English recipes in the fourteenth century. The crop was introduced into the USA in 1800.

Sugar beet, the most economically important member, was not used for sugar extraction until more recent times. The first sugar extraction factory was built in Kunern, Poland, in 1801. Rapid advances were then made in breeding for sugar content in both Germany and France, and now the cultivation of sugar beet can be fully mechanised, particularly since the introductions of pelleted seed and precision drilling. The main areas of production are the USSR, France, the USA, Germany and Poland.

In contrast to the biennial beets, spinach (*Spinacia oleracea* L.) is an annual green vegetable grown in cool moist climates. It is a native to south-western Asia and was cultivated by the Persians 2000 years ago, and is now widely grown throughout Europe, North and South America and the Far East. It was not known as a pot herb until Roman times and has spread beyond its native area since then. Other types of vegetable referred to as 'spinach' include the biennial spinach beet (*B. vulgaris*) and New Zealand spinach (*Tetragonia expansa* Murr.).

Legumes (Leguminosae)

Peas and beans have a world-wide distribution. Over 30 legume species are used as vegetables, with most grown for their seeds and/or pods, and a few for their tubers or tuberous roots.

The pea crop (*Pisum sativum* L.) constitutes one of the four most important seed legumes, with an estimated world production of about 14 million tonnes per annum. Neither the wild progenitor nor the early history of the pea crop is known, but carbonised pea seeds were found during excavations of Neolithic settlements (*c.* 7000 BC) in the Near East and in Europe, indicating that peas were cultivated as long ago as wheat and barley. It is almost certain that the crop has been grown by Man for several thousand years, but it was not until the sixteenth century that varieties were first described. Up to the 1950s varieties were selected for particular markets—for example, fresh market, dried peas and canning. Since that time, a new industry has developed for the production of immature peas for freezing. The criterion for maturity of such crops is strictly defined, with harvesting being controlled to a matter of hours. Harvesting time can be reduced by increasing the number of pods per node and by all the pods maturing more or less at the same time. The principal producers of green peas are the USA, the UK and France, and of dry peas, the USSR, China and France.

Beans (*Phaseolus* L. spp.) contain several quite distinct species that have contributed crop plants to agriculture in both the Old and New Worlds. They are usually consumed dry in tropical countries but in temperate countries, although some species are grown as dry beans (notably *P. vulgaris* L.), varieties have been developed for fresh pod consumption and for processing as frozen vegetables. Beans are an important source of protein and energy in human diets, particularly in tropical Africa and Central and South America, where they supplement the carbohydrate staple foods such as maize, plantain and cassava.

Phaseolus beans originated in Central America and were under domestication over 7000 years ago in Peru and Mexico. The common bean (*P. vulgaris*) reached Europe in the early sixteenth century and has since undergone important changes to its growth habit—for example, reductions in numbers of branches and leaves, and increases in pod and seed size and number of seeds per pod. China, Turkey, Italy and Spain are the main producers of green beans, and India, China and Brazil of dry beans.

Other economically important leguminous crops include the scarlet runner (*P. coccineus* L.), grown widely throughout temperate regions, and the Lima (*P. limensis* Macf.) and Mung (*P. aureus* Roxb.) beans, which tend to be grown in warmer regions. The broad bean (*Vicia faba* L.), which probably originated in south-west Asia, is essentially a temperate crop and is grown in Europe as a vegetable, where the seeds are used either fresh or frozen.

Tomatoes and Potatoes (Solanaceae)

Of the eight most important staple crops grown in the world, seven are grain crops and the other is a tuber group, the so-called white or Irish potato (*Solanum tuberosum* L.). Although this crop is produced in a greater quantity than any of the grain crops, it is less important for human nutrition, because of its relatively low dry-matter content (2% protein, 17% starch). The world production of potatoes in 1986 was over 300 million tonnes. They are cultivated mainly in temperate regions but can be grown in the tropics at higher altitudes, using short-day cultivars. The USSR and eastern Europe are the leading producers, contributing about 20% of world production.

The potato is native to the higher altitudes (>2000 m) of the Andes (between 10° N and 20° S) and has served as an important cultivated food since early civilisation. During the sixteenth century European explorers found potatoes growing from Chile to Columbia and introduced them to Spain in 1570. By the end of the 1500s, they were widely spread throughout Europe. The potato was introduced from Europe into North America in the early 1700s.

The tomato (*Lycopersicon esculentum* Mill.) is a warm-season perennial grown as an annual crop. In the USA it is ranked second only to potatoes in importance among vegetables. After the USA, the USSR, China, Italy and Turkey are major producers but there are few places where it is not grown and used in one form or another. In cooler temperate countries most production is under glass or polythene shelters.

While it is uncertain whether the tomato is native to Peru or Mexico, it was first domesticated in Mexico. It was not until the eighteenth century that it was accepted as a food in England and France, being previously thought to be poisonous. The spectacular increase in tomato cultivation came after 1920, with most of the increased production being processed into products such as juice, paste, sauce, and canned soups and fruit. Selection has occurred for a variety of fruit colours, shapes and sizes, among other traits.

1.2 MAJOR INVERTEBRATE PESTS OF VEGETABLE CROPS

Invertebrates and vertebrates damage vegetable crops, but, of the vertebrates, only a few species of birds and mammals are important pests. Invertebrates, on the other hand, are generally much more numerous in terms of both individuals and species, and, consequently, most pests of vegetable crops are invertebrate animals. Of these, the majority are insects, although a few smaller groups have some pest species—for example, nematodes, slugs, millepedes and mites (Jones and Jones, 1984). Pests mainly affect vegetable crops by feeding directly on the plants either by biting and devouring the tissue or by piercing it to feed on the sap. In most cases the immature stages cause damage, but adults can cause occasional damage. Symptoms vary considerably and all parts of plants may be affected by some pest or other. In some cases, whole plants may be eaten or killed, but damage is more frequently restricted to a particular part of the plant, such as the roots, stems, leaves, buds, flowers, fruits, etc. Damage done to vegetables near harvest is of obvious importance, since it has a direct effect on both the quality and the quantity of produce, but less conspicuous damage to roots, stems and leaves may impair growth and make plants more susceptible to disease. Pests may also have important indirect effects on plants by transmitting viruses and other micro-organisms that cause diseases and also by contaminating plants with excretions such as honeydew. Growth of sooty moulds on honeydew renders plants unacceptable for consumption and reduces the photosynthetic efficiency of the leaves.

The following section provides a brief introduction to the taxonomic orders that contain major pest species. More detailed accounts can be found under the chapters dealing with specific crop pests.

1.2.1 Nematodes (Nematoda)

Plant-parasitic nematodes are common in almost every environment, their distribution being limited by their dependence on water, varying from the films associated with soil particles to the sap of plants. Nevertheless nematodes can withstand long periods of drought in resting phases—for example, cysts—which can remain viable for many years until suitable conditions return. Most nematodes live in the soil, which imposes limits on their movement. Probably, the movement of nematodes is random until they come under the influence of gradients of substances from the rhizosphere. Many soil nematodes are polyphagous, but even the species with limited host ranges most likely test or invade plants indiscriminately and may become trapped in unsuitable roots. Host-specificity observed in some nematodes may depend on their inability to reproduce in unsatisfactory hosts rather than by host selection. Plant-parasitic species feed internally or externally on tissues, puncturing cells and extracting fluid contents. This injury causes discoloration, distortion and death of affected plants. Some species also transmit virus diseases.

Among the nematodes that attack vegetable crops, the stem nematode (*Ditylenchus dipsaci* (Kuhn) Filipjev), which is seed-borne, has a worldwide distribution and is polyphagous, attacking onions, carrots and beans. The pea cyst nematode (*Heterodera goettingiana* Liebscher) is more host-specific, feeding and developing only in legumes. Crop rotation and sanitation are important in the suppression of nematode populations. At least 17 species of root-knot nematode (*Meloidogyne* Goeldi spp.) occur in western Europe, including both native and other species introduced from the tropics, and are now established in glasshouses. They all attack roots and tubers, forming galls on susceptible plants such as cucumbers, tomatoes, lettuces, French beans and, occasionally, carrots, parsnips and beetroot. Attacks are worst in light soils and at high temperatures. The yellow potato cyst nematode (*Globodera rostochiensis* (Wollenweber) Behrens) is one of the most damaging pests of potatoes in temperate regions. The outstanding features of cyst nematodes are the enlargement of the female into a sub-spherical shape and the tanning of the body wall after death to produce a protective, horny cyst enclosing all the eggs. Such cysts aid the dispersal of the nematode by adhering to the soil or plant produce, which may be moved between fields.

1.2.2 Slugs (Mollusca: Gastropoda)

Slugs cause plant damage by rasping holes in roots, corms, tubers, stems, leaves, buds, flowers, seed capsules and fruits. Numbers of slugs in the soil increase with increasing organic matter, which improves water-holding capacity and provides alternative food when plants are absent. Slugs are most common in heavy soils, which hold moisture better than light soils. The distribution of slugs in the soil is largely determined by the distribution of aggregates which afford the slugs shelter. During the winter, slugs move down the soil profile, where they remain inactive but safe from lethal frosts. In spring and summer the activity of slugs increases considerably, especially in warm, moist conditions, when they feed on plants above and below ground, principally at night. Activity decreases in response to a drop in temperature and increases in response to humidity or rain. Slugs can be common during dry summers under crops which provide abundant shelter.

Vegetable crops are often seriously damaged by slugs. For instance, pea and bean seeds are injured soon after sowing, especially if they germinate slowly in cold weather; seedlings of brassicas and lettuces are severed at soil level both outdoors and under glass; and tubers of Jerusalem artichokes and potatoes are extensively holed, damage being usually worse on main-crop potatoes than early crops. Slug damage to potatoes can be lessened by growing cultivars less susceptible to attack and by lifting the main crops early.

Most control measures (e.g. poison baits and cultural methods) are not fully satisfactory, because only a proportion of the slug population is active on or near to the soil surface at any one time. Direct drilling (or minimal cultivation techniques) can lead to more damage from slugs.

1.2.3 Insects (Insecta)

Aphids (Homoptera: Aphididae)

Aphids belong to one of the more important families of economic insects, especially in temperate regions, where they are principally vectors of plant diseases. They are also among the most prolific of insects. Many show an alternation between a woody, winter (primary) host for egg laying and a herbaceous, summer (secondary) host on which they reproduce asexually (parthenogenetically and viviparously). The apterae or wingless adults reproduce rapidly, and the appearance of alatae or winged forms—which results from overcrowding or maturation of host plants or changes in daylight length and temperature—permits migration to new host plants (e.g. the willow-carrot aphid (*Cavariella aegopodii* Scopoli) is found on carrots in summer and on willows in winter). Some aphids—for example, the cabbage aphid (*Brevicoryne brassicae* L.)—remain on the same type of host plant (Cruciferae) throughout the year; some have no sexual phase, continuing throughout the year as viviparous females; and others—for example, the peach-potato aphid (*Myzus persicae* Sulzer) which has a sexual phase—can survive overwinter as apterae if the weather is mild and shelter is available.

Aphids can damage plants directly by removing plant sap, which contains essential food materials and plant growth promoters, or, more importantly, by transmitting viruses. The viruses they carry are either stylet-borne (non-persistent) or circulative (persistent) and, although the time required to feed on infected leaves to acquire these viruses varies together with their persistence in the aphid, both types can be transmitted very soon after viruliferous aphids start to feed on healthy plants. Control measures need to be applied either before or as quickly as possible after viruliferous aphids arrive on a crop, but in the case of direct damage, control can be delayed until the economic injury threshold is reached.

Aphids attack most vegetable and arable crops. The polyphagous species *M. persicae* is a major vector of potato and sugar beet viruses and is also found on turnips, swedes and spinach. Other species have a narrower host range. For example, the bean aphid (*Aphis fabae* Scopoli) is a serious pest of beans (*Vicia* spp.) and beet, while other species are restricted to a single summer host type, e.g. the potato aphids *Aulacorthum solani* Kaltenbach and *Macrosiphum euphorbiae* Thomas. *B. brassicae*, which remains on cruciferous plants, principally brassicas, throughout the year, overwinters on crops that are not cleared and destroyed. It is, therefore, inadvisable to grow brassicas close to overwintering seed crops. Unlike the lettuce aphid (*Nasonovia ribisnigri* Mosley), which feeds on leaves, the lettuce root aphid (*Pemphigus bursarius* L.) attacks the roots and causes lettuce plants to wilt and die. The overwintering woody host of *P. bursarius* is poplar.

Leaf and Plant Hoppers (Homoptera: Cicadellidae)

Leaf and plant hoppers are second in importance to aphids as virus vectors, especially in warm climates. They also transmit mycoplasmas, spiroplasmas and rickettsias. Leaf hoppers are very abundant in field crops, and although feeding is limited to certain crop plants, the adults are usually polyphagous. In the USA, leaf hoppers are found in such crops as potatoes and beet, where they can be important pests.

Capsid Bugs (Heteroptera: Miridae)

Many different species of capsids are common and widespread in Britain and northern Europe, mostly feeding on plants, although some are predatory. The nymphs and adults pierce plant tissues with their mouthparts, inject saliva and feed on sap. The saliva of the main pest species kills plant tissues, so that, following feeding, small ragged holes appear in young foliage. Buds and shoots may be killed and flowers developing from damaged buds may be deformed. Most damage is done in late spring and during the summer. Among the vegetable crops affected by capsid bugs are potatoes, runner beans and beet. Damage to crops is often found in fields surrounded by high hedgerows and, in Britain, is only occasionally sufficiently severe to warrant treatment, which can usually be confined to field margins only. On the continent of Europe the beet lace bug (*Piesma quadratum* Fieber) transmits beet leaf curl virus. The adult overwinters in the field surrounds, flying into the crop in early May.

Butterflies and Moths (Lepidoptera)

The order Lepidoptera contains many species of economic importance, especially in the tropics. The structure and habits of the adults and larvae are varied but relatively uncomplicated. The adults feed only on nectar and other sugary fluids or do not feed at all. Butterflies fly during the day and most moths at night, with some capable of long-distance flight. In contrast, the larvae are sluggish and are rarely found straying away from the food plants. Only the larvae cause crop damage, being voracious feeders. All larvae feed above ground except swift moth larvae (Hepialidae) and cutworms (Noctuidae), which live in the soil and cause damage to roots. Some noctuids feed above ground on leaves—for example, the cabbage moth (*Mamestra brassicae* L.) and the silver Y moth (*Autographa gamma* L.), which can badly damage cabbages, along with the cabbage white butterflies (*Pieris brassicae* L. and *P. rapae* L.).

Beetles (Coleoptera)

The order Coleoptera, which is one of the largest insect orders, contains more than 250000 known species, of which at least 5000 have been recorded in Britain and northern Europe. Many beetles are injurious to vegetable crops—for example, wireworms (Elateridae), pollen beetles (Nitidulidae), flea beetles (Chrysomelidae) and weevils (Curculionidae). Wireworms naturally inhabit permanent grassland, where they can develop large populations. When the grass is ploughed, being polyphagous, they can damage many arable crops, including *Phaseolus* beans, lettuces, tomatoes, carrots and potatoes. The life cycle lasts for 4–5 years and, like slugs, the larvae are more active at night and during damp weather, sheltering during daytime below the soil surface. The larvae, which take 3–4 years to mature, feed actively in spring, coincident with the seedling stage of many crops, and again in autumn, when most crops are mature. The type of injury caused by wireworms varies, but they burrow into potato tubers and carrots and also into the haulms of tomatoes. Crops with large seeds and thick stems, such as peas and beans, are not normally badly damaged and so can withstand a higher wireworm density than can potatoes, which are relatively susceptible. Although potatoes succeed in growing at almost any wireworm density, the tubers become riddled with holes.

Many species of flea beetle are pests of cruciferous crops, the adults attacking the cotyledons and stems of germinating seedlings in spring. The larvae feed on the roots and resemble wireworms in form and habits, except the larvae of *Phyllotreta nemorum* L., which are leaf miners. In contrast, the larvae of the cabbage stem flea beetle (*Psylliodes chrysocephala* L.) mine the stems and petioles of spring cabbages and kale sown in late summer. The pulse or seed beetles (Bruchidae) are pests of pods and seeds, particularly of the Leguminosae. The bean beetle (*Bruchus rufimanus* Boheman) is common in the UK in seeds of broad and field beans, and the pea beetle (*B. pisorum* L.) and the dried bean beetle (*Acanthoscelides obtectus* Say) are introduced annually in culinary peas and dwarf beans but fail to establish. The weevils, a very large and successful group, include many pest species. The larvae are mostly concealed feeders, either in the soil feeding on roots and nodules (e.g. pea and bean weevils, *Sitona* Germar spp.), in plant tissue (e.g. turnip gall weevil, *Ceutorhynchus pleurostigma* Marsham, and the cabbage stem weevil, *C. quadridens* Panzer), in flower buds or in seeds (e.g. cabbage seed weevil, *C. assimilis* Paykull). The Coleoptera also include predatory species (ground beetles and ladybirds) that feed on other insects and invertebrates and may, therefore, be beneficial.

Flies (Diptera)

Among the Diptera that attack vegetables is the pea midge (*Contarinia pisi* Winnertz), which attacks terminal blossoms and shoots and causes pod infection. The pea midge is widespread in Europe and attacks all varieties of peas. However, more important than the pea midge are the carrot fly (*Psila rosae* Fabricius) and the cabbage root fly (*Delia radicum* L.), which damage the roots of carrots and brassicas, respectively. Carrot fly rarely causes yield loss but its rust-coloured larval tunnels severely reduce crop marketability. Cabbage root fly causes damage to the roots of brassicas, thereby reducing water and nutrient uptakes, which disrupts the growth of the plants. Occasionally its larvae are found inside Brussels sprout buttons, making crops unacceptable for processing. The related turnip root fly (*D. floralis* Fallén) is common in Scotland, northern England and Scandinavia, and attacks turnips and cauliflower curds. The larvae of the onion fly (*D. antiqua* Meigen) tunnel into onion bulbs and can kill young onion and leek plants. Leatherjackets, the larvae of crane flies (principally *Tipula paludosa* Meigen), live in the soil and feed on the roots and stems of many different crops, and cause greatest damage in wetter areas. Affected plants turn yellow, wilt and die, and the symptoms often resemble those produced by cutworms and certain pathogenic fungi. Young brassicas and lettuces are susceptible to attack, and a few leatherjackets can cause considerable losses. Most damage is done to young plants in spring, but attacks may develop at other times during the growing season.

1.3 PEST CONTROL STRATEGIES

For rational management of crop pest populations, a knowledge of their life systems, which includes life cycles and interactions with the biotic (natural enemies and host plants) and the abiotic (weather and other physical components) factors in the agroecosystem, is essential. The basis for such a planned manipulation of a pest is to reduce its numbers to those which do not inflict unacceptable damage to host crops. 'Pest control' is a term used loosely to define this objective and implies a percentage reduction of the population, usually achieved by the application of pesticides to provide a satisfactory crop yield and quality of produce. The term 'economic threshold', which indicates the pest population density above which economic injury occurs (Matthews, 1984), is a convenient and helpful practical guide to the value of pesticidal treatment, but is also arbitrary, because the relationship between pest numbers and crop yield varies with, for instance, the degree of crop stress, crop growth stage and the economics of the treatment. In vegetable production, with cosmetic quality being critical to the value of the crop, the cost of pesticidal control may well exceed the cost of pest damage, simply to avoid the slightest damage or contamination by, for example, insect debris which could lead to crop rejection. Without the ability to predict those crops which will be significantly damaged, vegetable growers are usually compelled to resort to insurance-type treatments to meet the demand for increasingly high-quality produce. Some unnecessary use of pesticides is, therefore, inevitable unless practicable and reliable methods for pest forecasting are made available. Equally problematical in vegetable production is the implementation of integrated pest management, a strategy in which a variety of biological, chemical and cultural measures are combined to give stable, long-term, pest suppression. The integrated strategy is a problem, because most vegetable crops are of short-term duration and do not provide sufficient time for natural enemies to colonise before crops are harvested. Changes in the type of crop grown in a rotation may, nevertheless, help to avoid some pests, but may do little to discourage build-up of others. Vegetables, particularly those grown for processing, tend to be grown on a farm scale, and the extensive mechanisation required for large-scale production involves capital inputs and specialisation in a few crops. So, even on large farms, there is unlikely to be sufficiently large separation distances between crops necessary to counter pest mobility, particularly when rotations are short. Most horticultural systems are based on pesticide use, with a tendency to avoid long rotations and to concentrate production in defined areas for ease of harvesting, handling and storage. Rational control is a more practicable approach for pest suppression on vegetable crops than integrated pest management and is based on information on the crop, the weather and the pest and its natural enemies to determine the minimum amount of appropriate pesticide to produce the required level of control. This practice applies to those situations where 'timed' and 'supervised' control are used: in other words, where pesticidal treatments are applied on the basis of threshold activities or numbers of the pest species determined by monitoring either visually or from trap catches. Once the critical threshold has been reached in relationship to the stage of growth of the crop or its risk situation, a treatment is applied. The inclusion of cultural methods or resistant cultivars can contribute to the overall system, which can then be considered as truly integrated control. Most applied entomologists involved in crop protection agree that no single method of pest control will provide a complete solution to all pest problems and that total reliance on pesticides is not a wise strategy. The benefits from adopting a rational or integrated approach can reduce the amount of insecticide used, thereby reducing cost and improving efficiency.

The various components of pest management are described in the remaining part of this section. They can be used separately or together (integrated control), but unless a method can completely replace an existing pesticidal treatment, it has to be compatible with current pesticidal regimes, since soil and foliar applications are likely to remain the major methods of pest control for the foreseeable future (Finch, 1987).

1.3.1 Chemical Control

Chemical methods for the control of crop pests involve the application or use of chemicals in the pests' habitats within crop environments. The agents of chemical control include conventional pesticides, insect hormone analogues or growth regulators, and pheromones. The majority of pests which attack vegetable crops are currently controlled by chemicals, and their widespread use has to a great extent been highly successful. This success does not imply that all control measures are optimal: for example, control is often lacking for some soil-inhabiting insects and nematodes. In addition, some unfavourable cases of chemical use have led to environmental disturbances which have been sufficient to promote a drastic reappraisal of the future place of chemical control within insect pest management systems. Use of pesticides should, therefore, be determined on the basis of their potential advantages and disadvantages. The advantages of pesticides are: they are highly effective in reducing pest populations to the very low levels demanded by consumers, they work rapidly and economically, and they are readily available through well-organised supply systems. Pesticides can also be applied with relatively simple technology. Their potential disadvantages include pest resistance and insufficient selectivity. Lack of selectivity has generated environmental problems, particularly from persistent and broad-spectrum compounds which kill beneficial insects, thereby inducing secondary pest problems. Some compounds may also be highly toxic to Man. Alternatively, some compounds have a transient effectiveness or their performance may be weather-dependent, thereby necessitating repeated applications in many pest situations.

Development of pesticides over time demonstrates their progressive increase in potency against target species and reflects their increasingly specific modes of action. There has been a transition from the general enzyme inhibition of inorganic compounds to the specific disruption of metabolic processes, such as oxidative phosphorylation by dinitrophenol compounds and the specific inhibition of neurophysiological processes by most current insecticidal compounds (Graham-Bryce, 1987). However, such specific modes of action do not imply a greater species selectivity, as exemplified by the organochlorine insecticides, which are both stomach and contact poisons. By having no systemic action, DDT, for example, is ineffective against aphids and spider mites, and its injudicious use can cause resurgences of these pests by killing the predators that help to regulate them. With the decline of the organochlorine insecticides, the organophosphorus compounds became the predominant class of insecticides, although the pyrethroids have taken a large share of the total usage in recent years. The organophosphates cover a wide range of related compounds and there are few pest problems for which there is not at least one compound with appropriate insecticidal properties. The carbamates are a smaller class of general insecticides with a variety of properties that include general-purpose, contact chemicals, compounds with nematicidal and systemic properties, soil insecticides and others with specialist properties such as molluscicides (Corbett, 1978). The selective systemic aphicide pirimicarb has the advantage of controlling resistant strains and is of interest in insect pest management systems. The pyrethroid insecticides have a long history based on natural pyrethrin, extracted from flower heads of certain species of pyrethrum, but have only become of major importance in recent years with the discovery of synthetic pyrethroids (Elliot and Janes, 1978). These compounds are not only more active against many insect species than the natural pyrethroids, but are also safer to vertebrates. Bioresmethrin, for example, is 40 000 times more potent to houseflies than to rats on a weight-to-weight basis and is the safest insecticide in commercial use. Although the earlier synthetic pyrethroids are unstable in air and light, as are the natural compounds, the more recent ones are photostable and are highly active—for example, deltamethrin is the most potent insecticide yet discovered against many insect species.

Numerous, synthetic, active compounds have been found based on the structure of the juvenile hormone which regulates metamorphosis in arthropods. These compounds are known as juvenoids or juvenile hormone analogues, but only methoprene has been registered for practical use. Such compounds are limited by their physiological mode of action, i.e. they must be present during a critical phase of the life cycle, being ineffective against the larval stages, which often cause most damage to crops. Juvenile hormone analogues are also very labile. Of the other insect growth regulators (IGRs), the best-known are the urea insecticides, particularly diflubenzuron. IGRs inhibit insect cuticle formation, probably by disrupting chitin synthesis. Urea insecticides must be ingested and are not systemic, and late larval instars can cause substantial damage before they are eventually controlled. IGRs are also ineffective against adults, so their range of use is limited, but, being relatively selective, they are useful for control of Lepidoptera in insect pest management systems.

Chemicals that influence arthropod behaviour by attraction, repellency, food location, oviposition, mating and defensive strategies have been identified and are all potentially exploitable. The main interest has centred on sex pheromones, particularly those produced by females to attract distant males. Sex attractants are now used widely as lures in traps to monitor pest populations (Cammell and Way, 1987).

Application of pesticides is, in principle, a difficult task, since it is normally necessary to apply a blanket treatment over the crop area. However, the technology of application has maintained pace with the development of pesticides, and it is now possible to apply pesticides from the air, using volumes as low as 1 litre per hectare and from a variety of tractor-mounted or portable spraying and blowing machines and from a variety of hand-held packs, including aerosols, granule applicators, etc. (Matthews, 1979). Consequently, application is not a significant impediment to the use of pesticides.

Resistance to insecticides by crop pests is a serious disadvantage of chemical control and arises from the selection of less susceptible genotypes in a genetically heterogeneous population. The resistant individuals are not thought to arise as a result of insecticidal treatment. The insecticides are not mutagenic and resistance is assumed to be preadaptive, conferred by novel alleles which occur rarely in the untreated population by recurrent mutation (Sawicki and Denholm, 1984). Resistance occurs by decreased penetration, enhanced metabolic degradation or decreased sensitivity at the site of action. The rate at which resistance develops will depend on the frequency of resistant genes in the population, their dominance and their exposure to insecticides, as well as on the influences exerted by the ecology of the population—for example, reproductive potential and degree of outbreeding resulting from their dispersiveness and mixing with unselected populations. Resistance to one or more pesticides has been recorded in over 400 species from areas where chemical control has been applied intensively. Other cases of resistance have probably not been recorded because they have not given rise to serious problems (Georghiou, 1981). The development of resistant populations may lead to higher rates of chemical application in an attempt to maintain control, with all the attendant problems of chemical use, leading ultimately to the loss of the pesticide concerned as an effective control agent. Sometimes other chemicals in the same group can also be lost, and, in some cases, chemically unrelated compounds affected by the same resistance mechanisms can be included as well (cross-resistance). Resistance has also developed in some natural enemies to a wide range of insecticides (Croft and Brown, 1975), and there is much interest in exploiting resistant natural enemies in insect pest management because they could allow the use of the generally less expensive, broad-spectrum insecticides (Croft, 1981).

Pesticides do not provide a permanent control and have to be reapplied, usually every season. Although unfortunate from the users' point of view, it is ecologically advantageous, since any deleterious effect will eventually disappear as a result of chemical degradation. Pesticides are broken down by photodegradation and a variety of chemical reactions in the soil, particularly hydrolysis, but micro-organisms are also likely to play a major role.

Pesticides may be harmful to non-target species, either by direct contact or ingestion through a food chain or by indirect ecological effects following suppression of prey or host weed plants. As insect populations are regulated to some extent by their parasitoids and predators, it is desirable to choose insecticides that have a narrow spectrum of activity or to use them in such a way that they leave the beneficial species unharmed. While narrow-spectrum insecticides are desirable, in principle, such compounds are less likely to be as useful in a wide variety of situations as those with a broad spectrum. The relatively small number of market outlets for narrow-spectrum compounds normally makes them more expensive to develop and manufacture.

Although pesticides rarely provide a permanent solution to pest problems and have to be applied every season, their increasing use throughout the world demonstrates that the benefit they provide outweighs their cost.

1.3.2 Biological Control

Biological control uses living organisms to control pests, usually by introduction of pathogens, parasitoids or predators that can permanently reduce pest populations to below the economic injury threshold. Agents that cannot maintain sufficiently low population numbers to achieve this end can be reintroduced at intervals, but this approach is less satisfactory than the introduction that provides permanent control. Although biological control has mostly concentrated on exotic pests which reached their new habitat without the natural enemies that controlled them in their place of origin (Hall *et al.*, 1984), it can also be applied to either immigrant or native pests, using organisms from anywhere in the world, including virulent strains created or selected in the laboratory. A more recent activity associated with insect pest management has concerned manipulations of natural enemies already in the crop environment (Wratten, 1987).

Control by Viruses

The best-known insect viruses are the baculoviruses. They contain DNA and are of two types, nuclear polyhedral and granulosis viruses. Among other insect viruses, the cytoplasmic polynuclear viruses are of interest for biological control. Three characteristics influence the efficiency of viruses used for insect control: their high specificity, their relative instability and their slow speed of action. From the practical point of view, their specificity has a commercial disadvantage similar to that of selective insecticides, irrespective of ecological desirability. Insect viruses are sensitive to ultra-violet light, temperature and pH, and require to be formulated to protect them from sunlight to enhance their persistence on crop plants. The slow speed of action is more of a problem, as death may not occur for several days after ingestion. During the incubation period, larvae continue to feed and damage crop plants. Insect viruses are obligate parasites and can reproduce only within the living cells of hosts. Thus, all insect virus production methods involve rearing suitable larvae, infecting them and allowing each virus to reproduce before extracting it. Viruses can be applied in the same way as insecticides, but because they are infective only by ingestion, plants must be well covered by sprays (Entwistle and Evans, 1985).

Control by Bacteria

Of the hundreds of different species of bacteria associated with insects, those used in pest suppression are found within the Pseudomonadales, Enterobacteriaceae, Micrococcaceae and Bacillaceae, the latter family containing most spore-forming bacteria pathogenic to insects. The spore-forming bacilli are the most important pathogens, *Bacillus thuringiensis* Berliner (*B.t.*) being the most widely used. It is a facultative bacterium containing proteinaceous crystals which, upon ingestion, particularly by lepidopterous larvae, dissolve in the alkaline, reducing environment of the mid-gut, causing gut paralysis. Susceptible hosts may either be killed by the toxic crystals or be weakened to such an extent that the ingested spores penetrate the haemocoel and cause lethal septicaemia. The virulence of *B.t.* depends on whether the product contains the crystal (*B.t.* δ-endotoxin) formed during sporulation. Over 130 insect species in the Hymenoptera, Coleoptera, Diptera and Orthoptera are susceptible to *B.t.*, with some of the more serious vegetable pests included among the susceptible Lepidoptera (Bulla *et al.*, 1975). *B.t.* has a wider range of activity than have viruses, but a less wide range than have insecticides. It acts more quickly than viruses, although some damage may be caused to plants before the targeted insects are killed. It can persist for 10 days on cabbage leaves and for up to 4 months in the soil. Resistance to *B.t.* has been reported in a few species attacking field crops and stored products.

Control by Fungi

The fungi are the most numerous group of insect pathogens, with about 35 genera infecting insects. In general, the host range is broad, but some fungi are specific to only a few insect species. Infection is via germination of fungal spores on the insect cuticle, which is then penetrated. Since spore germination requires high humidity, environmental conditions affect the development of fungal epizootics more than those of bacteria and viruses, which enter insects orally (Burges and Hussey, 1971).

Control by Nematodes

Nematodes parasitising insects may be facultative or obligate, live internally or externally on their hosts and show narrow or broad host ranges. In general, insects attacked by nematodes are not killed. Nematodes which are found associated with most insect orders have a number of disadvantages as biological control agents (Poinar, 1971). The dependence of nematodes on moisture severely limits their distribution. Infective stages cannot survive at temperatures above 30 °C or in freezing water. Nematode propagation is difficult and insect hosts can respond to nematode invasion by encapsulating the parasites. So far, nematodes have not become a significant biological control agent for vegetable pests, although attempts have been made to evaluate them for cutworm control on lettuces (Theunissen and Fransen, 1984).

Control by Insects

Insects used as biological control agents have been responsible for more successes than any other group of controlling organisms. They can be used in one of three ways: (1) as exotic introductions into new environments (classical biological control); (2) to augment indigenous natural enemies (inundative control); and (3) to conserve and enhance indigenous natural enemies (Rosen, 1985). The practice of introducing exotic parasitoids or predators for pest suppression is more likely to be successful in ecologically circumscribed situations—for example, a vacant niche in the life system of a pest which might be filled by an introduced species; or the natural enemy in the niche is an inefficient regulator of the pest and can be displaced by a more efficient, introduced natural enemy. The first of these two situations is most commonly encountered with introduced species and the second situation with indigenous pests. The most important entomophagous agent used in mass-release programmes in field crops is the egg parasite *Trichogramma* Westwood spp., released annually against over ten lepidopterous pests, including species attacking vegetable crops. The controlled environment of glasshouses is favourable not only for pest development, but also for natural enemies. In Europe several glasshouse crop pests are controlled by parasitoids and predators (Hussey, 1985; Gould, 1987). As the degree of control achieved by a natural enemy is influenced by environmental factors, the level of crop damage caused by a target insect pest may well become unacceptable before control is effective. This problem has been recognised and 'complete control' may include the application of insecticides as necessary. However, timely release of artificially reared natural enemies can provide a good degree of control, to the extent that harvested produce is unblemished, particularly in glasshouses. The problem with the use of parasitoids and predators for biological control is not so much one of production as of discovering species with the correct attributes, such as good searching ability, appropriate degree of host-specificity, higher reproductive capacity than host or prey and adaptability over a wide range of environmental conditions. Nevertheless, biological control by parasitoids and predators has proved extremely effective and relatively inexpensive in glasshouses, but does have limitations for use on field vegetable crops, particularly when produce of high quality is demanded and when crops change from year to year. Problems also arise when there is a complex of pests associated with a crop, of which some will be successfully controlled with biological agents but most will require to be controlled with chemicals.

1.3.3 Genetic Control

There are many approaches to genetic control, all having the same principles and objectives—namely, to employ techniques that use the insect pest for its own destruction. Such techniques are known collectively as autocidal control, and the most frequently used is the sterile insect technique (SIT). Other forms of genetic control are, at present, of a more theoretical nature, and include the use of incompatible insect races or hybrid sterility (Boller, 1987).

Genetic control entails the mass production, sterilisation and release of large numbers of a pest that will mix and mate with wild populations in the field, leading to a reduction in fertility and, under certain circumstances, to population eradication. Unlike chemical control, genetic control is more efficient at low population levels. The release of sterile or genetically incompatible insects should, therefore, be preceded by a suppression programme based on insecticidal, biological or cultural methods. The characteristics of genetic control methods can be summarised as follows: they are species-specific and have no obvious ecologically deleterious side-effects, but, since they require technological support, they cannot be used by individual farmers. Genetic control is, therefore, often planned, organised and supplemented by government agencies, although there are exceptions where programmes have been conducted by private companies on a contract basis and at farm level—for example, onion fly (*Delia antiqua*) control in the Netherlands (Loosjes, 1976).

SIT is best considered as a strategic tool for long-term and large-scale pest suppression, being independent of crop-to-crop and season-to-season variations that dominate pest control at farm level. Current SIT programmes in the Americas, Africa and Japan operate on an international scale with the objective of eradication or as national, preventative, quarantine measures to protect major agricultural regions from invasion by new pests. There are three distinct phases during the operation of an SIT programme: (1) small-scale field and laboratory research to gather basic information on pest behaviour, ecology, rearing and sterilisation procedures; (2) large-scale feasibility studies to evaluate eradication and/or suppression techniques which should enable an economic appraisal to be made to provide a plan for the operational programme; and (3) operational programme use. Consequent on the complex organisation required for the application of SIT, the method cannot be expected to play other than an occasional very specialised role in vegetable crop pest management.

1.3.4 Breeding Resistant Crops

The breeding of varieties of a crop resistant to a pest is a genetic method of pest control applied to the crop rather than the pest. Resistance is based on one or more of the following mechanisms: antixenosis (factors affecting non-preference); antibiosis (factors affecting the performance of pests); and tolerance (factors affecting symptom development, growth and yield of plants).

The plant breeder is currently limited to the range of genetic variability which can be obtained from existing crop varieties, original centres of diversity of a crop and mutations produced by radiation or mutagenic chemicals. Consequently, it may not be possible to obtain resistance to a particular pest, either because resistance does not occur anywhere in the crop's genome or because it cannot be located or artificially produced. Genetically diverse strains of crop plants are screened for desirable characters, which are then incorporated by breeding into a hybrid genome expressing suitable agronomic characteristics. The new cultivars are then tested against the target pest and perhaps also against other pests. These tests should take place under a wide range of environmental conditions to ensure that crop yields are satisfactory in practice. Finally, enough seed must be produced for use. The genetic basis of crop resistance may be monogenic or oligogenic (single or few genes), polygenic (many genes) or occasionally cytoplasmic. Insects can sometimes break down oligogenic (vertical) resistance, which has led plant breeders to concentrate more on polygenic (horizontal) resistance, determined by many genes each having an individually small effect. The advantage of horizontal over vertical resistance is that it is likely to last longer, since a pest is unlikely to be able to modify its genotype to cope with many resistant genes. Clearly, vertical resistance carries a higher risk than does horizontal resistance of selecting adapted pest biotypes. However, vertical resistance is much easier to transfer to adapted varieties and is likely to provide a higher level of resistance. For these reasons, there is considerable interest in developing strategies for the management of vertical resistance.

Considerable efforts in breeding plants for pest resistance have been made in the USA, particularly into high-yielding varieties of the major world crops. Europe, on the other hand, has been slow to develop economically useful varieties resistant to insect pests, because they are seldom a key limiting factor in European crop production. Consequently, commercial varieties bred for resistance are few. Even the successful lettuce varieties Avon Defiance and Avon Crisp, which are almost totally resistant to lettuce root aphid (*Pemphigus bursarius*) attacks, were bred as varieties resistant to downy mildew. Varieties of swedes partially resistant to *Delia floralis* and *D. radicum* have also been bred in the UK and of melon resistant to *Aphis gossypii* Glover in France (Van Emden, 1987). The use of plant resistance in insect pest management is likely to concentrate on varieties not specifically bred for resistance, but identified as showing some resistance in comparison with other varieties. The combination of partial resistance with other components, such as chemical and biological methods of control, can play a unique role in insect pest management systems.

1.3.5 Cultural Control

Cultural control is the deliberate modification of a crop production system to make the crop habitat less favourable to a pest and more favourable to its natural enemies and/or the crop. Cultural control uses a wide range of techniques which are normally within the competence of individual growers to perform without the need for high technological input. Cultural control methods should be based on biological and ecological principles, and be economical. These techniques may be classified under the following headings: cultivation, timing of sowing and harvesting, crop rotation, sanitation, destruction of a pest's host plants, trap crops, water management and various other managerial practices (Coaker, 1987).

Cultivation is a traditional method of weed control and tilling the soil can also kill arthropods by mechanical injury, desiccation or exposure to predation. Direct drilling or minimum tillage can also have an inhibitory effect on some insects, but may favour survival of other pests (e.g. slugs). Cultivation does not discriminate between beneficial and pest species, so the timing of operations is critical to obtain the maximum effect on pest suppression.

Destruction of weed hosts, crop residues, scrub or shelter in which insects may harbour and from which they may move into crops involves work targeted solely at the control of insect pests, unlike other cultural control methods, which are targeted more directly at crop plants. In addition to harbouring pest species, wild plants in agroecosystems may also benefit natural enemies of pests, although the advantages of food and shelter offered to the natural enemies may to some extent be offset by similar advantages to the pest species.

Variation in sowing date can reduce or eliminate pest damage by growing crops when pests are inactive. Increasing seed rates may compensate for expected plant losses and may also help to reduce infestation by migrant pests—for example, aphids, which find dense plantings less attractive. Harvesting as soon as crops are mature can reduce damage. It can also remove pests before they emerge and perpetuate their population in the local area.

Crop rotation which attempts to separate pests and host plants in time and space is one of the oldest and most widespread farm practices. It is still one of the most effective controls for some nematode problems. Rotation is not likely to be effective against pest organisms that can survive for long periods in the soil without access to host plants or against mobile pests. Crop rotation normally reduces or delays pest attacks rather than completely preventing them, because, although important in single fields, it is a much less effective practice over larger cropping areas, where particular host crops are almost always likely to be growing somewhere.

Attempts to avoid pests by isolating crops from regularly infested sites have been frequently tried, particularly to prevent spread of insect-borne diseases, but because wild plants can be reservoirs of both vectors and the diseases they carry, this method has rarely proved successful on a regional scale.

Trap crops can be used to concentrate insect pests into small areas where they can then be killed by insecticides or by destroying the trap crops. Trap crops can be either an earlier planting of the crop to be protected or a 'preferred' host plant. Clearly, the destruction of the trap crop and its insect pests must be properly timed to avoid pest build-up.

Where crops are irrigated, manipulation of watering regimes can influence pests both favourably and unfavourably. Irrigation after dry conditions may cause pest incidence to rise dramatically, whereas excessive irrigation may wash pests off plants or drown them, and soil insects may be killed by colloidal particle pressure in saturated soil.

Intercropping or polyculture is one type of cultural system common in the tropics which may also provide potential for improved horticultural crop productivity in temperate climates. One of the advantages of intercropping compared with sole cropping is that insect pest attack is often less severe. There appear to be two reasons for this: first, polycultures provide better conditions for natural enemies by providing increased pollen and nectar sources, increased cover and alternative prey; and second, plant species growing in association with each other have a direct effect on the ability of insect herbivores to find and utilise their host plants. Intercropping also influences pest movement and reproductive behaviour. Compared with sole cropping, intercropping brassicas with various taxonomically unrelated plants reduced the infestation of *D. radicum*, *B. brassicae* and several lepidopterous pests. The incidence of carrot fly (*P. rosae*) was also reduced by intercropping with onions (Coaker, 1987).

Most cultural control measures have been developed empirically and require careful assessment of their economic benefits. For some measures, such as varying planting and harvesting times, the costs would essentially remain the same, since they would simply be used at different times. Cultural control methods are not, however, usually fully effective and are perceived in some circles as having the disadvantage of being preventative rather than curative. They should nevertheless be seen as a first-ditch defence against pest attacks in association with other techniques of pest control.

REFERENCES

Boller, E. F. (1987). Genetic control. In *Integrated Pest Management* (ed. A. J. Burn, T. H. Coaker and P. C. Jepson). Academic Press, London, pp. 161–187

Bulla, L. A., Rhodes, R. A. and St Julian, G. (1975). Bacteria as insect pathogens. *Ann. Rev. Microbiol.*, **29**, 163–190

Burges, H. D. and Hussey, N. W. (1971). Past achievements and future prospects. In *Microbial Control of Insects and Mites* (ed. H. D. Burges and N. W. Hussey). Academic Press, London, pp. 455–458

Cammell, M. E. and Way, M. J. (1987). Forecasting and monitoring. In *Integrated Pest Management* (ed. A. J. Burn, T. H. Coaker and P. C. Jepson). Academic Press, London, pp. 1–26

Coaker, T. H. (1987). Cultural control methods: The crop. In *Integrated Pest Management* (ed. A. J. Burn, T. H. Coaker and P. C. Jepson). Academic Press, London, pp. 69–88

Corbett, J. R. (1978). The future of pesticides and other methods of pest control. In *Applied Biology*, Vol. 3 (ed. T. H. Coaker). Academic Press, London, pp. 230–330

Croft, B.A. (1981). Use of crop protection chemicals for integrated pest control. *Phil. Trans. R. Soc. Lond.*, **B295**, 125–141

Croft, B.A. and Brown, A. W. A. (1975). *Ann. Rev. Entomol.*, **20**, 285–335

Elliott, M. and Janes, N. F. (1978). Synthetic pyrethroids—a new class of insecticide. *Chem. Soc. Rev.*, **7**, 473–505

Van Emden, H. F. (1987). Cultural methods: The plant. In *Integrated Pest Management* (ed. A. J. Burn, T. H. Coaker and P. C. Jepson). Academic Press, London, pp. 27–68

Entwistle, P. F. and Evans, H. F. (1985). Viral control. In *Comprehensive Insect Physiology, Biochemistry and Pharmacology*: Vol. 12, *Insect Control* (ed. G. A. Kerkut and L. I. Gilbert). Pergamon Press, Oxford, pp. 347–402

FAO Production Year Book (1987). *FAO Statistics Series* No. 70. FAO, Rome

Finch, S. (1980). Chemical attraction of plant-feeding insects to plants. In *Applied Biology*, Vol. 5 (ed. T. H. Coaker). Academic Press, London, pp. 67–143

Finch, S. (1987). Horticultural crops. In *Integrated Pest Management* (ed. A. J. Burn, T. H. Coaker and P. C. Jepson). Academic Press, London, pp. 257–293

Georghiou, G. P. (1981). *The Occurrence of Resistance to Pesticides in Arthropods*. FAO, Rome

Gould, H. J. (1987). Protected crops. In *Integrated Pest Management* (ed. A. J. Burn, T. H. Coaker and P. C. Jepson). Academic Press, London, pp. 403–424

Graham-Bryce, I. J. (1987). Chemical methods. In *Integrated Pest Management* (ed. A. J. Burn, T. H. Coaker and P. C. Jepson). Academic Press, London, pp. 113–159

Hall, R. W., Ehler, L. E. and Bisabri-Erhadi, B. (1984). Rate of success in classical biological control. *Entomol. Soc. Am. Bull.*, **25**, 280–282

Hussey, N. W. (1985). History of biological control in protected crops. In *Biological Pest Control: A Glasshouse Experience* (ed. N. W. Hussey and N. Scopes). Blandford Press, London, pp. 13–22

Jones, F. G. W. and Jones, M. G. (1984). *Pests of Field Crops*, 3rd edn. Edward Arnold, London

Loosjes, M. (1976). *Ecology and Genetic Control of the Onion Fly, Delia antiqua (Meigen)*. Centre for Agricultural Publishing and Documentation, Wageningen

Matthews, G. A. (1979). *Pesticide Application Methods*. Longman, London

Matthews, G. A. (1984). *Pest Management*. Longman, London

Peirce, L. C. (1987). *Vegetables: Characteristics, Production and Marketing*. John Wiley, New York

Poinar, G. O. (1971). Use of nematodes for microbial control of insects. In *Microbial Control of Insects and Mites* (ed. H. D. Burges and N. W. Hussey). Academic Press, London, pp. 181–203

Rosen, D. (1985). Biological control. In *Comprehensive Insect Physiology, Biochemistry and Pharmacology*: Vol. 12, *Insect Control* (ed. G. A. Kerkut and L. I. Gilbert). Pergamon Press, Oxford, pp. 413–464

Sawicki, R. M. and Denholm, I. (1984). Adaptations of insects to insecticides. In *Origins of Development and Adaptations*, CIBA Foundation Symposium 102. Pitman, London, pp. 152–166

Simmonds, N. W. (1976). *Evolution of Crop Plants*. Longman, London

Simpson, B. B. and Conner-Ogorzaly, M. (1986). *Economic Botany*. McGraw-Hill, New York

Theunissen, J. and Fransen, J. J. (1984). Biological control of cutworms in lettuce by *Neoaplectana bibionis*. *Med. Fac. Landbouww. Rijksuniv. Gent.*, **49**/3a, 771–776

Wratten, S. D. (1987). Native natural enemies. In *Integrated Pest Management* (ed. A. J. Burn, T. H. Coaker and P. C. Jepson). Academic Press, London, pp. 89–112

CHAPTER 2

Pests of Chenopodiaceous Crops

D. A. Cooke and A. M. Dewar

2.1 INTRODUCTION

Most of the chenopodiaceous vegetable and fodder crops, such as beet-root, fodder beet and mangolds, belong to the same species, *Beta vulgaris* L., and, as such, share many of the pests which attack sugar beet. Sugar beet crops are subject to attack by over 150 species of insects and mites across the world (Lange, 1987) as well as many nematode species and vertebrates such as skylarks and field mice. Fortunately, only a minority of this vast array of pests cause economic damage. The biology and control of sugar beet pests have been reviewed by several authors, including Benada *et al.* (1987), Cooke *et al.* (1989), Edwards and Heath (1964), Hill (1987), Jones and Dunning (1972), Jones and Jones (1984), Lange (1987), Lejealle (1982) and Whitney and Duffus (1986).

Most of the chenopodiaceous crops are grown for their roots and not their tops, the exceptions being spinach, spinach beet and chard. Consequently, damage to leaves is usually only important if it occurs early enough to remove a significant proportion of the young, photosynthesising leaf surface, thus reducing the yield and quality of the root. Once plants have become well established with a large rosette of leaves, they can compensate for most damage caused.

Many otherwise harmless arthropods are classed as pests because they cause significant damage to the roots of beet seedlings when the plants are at their most vulnerable. In many countries (e.g. the UK) where beet is sown to a stand (i.e. where most of the sown seeds are intended to produce plants which survive until harvest), damage can be devastating if not controlled. In contrast, where beet is still sown thickly and thinned by hand (e.g. in Eastern Europe), root pest attack may not be important or even noticed. In those countries, particularly in Western Europe, where the majority of the seed is pelleted, some protection is afforded to young seedlings by incorporating insecticides, such as methiocarb or carbofuran, in the pelleting material surrounding the seed. Naked seed may also be treated with insecticides but is more liable to suffer from phytotoxicity.

28

However, when seed treatments are applied as an insurance, serious damage can still occur if pest numbers are high and persistent.

2.2 APHIDS

2.2.1 *Myzus persicae* Sulzer: Peach Potato Aphid (Hemiptera: Aphididae)

M. persicae rarely reaches population levels which cause direct damage to sugar beet plants, but it is the most important vector of virus yellows disease, which is caused by beet yellows virus (BYV) and/or beet mild yellowing virus (BMYV). These viruses, if introduced to beet crops at an early stage, can cause yield losses of up to 50% and are, therefore, potentially among the most devastating diseases of sugar beet (Smith, 1986). *M. persicae* has developed resistance to many currently approved insecticides, making it an even more difficult pest to control (Dewar *et al.*, 1988).

A fuller description of the life cycle, distribution and control of *M. persicae* is given in Chapter 8.

2.2.2 *Aphis fabae* Scopoli: Black Bean Aphid (Hemiptera: Aphididae)

In contrast to *M. persicae*, *A. fabae* can achieve very dense populations in sugar beet and, sometimes, individual plants can be killed by direct feeding. In Europe, *A. fabae* does not introduce the yellowing viruses to crops, but it can spread viruses that have been introduced by other species. In the USA, however, particularly California, *A. fabae* has been implicated in the spread of viruses from nearby beet fields (Duffus, 1978). Yield losses in root crops caused by direct feeding of this species are negligible because of compensating growth by less affected plants and control is rarely cost-effective. In spinach and chard, leaf damage may cause substantial losses.

A fuller description of the life cycle, distribution and control of *A. fabae* is given in Chapter 6.

2.2.3 Other Aphid Species

Two potato aphids, *Macrosiphum euphorbiae* Thomas and *Aulacorthum solani* Kaltenbach, can occur in sugar beet crops. *M. euphorbiae* is quite common in most years but rarely reaches high numbers and is an inefficient vector of beet viruses. *A. solani* is uncommon in beet, but, when it colonises young seedlings at the two-leaf stage, it can cause curling of the cotyledons and severe distortion of the true leaves.

Fuller descriptions of both species are given in Chapter 8.

2.3 BEETLES

2.3.1 *Chaetocnema concinna* (Marsham): Mangold Flea Beetle (Coleoptera: Chrysomelidae)

The mangold flea beetle has a distinctly different colour pattern from flea beetles of the genus *Phyllotreta* Stephens (see page 98), although it is often confused with them, as both can be found in the same crop feeding upon weeds. Adult *C. concinna* are about 3 mm long, are black in colour with a bronze lustre and have rows of deep, puncture-like holes on the wing covers. They lack the distinct yellow stripe on each wing case of *Phyllotreta* spp. Like other flea beetles, *C. concinna* has hind legs with enlarged femora which allow them to jump when disturbed—hence their name. They feed on members of the Polygonaceae as well as chenopodiaceous hosts (Jones and Dunning, 1972).

Life Cycle

Adult beetles emerge from overwintering sites in grass, hedgerows, etc., during warm weather in March or April. Most damage to crops is caused by adults, following their dispersal from these sites in early spring. Eggs are laid in the soil near host plants and hatch into tiny, white larvae which feed on the seedling roots, apparently without causing injury. Pupation occurs in the soil after about 6 weeks and the next generation of adults emerges from July onwards. This generation does not cause significant damage, because the plants are much larger by this time.

Plant Damage

Characteristic circular pits that do not go right through the leaf are made by adult beetles on the young cotyledons or true leaves. As the leaves grow, these pits develop into holes, giving a lace-like appearance. Rapid growth of plants in warm weather can compensate for this early damage. However, when feeding is excessive, the plant may be defoliated and the growing point destroyed.

Distribution

Flea beetles are widespread across Europe and Asia—indeed wherever sugar beet is grown. In the USA and Canada, genera other than *Chae-*

tocnema, *Epitrix* Foudras and *Psyllioides* Berthold, cause damage.

Control

When damage is first noticed, sprays of persistent, contact insecticides such as gamma-HCH or some of the new pyrethroids (e.g. deltamethrin or cypermethrin) will limit further damage. Some control may also be given by granular carbamate insecticides (e.g. carbofuran) when they are applied to control other pests (Brock, 1983). However, these materials are usually too expensive to be used only for the control of flea beetles. Some control has been given by carbamate seed treatments such as carbosulfan and furathiocarb (Winder and Dewar, 1985).

2.3.2 *Atomaria linearis* Stephens: Pygmy Mangold Beetle (Coleoptera: Cryptophagidae)

Adult pygmy mangold beetles (or, simply, pygmy beetles) are about 2 mm long, elongate and dark-brown in colour. The adults' antennae have eleven segments, with the terminal three enlarged to form a club shape (Figure 2.1) (Bonnemaison and Lyon, 1968).

Figure 2.1 The pygmy mangold beetle, *Atomaria linearis*. © Broom's Barn Experimental Station

Life Cycle

Small, white, oval eggs are laid in the soil near beet plants from April onwards. They hatch soon afterwards and the larvae feed on the roots of sugar beet, reaching maturity in 3–6 weeks and then pupating in the soil. Second-generation adults emerge from these pupae over a long period and attack suitable hosts if they are available. These adults pass the summer and autumn in beet fields, sometimes in large numbers, and overwinter deep in the soil. In spring the following year, they re-emerge and migrate as soon as the weather becomes warm enough to fly (Cochrane and Thornhill, 1987).

Plant Damage

The adults which emerge in early spring cause the most serious damage to young beet seedlings. They do not migrate very far, so only affect crops which are grown continuously in the same field or in adjacent fields. The beetles chew characteristic pits in the young hypocotyls and roots (Figure 2.2), or small circular holes in the cotyledons and young heart leaves. The former damage is the more serious, as plants are often killed either outright or as a result of invasion by secondary fungal pathogens. Once plants are past the four-true-leaf stage, they can usually compensate for the damage and survive the attacks. Subsequent feeding by the next generation of larval or adult beetles is unimportant, except perhaps in newly sown seed crops or stecklings (Jones and Dunning, 1972).

Figure 2.2 Root damage caused by *Atomaria linearis*. © Broom's Barn Experimental Station

Distribution

The importance of this pest in the UK diminished in the 1930s when sugar company contracts prohibited the growing of beet in any field which had grown beet or brassica crops during the two previous years. Primarily, this action was to counteract the build-up of beet cyst nematode (see page 45). Pygmy beetle now causes damage only in areas where many beet crops are grown in close proximity or in close rotations—for example, in the Fens of eastern England. Elsewhere in Europe it is widespread and is regarded as a serious soil pest in Denmark, the Netherlands, Belgium, Ireland, Sweden and Eastern Europe (Dunning, 1972).

Control

Apart from widening the rotation and keeping beet fields far apart, control of pygmy mangold beetles can be achieved either by spraying broad-spectrum, persistent insecticides, such as gamma-HCH, on to the soil surface and incorporating them prior to drilling (Jones and Dunning, 1972), or by applying insecticidal granules (mainly carbamates such as aldicarb, carbofuran, carbosulfan, bendiocarb and benfuracarb) in the seed furrow at drilling (Thornhill, 1983; Winder, 1984). Partial control of a late attack may be achieved by spraying young plants with the same broad-spectrum sprays used prior to drilling. Insurance use of insecticides is not recommended except where damage is recurrent. Use of broad-spectrum materials can diminish the populations of beneficial soil-inhabiting insects and lead to outbreaks of other pests, such as aphids, at a later date. Cultural control methods include delayed sowing to avoid the main migration period, but this practice may incur a yield penalty (Draycott *et al.*, 1973).

2.3.3 *Agriotes lineatus* L., *Agriotes obscurus* L., *Agriotes sputator* L.: Wireworms (Coleoptera: Elateridae)

Wireworms are the larvae of click beetles, so called because of the sound they make when jumping. The adult beetles are not pests.

Description

Members of the genus *Agriotes* Eschscholtz, particularly *A. lineatus*, *A. obscurus* and *A. sputator*, are common pest species in many crops. The adults are elongate, black or brown to yellow in colour and about 6–12 mm long. The larvae are thin, yellow, smooth-bodied and about 25 mm long when fully grown. They have large, powerful, biting mouthparts and three pairs of short legs just behind the head.

Life Cycle

In May and June adult click beetles mate and the females lay eggs singly or in small clusters just below the soil surface of grassy or weedy ground. After about a month the eggs hatch into small, white larvae 1–1.5 mm long. These larvae grow slowly, spending some 3–5 years in the soil before maturing. During this time they feed on living and dead plant tissues and can damage crops severely. In July or August mature larvae burrow deeper in the soil (10–25 mm) and hollow out small cells within which they pupate. Adults may emerge within a month, but they usually overwinter underground unless disturbed, in which case they move upwards and overwinter elsewhere (Coghill, 1978).

Plant Damage

A wide variety of crops may be attacked by wireworms, including cereals, potatoes and many root vegetable crops. The severity of any damage depends on the vigour, size and density of the plants at the time of attack.

The greatest risk of attack is during the first two years after cultivation of grassland. Although populations of wireworms decrease quickly under cultivation, they may persist for three or four years at high densities, especially if the land is weedy.

Wireworms feed most actively in the spring, coinciding with the seedling stage of many crops. Feeding occurs also, but less actively, during September. Larvae migrate vertically in the soil throughout the year, being most numerous in the surface layers from March to May and again from July to September. At other times of the year they migrate to depths of up to 60 cm (Edwards and Heath, 1964).

Young beet plants following grass in the rotation are particularly susceptible to wireworm damage. They can be killed by small, subterranean wounds in the stems which cause wilting and death. Large, fully grown roots can be damaged by larvae feeding near the soil surface in the autumn (Coghill, 1978).

In potatoes, wireworms tunnel deeply into tubers, leaving fine round holes which, while not reducing the yield of the table crop, certainly affect its quality and marketability. Seed potatoes can also be attacked, but this seldom affects the growth of the plants, and damage to the developing tubers does not necessarily follow (Coghill, 1978).

In cereals, oats and wheat are more susceptible to wireworm attack than are barley and rye. Maize is at risk because of its low planting density. Typical signs of damage are wilting and dying plants with the stems just above the seed chewed and frayed.

Wireworms can be serious pests of many horticultural crops, including brassicas, French beans, lettuces, onions, strawberries and tomatoes—the last of these crops by boring into stem bases.

Some crops, including peas, beans and, particularly, linseed, seem to suffer little damage. However, other leguminous crops such as clover and lucerne, which are grown in a relatively uncultivated environment, favour the multiplication of wireworms.

Distribution

Wireworms are widespread throughout Europe, Asia and North America. The most common species in the USA and Canada belong to the genus *Limonius* Eschscholtz. In the UK they are most common on chalky soils where grass is an integral part of the rotation.

Control

Cultural practices which involve frequent and vigorous disturbance of the soil, such as deep ploughing, help to reduce wireworm populations by creating conditions which expose the eggs and larvae to desiccation and predation. Ploughing of grassland in February or March results in much greater decreases in wireworm numbers than ploughing in mid-summer, because more wireworms will be near the soil surface at the time of cultivation. Some alleviation of damage can be achieved by applying small amounts of nitrogen fertiliser and, as a result, encouraging rapid crop growth (Coghill, 1978).

Chemical control can be achieved in several crops. Sugar beet seedlings can be protected by applying insecticides, such as methiocarb, to seed pellets, or by incorporating persistent, contact chemicals such as gamma-HCH into the soil, or by applying insecticidal granules, such as carbofuran, to the seed furrow. The latter two soil treatments are necessary only where populations of wireworms are known to be very high—about 1.5 million per hectare (Jones and Dunning, 1972). Seedlings of other crops may be treated in similar ways, but gamma-HCH soil treatments should be avoided, as they taint red beets, carrots and developing potatoes for up to 18 months after application (Jones and Dunning, 1972). Potatoes are best protected by incorporating an insecticide into the soil during the season before the potatoes are to be planted, either with the previous crop or in the autumn. Some organophosphorus materials, such as phorate or fonofos, are recommended for wireworm control in the UK and the USA.

Wireworm control is wholly preventative. Once an attack is detected, control is almost impossible, as the pests are difficult, if not impossible, to reach with any chemical without disturbing the crop.

2.4 MOTHS

2.4.1 *Spodoptera exigua* (Hübner): Beet Armyworm
(Lepidoptera: Noctuidae)

The beet armyworm is the common name of the caterpillar stage in the life cycle of the small mottled willow moth. The caterpillars are a light-green colour when newly hatched from eggs, but vary in colour when fully grown from greenish-grey to blackish-brown with white dorsal and less distinctly coloured lateral stripes. Posterior to each caterpillar spiracle is a white-coloured dot with a ventral, broad, yellowish-green band (Benada *et al.*, 1987).

Life Cycle

Adult small mottled willow moths, which are grey to grey-brown in appearance, emerge from pupation cells in the soil at night during spring. Mating takes place within 24 h of emergence and egg-laying begins within 48 h of emergence. Eggs are laid *en masse* on the undersides of host plant leaves. Each egg mass has a fuzzy appearance because it is covered in fine hairs and scales from the body of the egg-laying adult female. About 500 eggs per female are deposited in batches of 50–150 over a 5 day period (Bass *et al.*, 1978). The eggs are ribbed, nearly spheroid, and range in colour from greenish-grey when freshly laid to cream, becoming very dark in colour just before hatching as the enclosed caterpillar head capsules become more prominent. Egg hatch occurs 3–4 days after laying. Newly hatched caterpillars are light green in colour and about 1 mm in length, with relatively large heads. They undergo five or six moults over a 15–25 day period, depending on temperature, and reach up to 32 mm in length before burrowing into the soil and pupating in pupal cells. Pupation lasts six or seven days. One generation can be produced in 21–24 days, and as many as eight generations per year occur regularly in Africa and Israel (Hill, 1987).

Plant Damage

Despite its name, the beet armyworm is a general herbivore, causing damage to a wide variety of crops such as sugar beet, alfalfa, cotton, upland rice, asparagus, tobacco, groundnuts, tomatoes, cabbages, potatoes, strawberries and many flowers. It is a major problem on onions in Asia and Mediterranean countries (Hill, 1987). Beet armyworms are gregarious feeders which systematically skeletonise host plant leaves.

Distribution

The beet armyworm is common in the United States, southern Europe, the Middle East, India, southern Asia, Japan, Australia and throughout Africa (Hill, 1987). In Canada it is an annual migrant from the USA.

Control

The sporadic nature of outbreaks of the beet armyworm make it a difficult pest to control, as infestations are usually serious by the time they become evident. Recommended insecticides include dichlorvos, endrin, trichlorphon and toxaphene. Other insecticides, used for cutworm control (see page 356), will also control beet armyworms (Hill, 1987). Methomyl and methamidophos have given good control in southern USA (Bass *et al.*, 1978). Certain cultural practices, such as spring ploughing and discing, help reduce damage to sugar beet crops.

2.5 FLIES

2.5.1 *Pegomya hyoscyami* (Panzer): Mangold Fly (Diptera: Anthomyiidae)

Adult mangold flies are light grey in colour and about the same size as common house flies. The larvae, known as beet leaf miners, damage sugar beet plants.

Life Cycle

In late April/early May adults emerge from the soil in fields which have previously grown a host crop and migrate to new beet crops. Small, white, cylindrical eggs are laid singly or sometimes in batches of between 2 and 10 on the undersurface of cotyledons and young leaves. Egg-laying continues for a month or more, and most eggs are laid by the end of May. After four or five days the eggs hatch and the larvae burrow into the leaves, causing characteristic mines between the upper and lower epidermis of each leaf. The larvae mature after 10–14 days, drop down to the soil and burrow beneath the surface to pupate. A second generation of flies emerges from the pupae after two or three weeks. These flies lay eggs which give rise to mangold flies mining in beet leaves during July. In some years, if temperatures are warm (>25 °C), a third generation of flies may be produced, but rarely does it cause significant damage. Many third-generation flies pupate and can cause an outbreak of damage early in the following year. However, the second and third generations are often attacked by parasites (Dunning, 1961; Jones and Dunning, 1972).

More eggs are laid per plant in crops with low plant populations. The number of eggs laid per plant where crops are drilled to a stand is therefore greater than in crops that will be hand-thinned, even though the same number of eggs are laid per hectare.

Plant Damage

The larvae cause characteristic leaf blisters (Figure 2.3) which prevent photosynthesis. The blisters eventually wither, giving the leaves a patchy, brown, scorched appearance. Severe attacks can reduce root yields by several tonnes per hectare, but the plants can usually survive considerable defoliation, provided that the terminal buds remain intact (Dunning, 1961).

Distribution

P. hyoscyami is widespread throughout Europe (Dunning, 1972), Asia, North Africa, Canada and the USA.

Control

The mangold fly is best controlled by using systemic organophosphorus insecticides such as dimethoate, after the eggs have hatched: the eggs are relatively tolerant of insecticides. In the UK treatment is recommended if

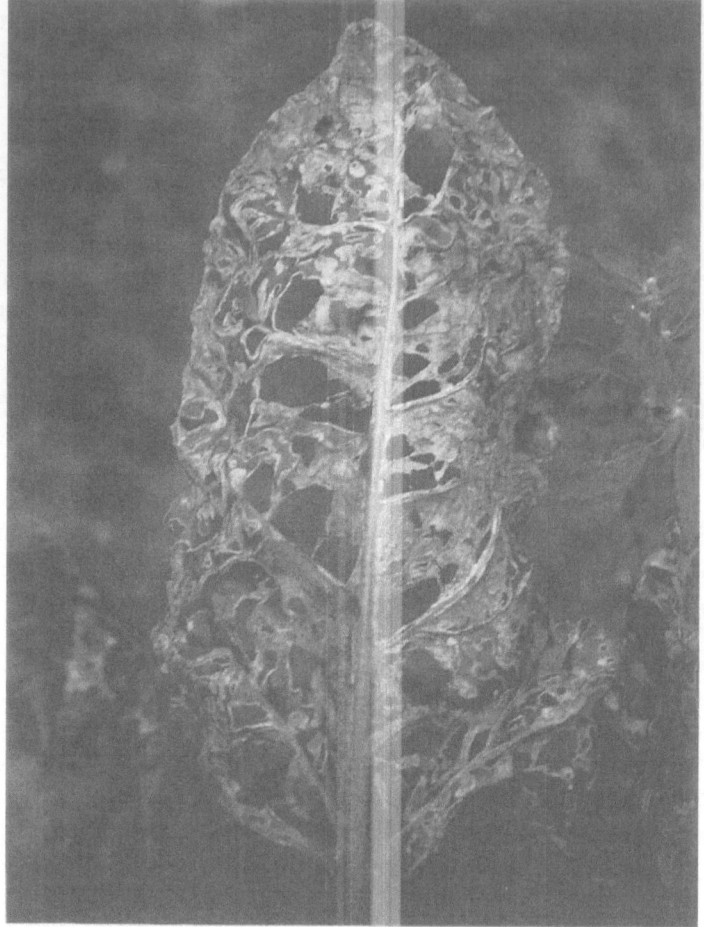

Figure 2.3   Blisters caused by larvae of the mangold fly, *Pegomya hyoscyami*. ©
Broom's Barn Experimental Station

the number of eggs and larvae per plant exceeds the square of the number
of true leaves. Crops which have grown past the eight-leaf stage are usually
not worth treating, because the plants can compensate for the damage. Use
of systemic organophosphorus insecticides to control the mangold fly
should be avoided if insecticide-resistant peach-potato aphids (*Myzus
persicae* Sulzer (see page 263)) are also present, since these insecticides are
now ineffective against resistant *M. persicae* morphs. Carbamate granular
pesticides applied to control other pests (e.g. pygmy mangold beetle) can
control first-generation *P. hyoscyami* larvae (Winder and Dunning, 1986).

2.5.2 *Tetanops myopaeformis* Röder: Sugar Beet Root Maggot (Diptera: Otitidae)

Description

Sugar beet root maggots are the larvae of small, dark-coloured flies. The maggots' bodies are whitish-creamy-coloured, pointed at one end, blunt at the other and about 15 mm in length when fully grown. Each maggot, while lacking head, legs and eyes, has two mouth hooks for feeding at the pointed end of the body.

Life Cycle

Adult flies, individually about the size of a housefly and each having a shiny, black-coloured body with dark-brown-coloured spots at the wing bases, emerge from the soil in previously planted sugar beet fields during April, May or June, depending on latitude. They migrate to nearby beet crops and deposit their eggs in the soil around developing roots. Up to 200 eggs per adult female fly are laid in batches of 6–12. In a few days, the eggs hatch into larvae which feed and develop on young sugar beet roots throughout early summer. The larvae are fully grown during August, when they begin to diapause. Usually, one generation only is produced per year, with the diapausing larvae overwintering in the soil before pupating in the spring (Whitney and Duffus, 1986).

Plant Damage

Sugar beet root maggots feed on the tap roots of sugar beet plants. If plants are attacked when young, maggots can eliminate whole plant stands. Larvae scrape tap root surfaces with their mouth hooks, causing irregular scars which later darken as sap becomes exposed to air. Grazing damage by larvae may also allow the secondary invasion of fungal pathogens. Yield losses can be considerable and have been estimated to be 481 000 t/year in the USA (Theurer *et al.*, 1982).

Distribution

The sugar beet root maggot is the key insect pest of sugar beet throughout the central plains and western, sugar-beet-producing, regions of Canada and the USA (Bechinski *et al.*, 1989). The maggot does not occur in other beet-growing areas of the world.

Control

The application of insecticides, such as terbufos, fensulfothion, diazinon, fonofos, carbofuran, chlorpyrifos, aldicarb and phorate, to sugar beet crops at sowing is standard practice for root maggot control (Yun and Sullivan, 1980; Bergen, 1984). This practice can, however, sometimes be wasteful if adult flies do not subsequently invade crops to lay eggs. The application of foliar insecticides at or near the peak of adult fly activity is more effective than insecticidal application at sowing time. Foliar sprays are, however, difficult to time accurately and their efficacy is affected by soil moisture content (Whitney and Duffus, 1986). To aid correct timing of foliar applications, coloured sticky traps have been used to monitor flight activity of adult flies, but thresholds vary from location to location as well as from year to year (Blickenstaff and Peckenpaugh, 1976). More practicable action guidelines based on the relationship between seasonal trap collections of adult *T. myopaeformis* and subsequent crop yield losses caused by larvae have been developed by Bechinski *et al.* (1989) in Idaho, USA; a cumulative catch of 34 flies per trap from the beginning of the growing season was adopted as the damage threshold, above which detectable root yield losses occurred; a minimum of 3 traps per field was recommended. Insecticides should be applied within 10 days of the damage threshold fly capture. As an alternative to insecticides, longer crop rotations help to reduce sugar beet root maggot populations.

2.6 SPRINGTAILS

2.6.1 *Onychiurus* Gervais spp.: White Blind Springtails (Collembola: Onychiuridae)

Springtails of many species occur in sugar beet crops, but only one or two can be regarded as pests. The root-feeding, entirely subterranean spring-tails *Onychiurus armatus* Tullberg are most serious. They are small, white, elongate insects about 1 mm long at the largest (Figure 2.4).

Figure 2.4 The springtail, *Onychiurus armatus*. © Broom's Barn Experimental Station

Life Cycle

Adults migrate from deep in the soil in the spring and, after mating, lay eggs in the top soil layers (upper 5 cm) (Brown, 1982). The eggs hatch into juveniles, which, in this primitive group of insects, resemble the parents, except that they are smaller. Both adults and juveniles feed on decaying organic matter, fungi and, when available, roots of young seedlings, including beet. One or two generations occur per year, depending on soil temperature (Brown, 1984). Onychiurids tend to migrate up and down the top 15 cm of the soil profile during the spring and summer in response to changes in the moisture content of the surface layers. They desiccate very quickly in dry conditions. In winter they migrate deeper in the soil to avoid frost.

Plant Damage

Feeding by onychiurids results in minute, rounded pits on developing hypocotyls or young roots. Feeding damage commonly occurs before emergence of the seedlings, when they are most vulnerable. Such damage can be fatal to plants during cold conditions when plant growth is slow or if feeding pits are numerous. The pits are also sites for invasion by pathogenic fungi, which may be the ultimate cause of plant death. Once plants have grown past the two-true-leaf stage, they can tolerate grazing by springtails (Jones and Dunning, 1972).

Distribution

Although *Onychiurus* spp. have a cosmopolitan distribution, damage is confined mainly to northern Europe. Springs in some years in northern Europe are cold, which slows development of beet seedlings and thereby prolongs the period when they are vulnerable to springtail attack. In the UK *Onychiurus* spp. occurs in organic or silty soils associated with the Fens, river estuaries such as Humber, Trent and Ouse, and the Yorkshire Wolds. Elsewhere in Europe they are a pest in Denmark, Belgium, Ireland, Germany, the Netherlands and, particularly, Sweden.

Control

Onychiurids have only recently been recognised as a pest: the recent adoption of wider seed spacings and extensive herbicide use has led to large concentrations of springtails around relatively isolated beet seedlings. Where hand-thinning and hand-weeding are still used to establish the crop, these organisms are not troublesome. When control is necessary, the same pesticides that are used to control pygmy beetles (i.e. gamma-HCH, various carbamates) will also control this pest. Some control can also be achieved by including insecticides such as tefluthrin or furathiocarb in the seed pellet (Winder and Dewar, 1988) or by incorporating organic matter such as farmyard manure into the soil to provide alternative food (Thornhill, 1983).

Other Springtails

Tullbergia Lubbock spp. and *Folsomia fimetaria* L. sometimes occur in association with onychiurids, but are not thought to damage sugar beet plants. *Sminthurus viridis* L. and *Bourletiella hortensis* Fitch are springtails which live at the soil surface. They are green or purple in colour and have globular bodies similar in size to the bodies of the onychiurids. They perforate plant foliage with their mouthparts to feed internally, but the damage caused is of no economic importance and control is unnecessary. Green springtails can be confused by growers with green aphids but are readily distinguished because, unlike aphids, springtails jump when disturbed (Lejealle, 1982).

2.7 SYMPHYLIDS

2.7.1 *Scutigerella immaculata* (Newport): Glasshouse Symphylid (Symphyla)

Symphylids, typified by the glasshouse symphylid, *S. immaculata*, are small, white, active arthropods about 5–10 mm long, with 12 pairs of legs (as adults) and long mobile antennae (Figure 2.5).

Figure 2.5 Adult glasshouse symphylid, *Scutigerella immaculata*. © Broom's Barn Experimental Station

Life Cycle

Symphylids lay their eggs in batches of 2–20 in the upper layers of soil throughout the year. The eggs are tended by the adults until they hatch, after about 3 weeks. Newly hatched symphylids have 6 legs only, but gain 2

with each moult until they have 24. Moults occur, depending on temperature, at intervals of 2–6 weeks. Young symphylids can easily be confused
with springtails or young millepedes, with which they frequently occur.
When symphylids have developed 12 pairs of legs, three further moults
occur (with no increase in the number of legs) before they become sexually
mature and begin to lay eggs. Further moults may take place throughout
adult life. Peak breeding time is in spring and early summer. Development
from egg to adult takes at least 3 months and some individuals can live for a
considerable time, up to 4 years in the laboratory (Binns *et al.*, 1981).

Symphylids migrate through the soil in cracks and fissures and are
associated particularly with heavier soils with a well-defined structure.
They live up to 2 m deep in the soil when the surface is cold, but migrate to
the upper layers in spring and early summer, just when young sugar beet
seedlings are becoming established. They will migrate back down the soil
profile if the surface becomes hot and dry.

Plant Damage

Symphylids can cause damage to a wide range of outdoor crops, including
asparagus, beans, brassicas, celery, cucumbers, parsley, peas, peppers,
potatoes, strawberries, sugar beet, anemones, roses and sweet-peas. In
glasshouses they are a pest of tomatoes, lettuces and chrysanthemums.

Symphylids cause plant damage by removing the hairs from young roots,
which results in wilting and stunting. This symptom can often be confused
with other disorders caused by waterlogging, acidity or excess salts. In
sugar beet there are often feeding pits which allow invasion by pathogenic
fungi. In tomatoes the leaves turn a bluish colour and plants become very
stunted.

Distribution

S. immaculata is widespread throughout Europe but is particularly noxious
to sugar beet grown in temperate areas, especially where the soil surface
layers are warm and moist.

Control

Pre-drilling treatment of soil with broad-spectrum sprays or granular
insecticides, as for pygmy beetles, can provide adequate protection in
sugar beet fields. Insecticidal seed treatments do not usually persist until
the time symphylids have migrated up to the soil surface.

Treatment of tomato crops must rely on a soil-drench of insecticides such
as diazinon or gamma-HCH, but timing of application is important to have
maximum effect. Trap plants can be used to determine the optimum time
for treatment (Binns *et al.*, 1981).

2.8 MILLEPEDES

2.8.1 *Blaniulus guttulatus* (Bosc): Spotted Snake Millepede (Diplopoda)
Brachydesmus superus Latzel: Flat Millepede (Diplopoda)

Two groups of millepedes commonly occur in sugar beet fields—snake millepedes, typified by the spotted snake millepede (*B. guttulatus*), and flat millepedes, an example of which is *B. superus*. The snake millepedes are more serious pests. Millepedes are root-feeding pests.

Description and Life Cycle

The snake millepede, *B. guttulatus*, has up to 60 segments and can measure 20 mm long as a full-grown adult. It is yellowish-white in colour, with a pair of bright orange-red spots on all segments except the first three or four and the last (Figure 2.6). It breeds in spring, summer and autumn, laying eggs in clusters in a cell in the soil. Immediately after hatching, young millepedes have only a few segments and three pairs of legs and at this stage they can be confused with springtails. As the millepedes grow, they gain more segments, the full number being reached after two or three years. Millepedes have up to 15 instar stages (Jones and Dunning, 1972).

Flat millepedes (e.g. *B. superus* and *Polydesmus angustus* Latzel) have fewer segments (up to 19 or 20), and are about 1.0–1.5 mm in breadth and 10–20 mm long. Body colour varies according to diet from pale purplish-white to dull red or brown. Eggs are laid in clusters of up to 300 in dome-shaped nests under stones. After hatching, new-born flat millepedes take 12 months to mature, moulting seven times in the process, usually in a chamber similar to the nest (Stephenson, 1960).

In some countries, notably in Eastern Europe, black millepedes (e.g. *Chromatoiulus unilineatus* C. L. Koch) can cause damage to sugar beet crops, but elsewhere they are rare.

Figure 2.6 The spotted snake millepede, *Blaniulus guttulatus*. © Broom's Barn Experimental Station

Plant Damage

As with springtails, millepedes are most injurious to sugar beet plants when there are no alternative sources of food, and in cold, slow-growing conditions. They feed, sometimes in very large numbers (up to 25 million per hectare), below soil level on the roots and stems of seedlings. Some feeding damage is superficial, but often large pits are gouged from young plant tissue, causing death directly or, as with pygmy beetles, by allowing the ingress of secondary fungal infections which can also lead to stunting of older plants. Because they are attracted to damaged tissue, millepedes are often associated with, and blamed for, damage caused by other organisms, such as wireworms or pygmy beetles.

Distribution

Millepedes occur in many soil types but are most numerous, and therefore most injurious, in peat, or silty soils, and are less common on sandy soils. In the UK they are prevalent in the Fens, the Yorkshire Wolds and the Trent and Humber estuaries. They are important in Northern Europe, particularly in Belgium, France, Ireland and Sweden.

Control

Insecticidal seed treatments are usually inadequate to control large numbers of millepedes, and pesticidal sprays or granules, as used for pygmy beetles, may be necessary in badly affected areas. Provision of alternative food, in the form of decaying organic matter such as farmyard manure, can alleviate the damage. It is virtually impossible to control damage by millepedes after seedlings have emerged, because pesticides cannot be incorporated into the soil at this stage and will not come into contact with the pest.

2.9 NEMATODES

2.9.1 *Heterodera schachtii* Schmidt: Beet Cyst Nematode (Tylenchida: Heteroderidae)

H. schachtii is the most important nematode pest of sugar beet, and occurs throughout the world's major beet-growing areas. Its host range includes several other crop species, most within the Chenopodiaceae and Cruciferae but some in other plant families. Following its recognition in the nineteenth century as a potentially devastating pest, it has been the subject of extensive and continuous investigation, recently reviewed by Cooke (1987).

Distribution

Infestations of *H. schachtii* have been recorded in all of the main continental land masses. Figure 2.7 illustrates the beet-growing areas in

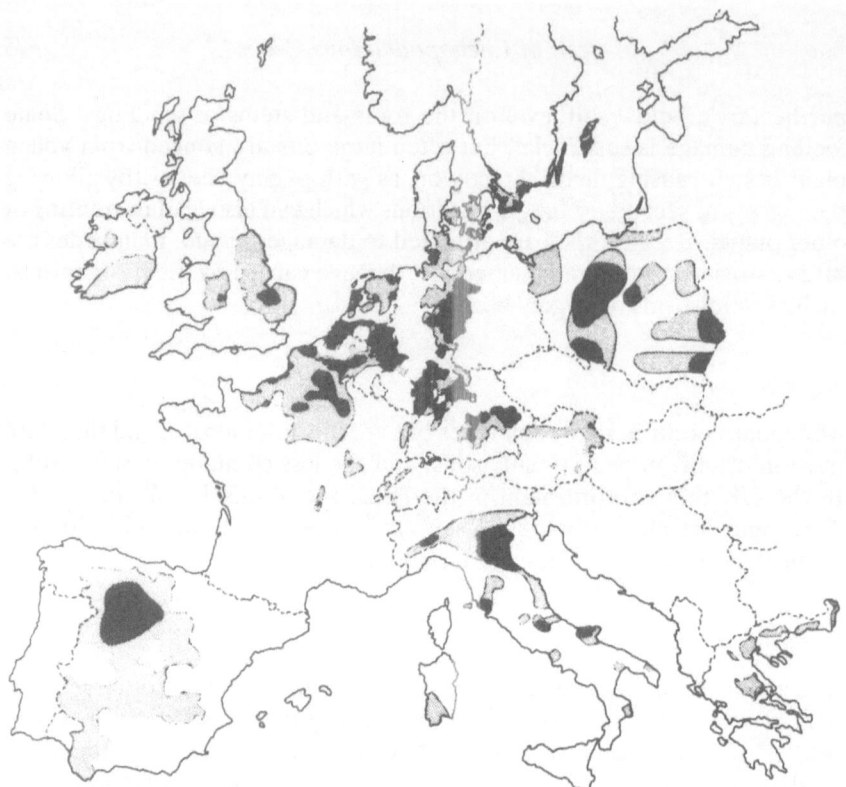

Figure 2.7 Distribution of *Heterodera schachtii* in western Europe: stippled, sugar-beet-growing areas; solid, areas where detectable populations of *H. schachtii* commonly occur

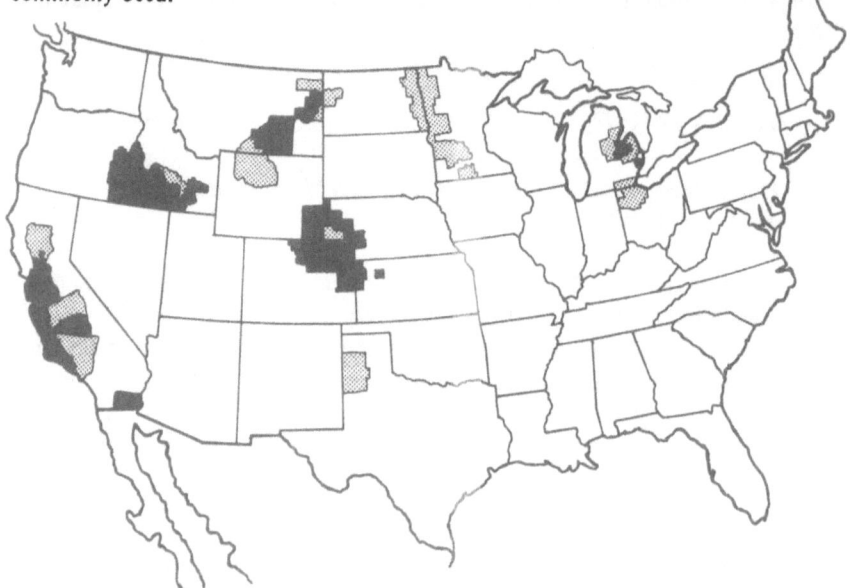

Figure 2.8 Distribution of *Heterodera schachtii* in the USA: stippled, sugar-beet-growing areas; solid, areas where detectable populations of *H. schachtii* commonly occur. Redrawn from Miller (1987)

western Europe where detectable populations commonly occur. Figure 2.8 shows the beet-growing counties of the USA in which infestations of *H. schachtii* have recently been recorded (Miller, 1987). Infested brassica and red beet crops have also been found in Florida, New York and Ohio.

Description

The cyst of *H. schachtii*, which comprises up to 600 eggs protected by the cuticle of the dead female, is brown in colour and lemon-shaped, with a prominent neck anteriorly and a vulval cone posteriorly. The second-stage juvenile (which hatches from the egg) is vermiform, with a rounded head and a pointed tail (Figure 2.9, L). The third- and fourth-stage juveniles occur only within plant roots and are of little diagnostic value. The adult male is vermiform, with a well-developed stylet and curved copulatory spicules (Figure 2.9, C, G, H). The adult female is white in colour and lemon-shaped. The female body surface is covered with a white waxy material called the sub-crystalline layer which persists until the young cyst stage (Figure 2.9, A). The vulval cone may be covered with a gelatinous egg sac (Figure 2.9, B) which contains a few eggs, although most are retained in the body.

Life Cycle

The life cycle of *H. schachtii* is shown diagrammatically in Figure 2.10. The cysts contain several eggs, each egg enclosing a second-stage juvenile (the first moult having already taken place). Unhatched juveniles can remain infective for many years. Each year, even in the absence of a host crop, about half of them emerge from the eggs, using their stylets to cut a slit in the eggshell, but die or become uninfective if they are unable to find weed hosts such as shepherd's purse (*Capsella bursa-pastoris* (L.)) or fat hen (*Chenopodium album* (L.)) to sustain them. When host crops are grown, their roots produce exudates which further stimulate hatch. Once in the soil, the juveniles aggregate round host plant roots (having been attracted along a concentration gradient of root exudates) and, using their stylets, they penetrate the epidermal cells. They migrate through the cortical cells to their permanent feeding sites adjacent to the vascular cylinder. Saliva injected into the vascular cells through the stylet induces cell wall break-down and the subsequent formation of 'transfer cells' which act as food sources for the nematodes (Wyss and Zunke, 1986). Development through third- and fourth-stage juveniles to adult males or females continues, with sex being determined by the environment: the male:female ratio is greater where food is in short supply (e.g. in small roots, in resistant cultivars and where crowding occurs: Müller, 1985). Males escape into the soil and are attracted to the exposed bodies of the females which have swelled and split the root cortex. Each female may be mated by several males, each of which

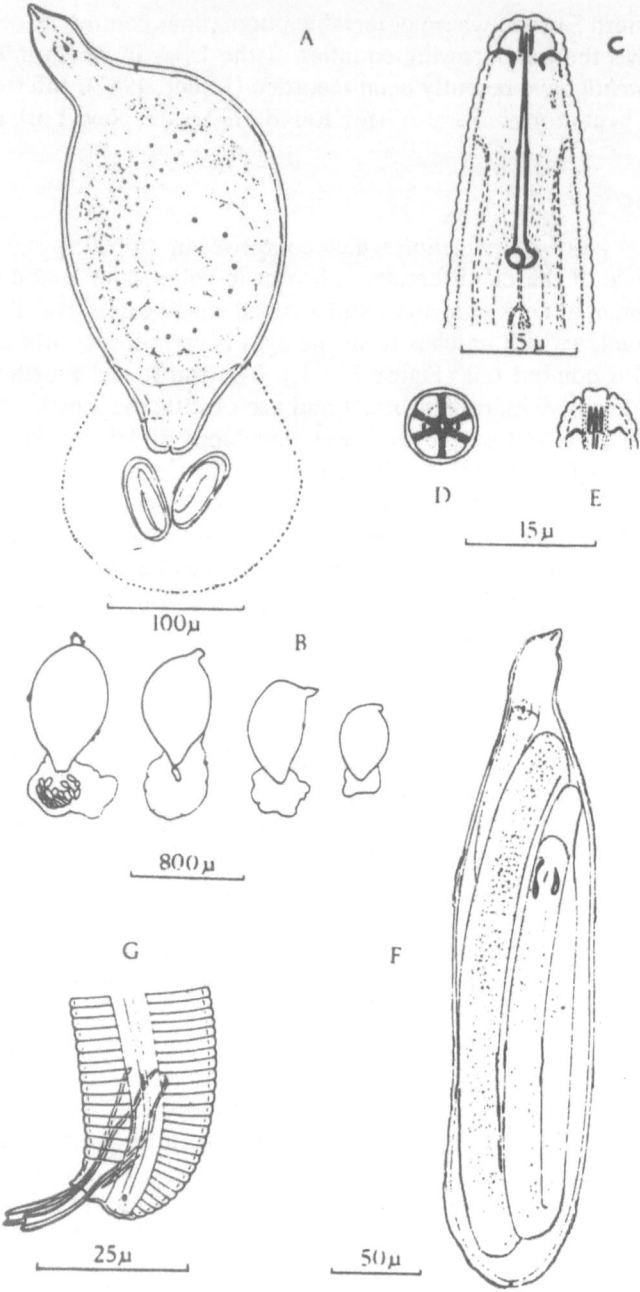

Figure 2.9 Morphology of *Heterodera schachtii*: A, adult female with egg sac; B, cysts with egg sacs; C, head of male, showing anterior and posterior cephalids; D, end-on view of male head; E, dorsoventral view of male head, showing amphid

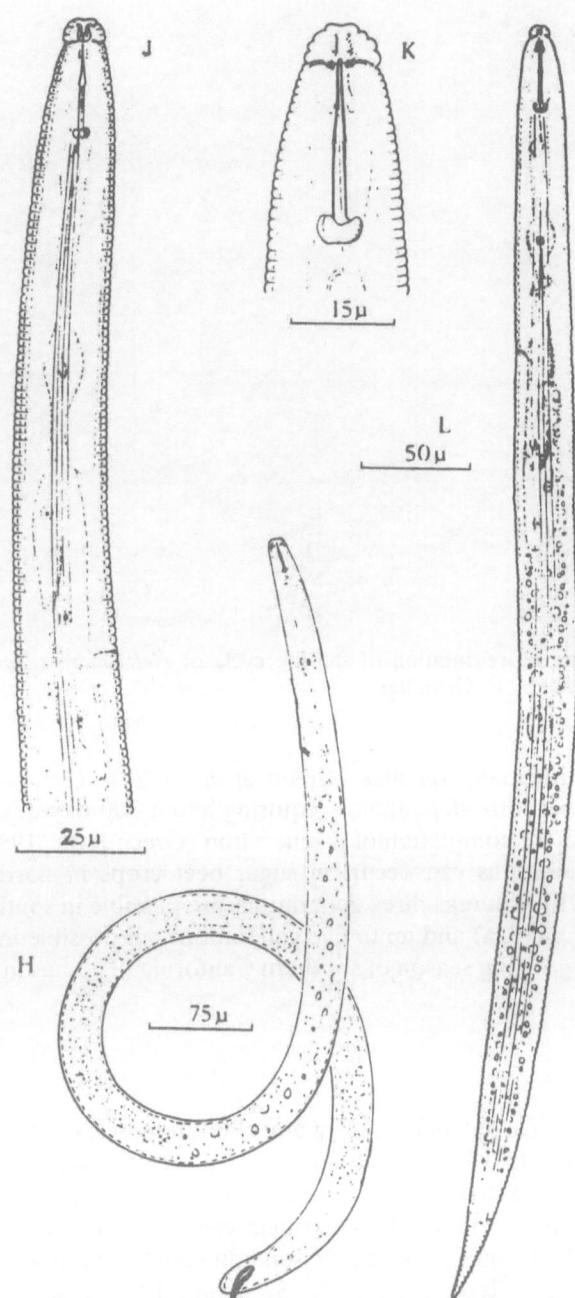

openings; F, fourth-stage male moulting; G, male tail; H, male; J, male oesophageal region; K, head of second-stage juvenile; L, second-stage, infective juvenile. From Franklin (1972)

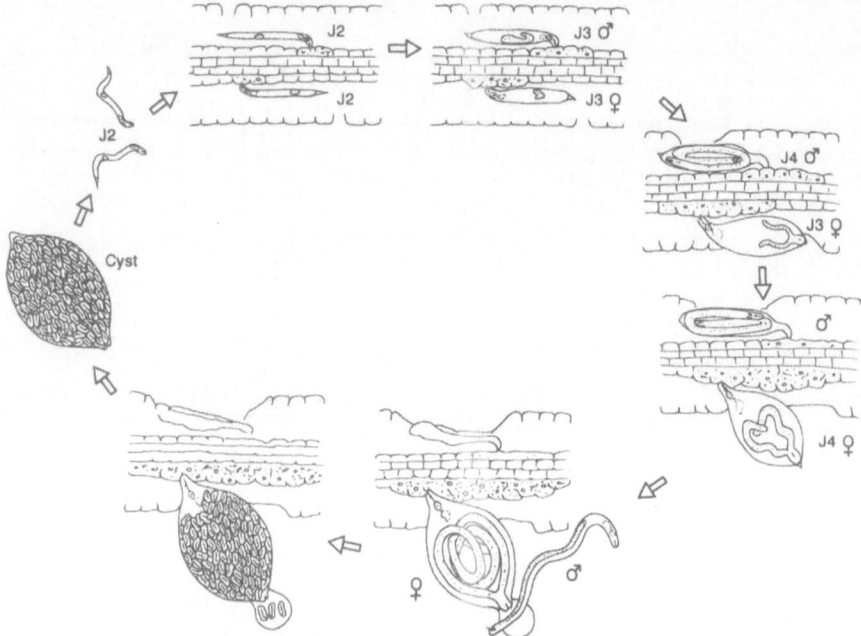

Figure 2.10 Schematic representation of the life cycle of *Heterodera schachtii*.
Redrawn from a diagram by F. Grundler

can in turn mate with many females (Green *et al.*, 1970). The rate of
development is temperature-dependent, requiring about 300 day-degrees
above a base of 10°C for completion of a generation (Greco *et al.*, 1982a).
Therefore, two generations can occur on sugar beet crops in northern
Europe (Müller, 1979), whereas three generations are possible in southern
Europe (Greco *et al.*, 1982a) and up to five generations are possible in the
warm soils and long growing season of southern California (Thomason and
Fife, 1962).

Plant Damage

Heavy infestations of *H. schachtii* result in a proliferation of lateral roots,
an increased tendency to wilt and stunted, poorly yielding plants. The
relationship between nematode populations and yield losses of crops varies
with site and season: for example, the sugar beet yield losses caused by an
initial population of 10 eggs/g of soil in separate field experiments in
California, England and Italy were 64%, 37% and 19%, respectively
(Cooke and Thomason, 1979; Cooke, 1984; Greco *et al.*, 1982b). The
damage caused to the root system of cabbage crops results in smaller, less
compact heads and greatly reduced yields (Abawi and Mai, 1980). The
yields of Brussels sprouts, cauliflowers and other vegetable crops can also
be severely decreased. Lesions produced when roots are invaded by

juvenile nematodes act as entry sites for secondary pathogens, and interactions with several species of soil fungi have been demonstrated.

Control

There are several approaches to managing populations of *H. schachtii* and minimising yield losses. Some cultural practices decrease the rate of nematode spread and reduce yield losses in infested fields. Soil or organic matter containing viable cysts (e.g. from sugar beet factories, from sewage farms, beneath cleaner/loaders or adhering to machinery or animals) should not be allowed to contaminate uninfested fields. Cattle fed on tops and root pieces from plants grown in infested fields should not be moved on to uninfested fields, because cysts can remain viable during passage through the animals' digestive tracts (Kontaxis *et al.*, 1976). Crops grown in infested fields should be sown as early as possible when temperatures are too low for rapid hatch and invasion but high enough for germination and seedling establishment. Field experiments have shown that yield losses are smaller on earlier-sown crops (Weischer and Steudel, 1972).

A wide rotation of host crops is the most effective way to keep down cyst nematode populations. Numbers of viable eggs decrease by about 50% per year under non-host crops. The increase rate under host crops is difficult to predict, because it is density-dependent and related to several environmental factors. A rotation policy based on the population level in soil samples has been developed in the Netherlands (Table 2.1) but a more general recommendation for sugar beet is to restrict cropping with host plant species to once every 3 years in fields where the cyst nematodes cannot be detected and to widen the rotation to 1 in 5 years in heavily infested fields. The most important host crops are within the Chenopodiaceae (e.g. sugar beet, fodder beet, mangolds, red beet) and Cruciferae (e.g. rape, turnip, swede, mustard, broccoli, Brussels sprouts, cabbage, cauliflower, kale). A more complete host list is given by Steele (1965).

Nematode-resistant green manure crops (cultivars of oil radish or white mustard) are grown extensively in parts of Europe. They are typically sown shortly after cereal harvest and ploughed in after at least 2 months' growth. These crops not only maintain soil organic matter content and improve fertility, but also stimulate hatch of, and root invasion by, *H. schachtii* while preventing completion of the life cycle, thereby reducing the numbers of viable eggs (Steudel *et al.*, 1985). Attempts to produce nematode-resistant sugar beet varieties by hybridisation with highly resistant species such as *B. patellaris* Moq. or *B. procumbens* Chr. Sm. have had some success, although a commercial cultivar is still years away (Yu, 1984; Heijbroek *et al.*, 1983; Lange *et al.*, 1990). Attempts are being made to produce virulent isolates of parasitic fungi (e.g. *Verticillium chlamydosporium* Goddard (Ascomycetes: Hypocreales)) which can be added to nematode-infested soil.

Table 2.1 Advisory scheme used in the Netherlands for growing sugar beet in soils with different levels of *H. schachtii* infestation (after Heijbroek, 1973)

Eggs + juveniles/ml soil			
Sand, loam, peat	*Clay*	*Degree of infestation*	*Advice*
0	0	Uninfested	No beet cyst nematodes found
<1	<1.5	Very light	At present no significant damage is expected
1–3	1.5–4	Light	At the most, a slight amount of damage is anticipated
3–6	4–7	Moderate	There is a risk that some damage will occur. Delay growing sugar beet for 1–2 years. No delay is necessary if chemical control is used
6–15	7–20	Quite severe	There is a great risk of significant damage. Delay growing sugar beet for 3–4 years, then take another soil sample. If chemical control is used, a 1 year delay is desirable
15–30	20–40	Severe	There is a great risk of substantial damage. Delay growing sugar beet for 4–5 years, then take another soil sample. If chemical control is used, a 2 year delay is desirable
>30	>40	Very severe	Serious damage is expected. Delay growing sugar beet for 5–7 years, then take another soil sample. If chemical control is used, a 3–4 year delay is desirable

Nematicides are indispensable in some situations, although they should not be used simply as insurance treatments. Soil fumigants such as 1,3-dichloropropene, 1,3-dichloropropene–1,2-dichloropropane mixture and sodium *N*-methyl-dithiocarbamate are expensive and not very effective in heavy or peaty soils. However, they are able to control a wide range of pests and weeds and can be useful as once-in-a-rotation treatments. They must be injected into well-drained soils, in approximately seed-bed condition, at the correct temperature (e.g. above 5 °C for 1,3-dichloropropene). Sealing the soil surface (e.g. by smearing with a powered roller) can improve control by preventing the fumigant from escaping too rapidly from the top few centimetres. Non-fumigant nematicides, including organophosphates (e.g. fenamiphos) and carbamates (aldicarb, oxamyl, carbofuran), are cheaper, easier to apply (usually as granules) and less phytotoxic than fumigants. These materials do not necessarily kill nematodes but adversely affect their movement and behaviour. Application methods should ensure a uniform distribution of active ingredient in the root zone.

2.9.2 *Trichodorus* **Cobb spp.,** *Paratrichodorus* **Siddiqi spp.: Stubby Root Nematodes (Enoplida: Trichodoridae)**

Stubby root nematodes are root ectoparasites with wide host ranges. They are important pests of several crops, causing yield losses either by killing roots during the feeding process or by transmitting viral diseases. Various aspects of their biology, pathogenicity and control have been reviewed by Winfield and Cooke (1975), Thomason and McKenry (1975), Wyss (1975) and Taylor and Brown (1981).

Distribution

Stubby root nematodes are widespread in North America and Europe. They have been recorded from all the main continental land masses and from many islands, especially in the West Indies and Australasia. Several surveys have shown that most species are associated with coarse, sandy soils and occur more infrequently in heavier soils (see, e.g., Cooper, 1971). The main exception to this distribution pattern seems to be *T. primitivus* (de Man) Micoletzky, which is found in a wide range of soils, including clays (Cooper, 1971). Stubby root nematodes are found more frequently in deeper soil layers—often being more numerous below 20 cm depth. This distribution of stubby root nematodes in the soil occurs possibly because of their susceptibility both to mechanical injury (Bor and Kuiper, 1966), which may occur during ploughing and cultivation, and to dry soil conditions (Wyss, 1970; Cooke and Draycott, 1971), which often occur in the surface layers of the free-draining sandy soils in which the nematodes are most prevalent.

Description

There are over 30 species of *Trichodorus* and over 20 of *Paratrichodorus*. Adults vary in size from 0.5 mm to just over 1.5 mm long, and females, although vermiform, often appear rather plump and cigar-shaped (Figure 2.11). They all possess a characteristically shaped, elongate, ventrally curved mouth stylet (onchiostyle). *Paratrichodorus* spp. are distinguished from *Trichodorus* spp. by their more swollen cuticles (especially after microscopic fixation), the presence of caudal alae in males and the absence of lateral pores near the vulvae of females.

Life Cycle

In some species (e.g. *P. christiei* (Allen) Siddiqi, *P. allius* (Jensen) Siddiqi, *P. teres* (Hooper) Siddiqi), males are rare and reproduction is partheno-genetic, whereas in others (*P. pachydermus* (Seinhorst) Siddiqi, *T. primitivus*) males are common and sexual reproduction occurs. Eggs

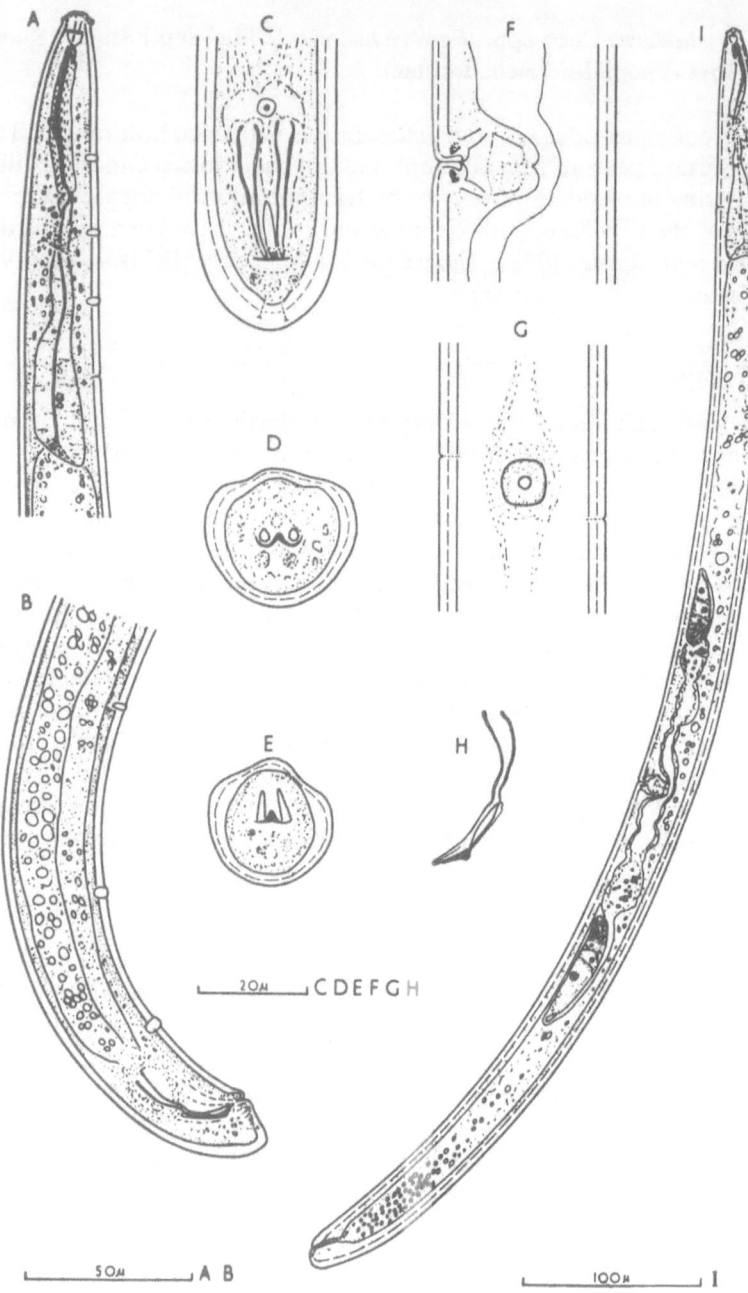

Figure 2.11 Morphology of *Trichodorus viruliferus*: A, anterior end, lateral, male; B, tail end, lateral, male; C, tail, ventral, male; D, optical section through proximal end of male gubernaculum; E, optical section through distal end of male spicules, showing position of gubernaculum keel; F, vulva and vagina, lateral; G, vulva, ventral; H, spicule and gubernaculum, male; I, female, lateral. From Hooper (1976)

are deposited in the soil, usually in the vicinity of host plant roots, and the first-stage juveniles hatch, not as a result of stylet action, but merely because of the pressure exerted on the eggshells by the moving nematodes (Wyss, 1974). There are four juvenile stages, each separated by a moult. Feeding and reproduction take place throughout the growing season of the host plant and there is no resistant stage in the life cycle which is capable of remaining dormant in the soil, so populations consist of a mixture of adults and juveniles. Estimates of the time taken to complete the life cycle vary from 16–17 days for *T. christiei* (Allen) Siddiqi at 30°C (Rohde and Jenkins, 1957a) to 45 days for *T. viruliferus* Hooper at 15–20°C (Pitcher and McNamara, 1970).

Plant Damage

Stubby root nematodes can directly affect plant growth as a result of damage to the roots during feeding. They usually aggregate near the tips of growing roots and feed on the epidermal cells or root hairs. Up to 300 nematodes have been recorded around a single root tip (Pitcher, 1967). The feeding cycle consists of five phases: exploration of the cell wall, perforation of the cell wall by repeated stylet thrusts, salivation, ingestion of partially digested cell contents through a feeding tube probably formed from hardened saliva, and withdrawal (Wyss, 1975). Feeding periods at a single site rarely exceed 6 min, after which the nematode moves to a new cell (or occasionally a different site on the same cell). Feeding damage causes a superficial browning and collapse of epidermal cells. Root growth stops, resulting in the stubby root symptoms first described by Christie and Perry (1951). Stubby root nematodes have wide host ranges, and damage has been reported in several crops, including sugar beet, cabbage, cauliflower, celery, soybean, tobacco, lucerne, maize, wheat, fig, onion, cranberry, sugar cane, tomato and various grasses. The damage to sugar beet (known as Docking disorder in England and t-disease in the Netherlands) has been extensively studied. It is restricted to light sandy soils (which grow about 30000 ha of sugar beet each year in England) and yield losses are most severe following wet springs, when soil conditions favour nematode activity (Cooke, 1973). Stubby root symptoms can be seen on sugar beet seedlings, and severely damaged plants remain stunted, yield poorly and often have fangy, misshapen roots (Figure 2.12).

Apart from direct plant damage, stubby root nematodes are also virus vectors. The two viruses transmitted by these nematodes are the only members of the tobra (*tob*acco *ra*ttle) group, which have straight tubular particles and wide host ranges: tobacco rattle virus (TRV) and pea early browning virus (PEBV). The nematodes acquire the viruses during feeding when the cytoplasmic contents of an infected cell are ingested. Viral particles are retained on the cuticular lining of the pharynx and the oesophagus. These surfaces are shed during moulting and infectivity

Figure 2.12   Fangy sugar beet roots, the result of damage at the seedling stage by stubby root nematodes *Trichodorus* spp. and/or *Paratrichodorus* spp. © Broom's Barn Experimental Station

probably does not persist from one developmental stage of the nematode to the next. However, adults can remain infective for over a year (van Hoof, 1970). Virus transmission probably takes place during the initial stages of feeding, and infection is more likely to occur when the host plant cell is only slightly damaged and cytoplasmic streaming is not prevented (e.g. when feeding is interrupted). Fourteen species of stubby root nematodes have been shown to transmit viruses (Lamberti and Roca, 1987) which can infect hundreds of species of plants. The crop species which can suffer important yield loss or quality impairment as a result of infection with a tobra virus include potatoes (in which TRV is one of the causes of spraing, or corky ring spot), bulbous ornamentals and peas (early browning disease caused by PEBV). Many other crops (e.g. sugar beet) can be affected and show symptoms but suffer little or no yield loss.

Control

Stubby root nematodes have wide host ranges, and although several studies have revealed large differences in host efficiency (see, e.g., Rohde and Jenkins, 1957b), the practical limitations on rotational policy which apply on most farms mean that crop rotation is rarely used as a means of controlling populations. The use of nematicides remains the most reliable method of avoiding serious damage, and the choice of materials and rates will depend upon whether the application is aimed at minimising direct damage to the roots or at preventing virus transmission. Work on sugar beet crops growing in light soils in England which are subject to Docking disorder has confirmed that a seed-furrow application of a granular pesticide (e.g. aldicarb at 0.5–1.0 kg a.i./ha, carbofuran or carbosulfan at 0.6 kg a.i./ha) can produce large yield increases, especially following wet springs (Cooke *et al.*, 1974; Maughan *et al.*, 1984; Cooke, 1989). Higher rates of nematicide application or the use of soil fumigants (see sub-section 2.9.3) may be necessary to prevent virus transmission, because infection can result from the feeding of a single viruliferous nematode. Effective treatment of topsoil may still be insufficient to prevent virus transmission, because infective nematodes can migrate upwards from deeper horizons where control measures are ineffective (Weingartner *et al.*, 1980).

2.9.3 *Longidorus* (Micoletzky) Filipjev spp.: Needle Nematodes (Dorylaimida: Longidoridae)

Needle nematodes are root ectoparasites with wide host ranges. Like stubby root nematodes, they affect crop yield and quality either by damaging the root system directly during feeding or by transmitting virus diseases. Different aspects of their biology and control have been reviewed by Taylor (1971, 1978), Cohn (1975), Wyss (1975, 1981) and Taylor and Brown (1981).

Distribution

Needle nematodes are found throughout the world, although the distribution of individual species is influenced by climate, soil type and cropping regime. For example, *L. africanus* Merny typically inhabits tropical and subtropical regions, whereas *L. elongatus* (de Man) Thorne and Swanger is most common in temperate climates. In Britain *L. macrosoma* Hooper is found mainly in the south and Midlands, usually on heavy soil; *L. attenuatus* Hooper is largely restricted to light, sandy, well-drained soils, especially in East Anglia; *L. goodeyi* Hooper and *L. leptocephalus* Hooper are widely distributed, usually on heavy soils and often under grass; *L. profundorum* Hooper is only common in the south-east, often in orchard

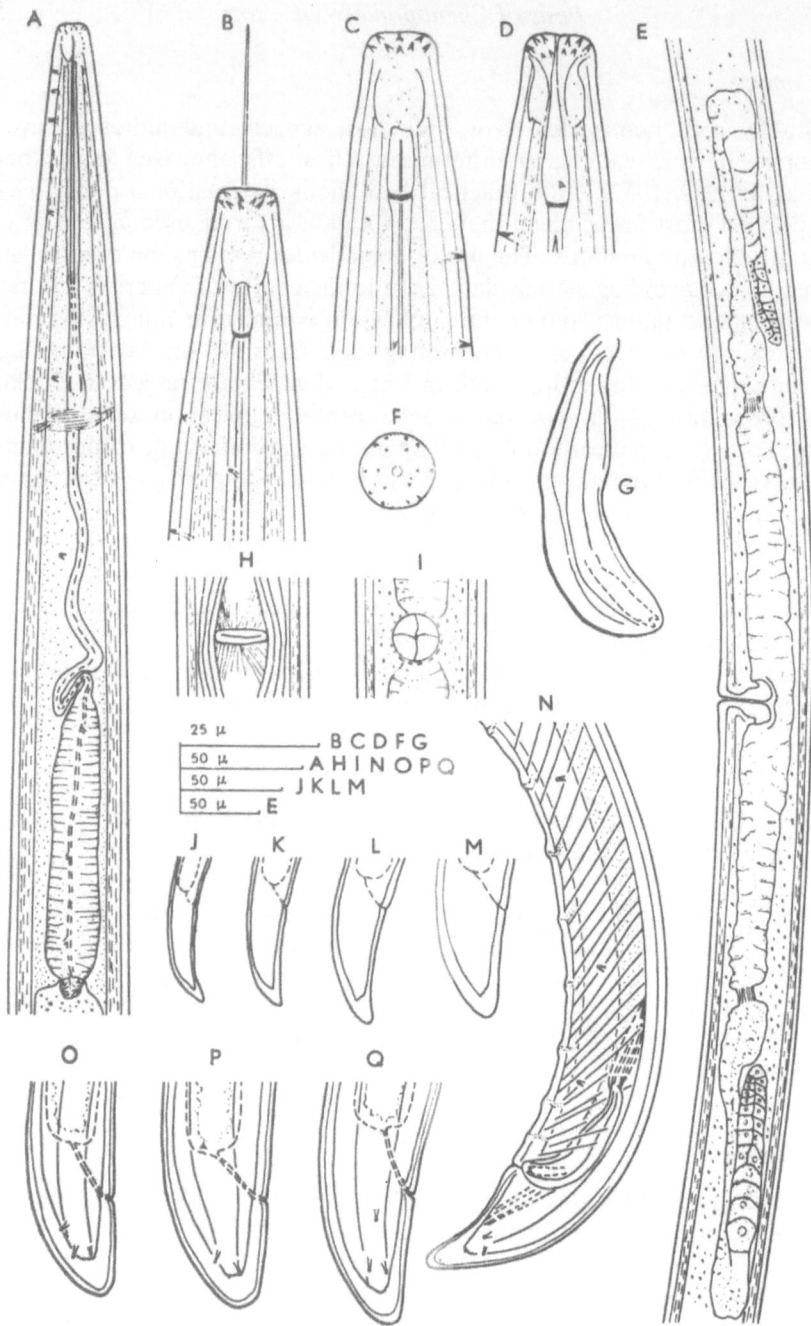

Figure 2.13 Morphology of *Longidorus elongatus*: A, oesophageal region, dorso-lateral; B, head, lateral; C, female head, lateral; D, female head, ventral; E, female gonads, lateral; F, lip region, *en face*; G, male spicules and lateral guiding piece; H, vulva, ventral; I, vagina, ventral; J, juvenile tail, lateral, J1; K, juvenile tail, lateral, J2; L, juvenile tail, lateral, J3; M, juvenile tail, lateral, J4; N, male tail, lateral; O–Q, female tails, lateral. From Hooper (1973)

soil; and *L. elongatus* has been recorded from most areas, although it is particularly prevalent in Scotland and the north-east of England in sandy to loamy soils.

Description

There are now over 60 valid species of needle nematodes. Although they are rather slender, they are among the largest of plant parasitic nematodes, with adults varying in length from 1.5 mm to 13 mm. Most European species are between 5 mm and 10 mm long. Indeed, they can sometimes be seen with the naked eye, attached to host plant roots, where they resemble short lengths of white thread. They are characterised by the presence of a long, hollow mouth spear (odontostyle) with which they puncture host plant roots and feed on cell contents (Figure 2.13). They are rather slow-moving and assume a characteristic C-shape when killed by gentle heat.

Life Cycle

In populations of many species (e.g. *L. elongatus*, *L. attenuatus* and *L. leptocephalus*), males are rare and reproduction seems to be partheno-genetic. Most species have long life cycles and adults may live for several years, particularly in temperate climates. The estimated times for completion of a generation of *L. elongatus* in field conditions were 2 years in Scotland and 1.3 years in the warmer soils of southern England (Griffiths and Trudgill, 1983). Much shorter generation times with, as a result, the completion of several generations in a season, were indicated for *L. africanus* in the Imperial Valley, California, where soil temperatures often exceed 28 °C (Kolodge *et al.*, 1987). Egg-laying cycles usually coincide with periods of active root growth of host plants (Flegg, 1968; Thomas, 1969). During each egg-laying cycle, females of *L. elongatus* may produce up to 20 eggs, but high rates of mortality result in population increases of only 2–4-fold per annum, even on good hosts (Thomas, 1969). Presumably because of their long life cycle, needle nematodes are associated with relatively undisturbed soil conditions and are particularly prevalent in orchards, woodland, grass leys, hedgerows and a variety of perennial crops (e.g. hops, grapes and raspberries).

Plant Damage

Longidorus spp. feed on root tips, forming galls in many herbaceous plants. The long stylet resembles a hypodermic needle which can be inserted deeply into the root tissue and which, in thin roots, can penetrate the stele. Because of this deep stylet penetration, feeding periods are typically much longer than the times taken to feed by most other ectoparasitic nematodes and can last for several hours, or even days. In a detailed investigation of the feeding of *L. caespiticola* Hooper on ryegrass (*Lolium perenne* L.) there was indirect evidence of salivation leading to the characteristic swelling of the root tip (Towle and Doncaster, 1978). The stylet tip remains in a single cell throughout ingestion, although food may also be derived from adjacent cells. Symptoms of direct damage are remarkably similar on a range of plants: root systems are reduced; lateral and tap roots may be stunted; and galls are apparent on root tips, especially younger roots. A necrotic spot is often present at the periphery of galls. Damage to several crops by *Longidorus* spp. has been proved or assumed on the basis of consistent association. Stunting of sugar beet plants (known in England as Docking disorder) on sandy or peaty soils can be caused by *L. attenuatus* or *L. elongatus* (Whitehead and Hooper, 1970) and probably by other species of needle nematodes, as well as by stubby root nematodes (q.v.). There are reports of damage to carrots by *L. elongatus* (Whitehead and Hooper, 1970); to onions by *L. vineacola* Sturhan and Weischer (Cohn and Krikun, 1966; Williams *et al.*, 1981); to celery by *L. vineacola* (Cohn and Ausher, 1971); to lettuce by *L. africanus* (Radewald *et al.*, 1969); and to several cereal and soft fruit crops.

The viruses transmitted by *Longidorus* spp. belong to the nepo group (*ne*matode-transmitted viruses with *po*lyhedral particles), which can also be transmitted by *Xiphinema* Cobb spp. and by *Paralongidorus maximus* (Bütschlii) Siddiqi. Viruses are acquired during feeding and particles are retained as a single layer on the inner surface of the odontostyle or between the odontostyle and the guiding sheath. Transmission occurs when the particles become detached, probably when saliva passes through the odontostyle, and are inoculated into a new plant cell. The sites of virus retention are shed during moulting, so infectivity does not persist from one developmental stage to the next. Most records of virus transmission by *Longidorus* spp. are from Europe, and diseases are caused on soft fruits (raspberry ringspot virus), artichoke (artichoke Italian latent virus), sugar beet (tomato black ring virus) and a range of other crops (Taylor and Brown, 1981).

Control

Most species of *Longidorus* appear to have wide host ranges and can survive for months, if not years, without feeding, so crop rotation is rarely a useful way to manage populations. However, because weeds and grasses are good hosts of many species, a sound weed control policy can help to restrict population development. Where prevention of direct damage to roots is required, relatively low rates of granular pesticides (e.g. aldicarb at 0.5–1.0 kg a.i./ha, carbofuran or carbosulfan at 0.6 kg a.i./ha) can give adequate protection. These materials, applied in the seed furrow, have been shown to increase yields of sugar beet crops at risk from Docking disorder in years with wet springs, and are widely used commercially to minimise damage by needle and stubby root nematodes as well as by a range of other soil pests (Cooke *et al.*, 1974, 1985; Dewar and Cooke, 1986). Soil fumigants such as 1,3-dichloropropene or 1,3-dichloropropene– 1,2-dichloropropane mixture have also decreased populations of needle nematodes and consequently increased yields of sugar beet. Fumigants have been applied either at relatively low rates (16–132 l of product/ha) as row treatments shortly before drilling (Cooke *et al.*, 1974) or at much higher rates (up to 410 l/ha) as overall soil treatments in the autumn prior to drilling. Overall fumigation gives the added advantage of persistent control for 2 or more years after treatment, because of the slow rate of subsequent population increase of needle nematodes (Whitehead *et al.*, 1970; Cooke and Hull, 1972; Cooke and Holden, 1975). The expense and inconvenience of fumigant application have precluded its use on sugar beet, where granules have proved a satisfactory alternative. However, if a high degree of control is required in other crops (e.g. to prevent virus transmission), there may be no alternative to the use of fumigants.

2.9.4 *Meloidogyne* Goeldi spp.: Root-knot Nematodes (Tylenchida: Meloidogynidae)

Root-knot nematodes are obligate, sedentary endoparasites of higher plants. They constitute the most economically important group of plant parasitic nematodes, attacking a wide range of agricultural and horticultural crops, often with devastating effects on yield or quality. They are primarily thought of as pests of the tropics, where they affect not only staple food crops (e.g. rice, yam, potato), but also cash crops (e.g. tobacco, coffee) upon which national economies may depend. Many species which are widespread in the tropics (e.g. *M. incognita* (Kofoid and White) Chitwood, *M. javanica* (Treub) Chitwood, *M. arenaria* (Neal) Chitwood) can also attack vegetables, fruits and ornamental crops grown in Mediterranean and subtropical countries or even, under glass, in temperate climates. Other species (especially *M. hapla* Chitwood, *M. naasi* Franklin and *M. artiellia* Franklin) are widespread and sometimes damaging outdoors in temperate climates. There is a vast amount of literature on this genus, but general accounts are given by Franklin (1978) and in a conference report edited by Lamberti and Taylor (1979). A full review of *Meloidogyne* taxonomy is given by Jepson (1987). In general, the following account is restricted to temperate species.

Distribution

There are over 50 species of *Meloidogyne*. The most economically impor-

tant (*M. incognita* and *M. javanica*) are widespread in tropical regions of
Central and South America, Africa, Asia and Australasia, as well as in
subtropical and Mediterranean climates. In temperate climates, reports of
damage by *M. hapla* (northern root-knot nematode) have come from most
European countries, North America, the USSR, Japan, Australia and New
Zealand. *M. naasi* (cereal root-knot nematode) has been reported most
frequently from north-west Europe and North America, although it is also
known to affect oats in southern Chile. *M. artiellia* is also widely scattered
throughout northern Europe, but other temperate species (e.g. *M. litoralis*
Elmiligy, *M. ovalis* Riffle) have more limited distributions.

Description

Root-knot nematodes (Figure 2.14) are characterised by white, swollen
females, which are very variable in length (300–425 μm) and have a short
'neck', often bent at an angle to the main body axis. The mouth spear is
short, with well-developed basal knobs. The features of the posterior end
of the female (vulva, anus, tail tip, phasmids and cuticular striations) form
the perineal pattern, which is used in species identification. Males are
vermiform and up to 2 mm long. The slender, second-stage juveniles which
emerge from the eggs can occur in the soil, whereas third- and fourth-stage
juveniles occur only within host plant roots.

Life Cycle

The life cycles of root-knot and cyst nematodes are similar. Eggs are
usually present in the soil as egg masses, each mass protected by a
gelatinous matrix. The first moult to second juvenile stage takes place in
the egg. Hatching usually occurs readily in water and does not depend, as is
the case with some cyst nematodes, on stimulation by root exudates
(although exudates may increase the proportion of hatching eggs). There is
some evidence of a partial diapause in at least one species, so that egg
hatch does not always occur, even in favourable conditions (de Guiran,
1974). Second-stage juveniles invade host plant roots, usually just behind
the root tip, where there is intense meristematic activity, and move
through the cortical cells to their feeding sites close to the pericycle.
Secretions from the oesophageal glands injected into host cells induce the
formation of giant cells which usually develop from undifferentiated cells
of the pericycle. These multinucleate feeding cells arise as the result of a
failure of cytokinesis and not (like cyst nematodes) as the result of cell wall
breakdown (Jones, 1981). The second-stage juveniles swell and at the end
of this stage the sexes can usually be differentiated. Three further moults
occur quite rapidly. Third- and fourth-stage juvenile females do not feed,
but after the final moult to adult females feeding recommences, further
growth occurs and egg production starts. Males also do not feed after the

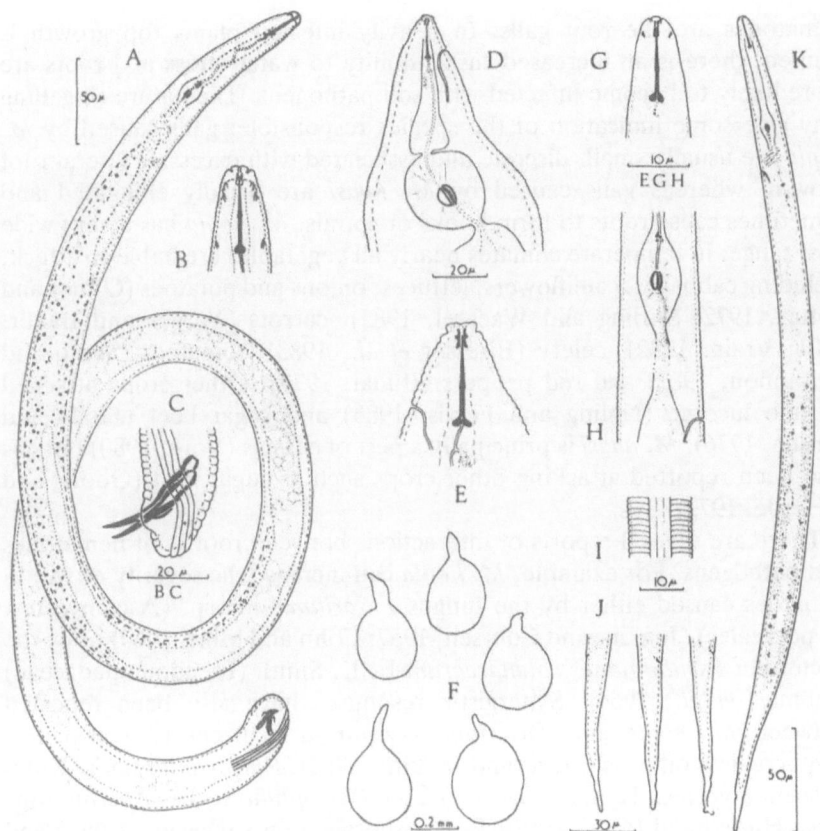

Figure 2.14 Morphology of *Meloidogyne naasi*: A, adult male; B, male head, dorsoventral; C, male tail; D, female head and neck; E, female anterior part of neck, lateral; F, adult females; G, juvenile head end, dorsoventral; H, juvenile median bulb region; I, lateral field of juvenile; J, juvenile tails; K, juvenile, infective second stage. From Franklin (1973)

second juvenile stage, but after development through third and fourth juvenile stages the vermiform males escape into the soil. In some species sexual reproduction occurs but, in most species, males are rare and, even if numerous, they appear to be functionless and reproduction is nearly always parthenogenetic. Females lay several hundred eggs, with a maximum of about 3000 having been recorded. The rate of development is temperature-dependent. Tropical and subtropical species may complete several generations in the course of a year, but in cooler climates fewer generations occur—for example, *M. naasi* completes only one generation per year on cereals.

Plant Damage

In attacked plants the most obvious symptoms of damage by root-knot nematodes are the root galls. In heavily infested plants top growth is stunted, there is an increased susceptibility to water stress and roots are more likely to become infected with soil pathogens. The nature of galling may give some indication of the species responsible: galls caused by *M. hapla* are usually small, discrete and associated with excessive lateral root growth, whereas galls caused by *M. naasi* are usually elongated and sometimes cause roots to form hooks or spirals. *M. hapla* has a very wide host range: in temperate climates nearly all vegetables are liable to attack, including cabbages, cauliflowers, lettuces, onions and potatoes (Olthof and Potter, 1972; Stirling and Wachtel, 1985), carrots (Yarger and Baker, 1981; Vrain, 1982), celery (Bisessar *et al.*, 1983), tomatoes (Santo and O'Bannon, 1982) and red peppers (Budai, 1979). Other crops attacked include lucerne (Noling and Ferris, 1985) and sugar beet (Jatala and Jensen, 1976). *M. naasi* is principally a pest of cereals (York, 1980) but has also been reported attacking other crops such as sugar beet (Gooris and d'Herde, 1972).

There are several reports of interactions between root-knot nematodes and pathogens. For example, *M. hapla* can increase the severity of wilt in tomatoes caused either by the fungus *Fusarium* Link sp. (Ascomycetes: Hypocreales) (Jenkins and Coursen, 1957; Cohn and Minz, 1960) or by the bacterium *Pseudomonas solanacearum* E. K. Smith (Pseudomonadaceae) (Libman *et al.*, 1964). Synergistic responses have also been reported between *M. hapla* and *Fusarium oxysporum* Schlecht (Ascomycetes: Hypocreales) on peas (Davis and Jenkins, 1963), *Rhizoctonia solani* Kühn (Basidiomycetes: Hymenomycetes) and *Phytophthora megasperma* var. *sojae* Hildebrand (Oomycetes: Peronosporales) on soybeans (Taylor and Wyllie, 1959; Wyllie and Taylor, 1960) and *Aspergillus flavus* Link (Ascomycetes: Eurotiales) on peanuts (Minton *et al.*, 1969).

Control

Several control methods have been investigated. As most species of root-knot nematode have wide host ranges, and variations in pathogenicity occur between populations of the same species, devising a suitable crop rotation policy may be difficult. Because cereals are usually poor hosts of *M. hapla*, frequent cereal cultivation has been used to prevent the build-up of this species on arable land in the Netherlands (Hijink and Kuiper, 1964). Conversely, *M. naasi*, found principally on wheat and barley, has been controlled by growing potatoes for 2 years (Allen *et al.*, 1970). The sugar beet crop is also a poor host of *M. naasi* and causes populations to decrease, although it can suffer severe damage (Gooris and d'Herde, 1972). Oats have been used as a trap crop to be invaded by the nematode, which was then unable to complete its life cycle (Gooris and d'Herde, 1972). Resistant cultivars of host crops have been bred: for example, a lucerne variety resistant to *M. hapla* has recently been registered in

California (Lehman *et al.*, 1983) and a genotype of carrot resistant to the same species has also been found (Frese, 1983). Cultural techniques which can decrease the extent of damage in infested fields include removing residues of previous crops, controlling weed hosts and planting crops early when soil temperatures are below the optimum for plant invasion and nematode reproduction (e.g. <18°C for *M. incognita*). The use of nematicides is probably the most reliable method of control in heavily infested fields. A number of carbamate and organophosphate nematicides have controlled *M. hapla* on peanuts (Phipps, 1987). Soil fumigants, especially 1,3-dichloropropene applied by conventional soil-injection techniques, are extensively used to control *M. hapla* and *M. chitwoodi* Golden, O'Bannon, Santo and Finley on potatoes in the north-western United States, although metham sodium (sodium *N*-methyl-dithiocarbamate) applied in irrigation water onto a lucerne crop or wheat stubble during the autumn before planting the potato crop also controlled the nematodes and decreased the proportion of infested tubers (Santo and Qualls, 1984). Oxamyl applied as a foliar spray, a soil drench or granules has been shown to decrease root galling on tomatoes, although timing of application can significantly alter the effectiveness of treatment (Stephan and Trudgill, 1983). Whether or not to use a nematicidal chemical to control root-knot nematodes, and the choice of chemical and application method, will depend on cost, statutory restrictions and the perceived effectiveness of the available chemical treatments. However, their use should be considered only as part of a pest management scheme which incorporates alternative methods of population control.

2.9.5 *Nacobbus aberrans* (Thorne) Thorne and Allen: False Root-knot Nematode (Tylenchida: Pratylenchidae)

The false root-knot nematode is an obligate parasite of higher plants which, as its name suggests, causes root galling similar to the galling caused by root-knot nematodes (see page 61). Recent general accounts of different aspects of its distribution, biology and control are given by Stone and Burrows (1985), Inserra *et al.* (1985) and in an OEPP bulletin (Anon., 1984). Sher (1970) synonymised *N. serendipiticus* Franklin, *N. batatiformis* Thorne and Schuster and *N. serendipiticus bolivianus* Lordello, Zamith and Boock with *N. aberrans*.

Distribution

N. aberrans appears to be native to South, Central and North America, having been recorded from Chile, Bolivia, Ecuador, Argentina, Peru, Mexico and some western states of the USA. Infested glasshouse tomatoes found in England (Franklin, 1959) and the Netherlands (de Bruijn and

Stemerding, 1968) almost certainly resulted from isolated introductions of infected root material.

Description

Mature females of *N. aberrans* have white-cream-coloured, swollen bodies which are 0.7–1.9 mm long, are typically oval in shape, and taper both anteriorly and posteriorly. Immature females are elongated and each has a robust, feeding stylet and a posteriorly situated vulva. Each male has a short, arcuate tail enveloped by a bursa which arises close behind the heads of the copulatory spicules. Second-, third- and fourth-stage juveniles are all vermiform with heavily sclerotised head skeletons and robust stylets, typical of members of the family Pratylenchidae.

Life Cycle

False root-knot nematode second-stage juveniles, which emerge from the eggs, move through the soil before penetrating host plant roots. After invasion, they move intracellularly and may cause swellings of the root tips or along the root axes, large numbers of juveniles often occurring in the same swelling. Juveniles, males and immature females can leave the swellings and move through adjacent root tissues. These stages are also able to leave the roots altogether and reinvade at different sites. Mature females, by contrast, adopt a sedentary, endoparasitic habit and their feeding causes hypertrophy and hyperplasia of host tissues, resulting in the formation of true galls, containing syncytia. Fertilisation of the females by the males probably occurs after the females have become sedentary. Each female exudes a gelatinous matrix into which it discharges the fertilised eggs. The complete life cycle (including embryogenic development) takes about 48 days at 25 °C (Inserra *et al.*, 1983).

Plant Damage

Invasion of, and migration within, host plant roots by juvenile stages of the false root-knot nematode cause extensive damage to the cortical parenchyma and the stelar area, reducing nutrient and water uptakes. Root function is further impaired by the formation of galls, which tend to bear numerous small rootlets. Severely affected plants often appear stunted and yellow in colour, and large yield losses can occur, especially on coarse-textured soils, e.g. *N. aberrans* on sugar beet crops in the states of Nebraska, Colorado, Kansas, Montana and Wyoming in the USA can cause yield losses of over 20% (Altman and Thomason, 1971).

   N. aberrans has a wide host range including members of the Cactaceae, Chenopodiaceae, Cruciferae, Zygophyllaceae, Cucurbitaceae, Solanaceae, Leguminosae and Umbelliferae. Several important crops such as sugar beet, cabbage, Brussels sprout, turnip, lettuce, cucumber, pea, carrot, tomato and potato are hosts (Stone and Burrows, 1985). Physiological races probably exist within the species. In the USA, the sugar beet race only is thought to occur. This race can reproduce on kochia, tomato and sugar beet, but not on pepper and potato (Inserra *et al.*, 1984).

Control

Although *N. aberrans* has a wide host range, there are several, important, non-host crops. Crop rotation can successfully be used, therefore, as a control measure. Rotation with non-host crops (e.g. cereals, lucerne, potatoes and onions) with four- to six-year intervals between host crops effectively controls populations of the sugar beet race in the USA (Altman and Thomason, 1971). Because the nematode typically occurs on light, sandy soils, fumigation is an effective method of reducing populations. Carbamate nematicides also showed promise in trials on potato in Peru, South America (Cornejo Quiroz, 1977).

2.9.6 *Ditylenchus dipsaci* (Kühn) Filipjev: Stem Nematode (Tylenchida: Tylenchidae)

Stem nematodes are migratory endoparasites which feed on parenchymatous tissue in stems and bulbs, causing swelling and distortion of infested areas (Hooper, 1972). They are particularly important pests of monocotyledonous crops (e.g. oats, onions) and are discussed in detail in Chapter 7. They can also attack crops in many dicotyledonous families including the Chenopodiaceae, and are recognised pests of mangolds and sugar beet. Stem nematodes invade seedlings, causing malformation of cotyledons, and can kill the growing points, resulting in multiple-crowned plants with many small leaves. In the autumn the crowns may be invaded, causing a canker which soon contains various secondary pathogens. Chemical control measures are rarely warranted to prevent damage to chenopodiaceous crops, but in Switzerland stem nematode is considered a major pest of sugar beet, requiring prophylactic, nematicidal treatment (Graf and Meyer, 1973).

2.10 ACKNOWLEDGEMENTS

Figures 2.9, 2.11, 2.13 and 2.14 are reproduced by kind permission of Dr R. Muller, Commonwealth Institute of Parasitology.

REFERENCES

Abawi, G. S. and Mai, W. F. (1980). Effect of initial population densities of *Heterodera schachtii* on yield of cabbage and table beets in New York State. *Phytopathology*, **70**, 481–485

Allen, M. W., Hart, W. H. and Baghott, K. (1970). Crop rotation controls barley root-knot nematode at Tulelake. *Calif. Agric.*, **24**, 4–5

Altman, J. and Thomason, I. (1971). Nematodes and their control. In *Advances in Sugarbeet Production* (ed. R. T. Johnson, J. T. Alexander, G. E. Rush and G. R. Hawkes). Iowa State University Press, pp. 335–370

Anon. (1984). *Nacobbus aberrans* (Thorne) Thorne and Allen (*sensu lato*). *Bull. OEPP*, **14**, 61–65

Bass, M. H., Cobb, P. P. and Higgins, D. (1978). *Beet Armyworm. Biology and Control.* Report of the Agricultural Experiment Station, Auburn University, Alabama. Leaflet 94

Bechinski, E. J., McNeal, C. D. and Gallian, J. J. (1989). Development of action thresholds for the sugarbeet root maggot (Diptera: Otitidae). *J. Econ. Ent.*, **82**, 608–615

Benada, J., Sedivy, J. and Spacek, J. (1987). *Atlas of Diseases and Pests in Beet.* Elsevier, Amsterdam, Oxford, New York and Tokyo

Bergen, P. (1984). Guide to chemical weed and insect control in sugarbeets. Alberta Agriculture Agdex 171/605-1, Edmonton, Alberta

Binns, E. S., Brown, R. A. and Dunning, R. A. (1981). *Symphylids.* MAFF Leaflet No. 484. HMSO, London

Bisessar, S., Rinne, R. J. and Potter, J. W. (1983). Effects of heavy metals and *Meloidogyne hapla* on celery grown on organic soil near a nickel refinery. *Pl. Dis.*, **67**, 11–14

Blickenstaff, C. C. and Peckenpaugh, R. E. (1976). Sticky stake traps for monitoring fly populations of the sugarbeet root maggot and predicting maggot populations and damage ratings. *J. Am. Soc. Sugar Beet Technol.*, **19**, 112–117

Bonnemaison, L. and Lyon, J. P. (1968). L'atomaire de la betterave (*Atomaria linearis* Steph.), biologie et méthodes de lutte. *Annls Epiphyt.*, **18**, 401–450

Bor, N. A. and Kuiper, K. (1966). Gevoeligheid van *Trichodorus teres* en *T. pachydermus* voor uitwendige invloeden. *Meded. Fac. Landbouwwet., Rijksuniv. Gent*, **31**, 609–616

Brock, A. M. (1983). *Flea Beetles.* MAFF Leaflet No. 109. HMSO, London

Brown, R. A. (1982). *The Ecology of Soil-inhabiting Pests of Sugar-beet, with Particular Reference to Onychiurus armatus.* PhD Thesis, University of Newcastle upon Tyne

Brown, R. A. (1984). The soil pest complex. *Br. Sugar Beet Rev.*, **52**(1), 31–32

Budai, C. (1979). Spread of and damage caused by the root-knot nematode, *Meloidogyne hapla* Chitwood, in the red pepper growing area of Szeged. *Acta Phytopathol. Acad. Sci. Hung.*, **14**, 543–548

Christie, J. R. and Perry, V. G. (1951). A root disease of plants caused by a nematode of the genus *Trichodorus. Science, N.Y.*, **113**, 491–493

Cochrane, J. and Thornhill, W. A. (1987). Variation in annual and regional damage to sugar beet by pygmy beetle (*Atomaria linearis*). *Ann. Appl. Biol.*, **110**, 231–238

Coghill, K. J. (1978). *Wireworms.* MAFF Leaflet No. 199. HMSO, London

Cohn, E. (1975). Relations between *Xiphinema* and *Longidorus* and their host plants. In *Nematode Vectors of Plant Viruses* (ed. F. Lamberti, C. E. Taylor and J. W. Seinhorst). Plenum Press, London, pp. 365–386

Cohn, E. and Ausher, R. (1971). Seasonal occurrence of *Longidorus vineacola* on celery in Israel and its control. *Israel J. Agric. Res.*, **21**, 23–25

Cohn, E. and Krikun, J. (1966). A disease of onion associated with an ectoparasitic nematode of the genus *Longidorus. Pl. Dis. Reptr.*, **50**, 711–712

Cohn, E. and Minz, G. (1960). Nematodes and resistance to *Fusarium* wilt in tomatoes. *Hassadeh*, **40**, 1347–1349

Cooke, D. A. (1973). The effect of plant parasitic nematodes, rainfall and other factors on Docking disorder of sugar beet. *Pl. Pathol.*, **22**, 161–170

Cooke, D. A. (1984). The relationship between numbers of *Heterodera schachtii* and sugar beet yields on a mineral soil, 1978–81. *Ann. Appl. Biol.*, **104**, 121–129

Cooke, D. A. (1987). Beet cyst nematode (*Heterodera schachtii* Schmidt) and its control on sugar beet. *Agric. Zool. Rev.*, **2**, 135–183

Cooke, D. A. (1989). Damage to sugar-beet crops by ectoparasitic nematodes, and its control by soil-applied granular pesticides. *Crop Prot.*, **8**, 63–70

Cooke, D. A., Bromilow, R. H. and Nicholls, P. H. (1985). The extent and efficacy of granular pesticide usage to control ectoparasitic nematodes on sugar beet. *Crop Prot.*, **4**, 446–457

Cooke, D. A., Dewar, A. M. and Asher, M. J. C. (1989). Pests and diseases of sugar beet. In: *Pest and Disease Control Handbook* (ed. N. Scopes and L. Stables). British Crop Protection Council, Thornton Heath, pp. 241–259

Cooke, D. A. and Draycott, A. P. (1971). The effects of soil fumigation and nitrogen fertilizers on nematodes and sugar beet in sandy soils. *Ann. Appl. Biol.*, **69**, 253–264

Cooke, D. A., Dunning, R. A. and Winder, G. H. (1974). The effect of nematicides, applied to the seed rows in spring, on growth and yield of sugar beet in Docking-disorder-affected fields. *Ann. Appl. Biol.*, **76**, 289–298

Cooke, D. A. and Holden, M. R. (1975). A comparison of two methods of overall fumigation for the control of Docking disorder of sugar beet. *Pl. Pathol.*, **24**, 134–139

Cooke, D. A. and Hull, R. (1972). The effect of soil fumigation with D-D on the yields of sugar beet and other crops. *Ann. Appl. Biol.*, **71**, 59–67

Cooke, D. A. and Thomason, I. J. (1979). The relationship between population density of *Heterodera schachtii*, soil temperature and sugar beet yields. *J. Nematol.*, **11**, 124–128

Cooper, J. I. (1971). The distribution in Scotland of tobacco rattle virus and its nematode vectors in relation to soil type. *Pl. Pathol.*, **20**, 52–58

Cornejo Quiroz, W. (1977). Control quimico de *Nacobbus aberrans* y *Globodera* spp. *Nematropica*, **7**, 6

Davis, R. A. and Jenkins, W. R. (1963). Effects of *Meloidogyne* spp. and *Tylenchorhynchus claytoni* on pea wilt incited by *Fusarium oxysporum* f. *pisi* race 1. *Phytopathology*, **53**, 745

de Bruijn, N. and Stemerding, S. (1968). *Nacobbus serendipiticus*, a plant parasitic nematode new to the Netherlands. *Neth. J. Pl. Path.*, **74**, 227–228

de Guiran, G. (1974). Partial diapause within the egg masses of *Meloidogyne incognita*. In *Proc. Twelfth Int. Nematol. Symp. Granada 1974*, pp. 37–39

Dewar, A. M. and Cooke, D. A. (1986). Recent developments in the control of nematode and soil-arthropod pests of sugar beet. *Asp. Appl. Biol.*, **13**, 89–99

Dewar, A. M., Devonshire, A. and Ffrench-Constant, R. (1988). The rise and rise of the resistant aphid. *Br. Sugar Beet Rev.*, **56**(1), 41–43

Draycott, A. P., Webb, D. J. and Wright, E. M. (1973). The effect of time of sowing and harvesting on growth, yield and nitrogen fertilizer requirement of sugar beet. *J. Agric. Sci., Camb.*, **81**, 267–275

Duffus, J. E. (1978). The impact of yellows control on California sugarbeets. *J. Am. Soc. Sugar Beet Technol.*, **20**, 1–5

Dunning, R. A. (1961). Mangold fly incidence, economic importance and control. *Pl. Pathol.*, **10**, 1–9

Dunning, R. A. (1972). Sugar beet pest and disease incidence and damage, and pesticide usage. Report of an IIRB enquiry. *IIRB*, **6**, 19–34

Edwards, C. A. and Heath, G. W. (1964). *The Principles of Agricultural Entomology*. Chapman and Hall, London

Flegg, J. J. M. (1968). Life cycle studies of some *Xiphinema* and *Longidorus* species in south-eastern England. *Nematologica*, **14**, 197–210

Franklin, M. T. (1959). *Nacobbus serendipiticus* n. sp., a root galling nematode from tomatoes in England. *Nematologica*, **4**, 286–293

Franklin, M. T. (1972). *Heterodera schachti. Commonwealth Institute of Helminthology Descriptions of Plant Parasitic Nematodes.* Set 1, No. 1. Commonwealth Agricultural Bureaux

Franklin, M. T. (1973). *Meloidogyne naasi. Commonwealth Institute of Helminthology Descriptions of Plant Parasitic Nematodes.* Set 2, No. 19. Commonwealth Agricultural Bureaux

Franklin, M. T. (1978). *Meloidogyne.* In *Plant Nematology* (ed. J. F. Southey). HMSO, London, pp. 98–124

Frese, L. (1983). Resistenz der Wildmöhre *Daucus carota* ssp. *hispanicus* gegen der Wurzelgallnematoden *Meloidogyne hapla. Gartenbauwissenschaft*, **48**, 259–265

Gooris, J. and d'Herde, C. J. (1972). Mode d'hivernage de *Meloidogyne naasi* Franklin dans le sol et lutte par rotation culturale. *Rev. Agric., Brux.*, **25**, 659–664

Graf, A. and Meyer, H. (1973). Bedeutung des Rübenkopfälchens (*Ditylenchus dipsaci*) in der Schweiz und seine Bekämpfungsmöglichkeiten. *IIRB*, **6**, 117–126

Greco, N., Brandonisio, A. and De Marinis, G. (1982a). Investigation on the biology of *Heterodera schachtii* in Italy. *Nematol. Medit.*, **10**, 201–214

Greco, N., Brandonisio, A. and De Marinis, G. (1982b). Tolerance limit of the sugar beet to *Heterodera schachtii. J. Nematol.*, **14**, 199–202

Green, C. D., Greet, D. N. and Jones, F. G. W. (1970). The influence of multiple mating on the reproduction and genetics of *Heterodera rostochiensis* and *H. schachtii. Nematologica*, **16**, 309–326

Griffiths, B. S. and Trudgill, D. L. (1983). A comparison of the generation times of and gall formation by *Xiphinema diversicaudatum* and *Longidorus elongatus* on a good and a poor host. *Nematologica*, **29**, 78–87

Heijbroek, W. (1973). Forecasting incidence of and issuing warnings about nematodes, especially *Heterodera schachtii* and *Ditylenchus dipsaci. IIRB*, **6**, 76–86

Heijbroek, W., Roelands, A. J. and De Jong, J. H. (1983). Transfer of resistance to beet cyst nematode from *Beta patellaris* to sugar beet. *Euphytica*, **32**, 287–298

Hijink, M. J. and Kuiper, K. (1964). Crop rotation effects in Leguminosae due to *Meloidogyne hapla. Nematologica*, **10**, 64

Hill, D. S. (1987). *Agricultural Insect Pests of Temperate Regions and Their Control.* Cambridge University Press, Cambridge

Hooper, D. J. (1972). *Ditylenchus dipsaci. Commonwealth Institute of Helminthology. Descriptions of Plant-parasitic Nematodes.* Set 1, No. 14. Commonwealth Agricultural Bureaux

Hooper, D. J. (1973). *Longidorus elongatus. Commonwealth Institute of Helminthology Descriptions of Plant Parasitic Nematodes.* Set 2, No. 30. Commonwealth Agricultural Bureaux

Hooper, D. J. (1976). *Trichodorus viruliferus. Commonwealth Institute of Helminthology Descriptions of Plant Parasitic Nematodes.* Set 6, No. 86. Commonwealth Agricultural Bureaux

Inserra, R. N., de Vito, M. and Ferris, H. (1984). Influence of *Nacobbus aberrans* densities on growth of sugarbeet and kochia in pots. *J. Nematol.*, **16**, 393–395

Inserra, R. N., Griffin, G. D. and Anderson, J. L. (1985). The false root-knot nematode *Nacobbus aberrans. Res. Bull. Utah St. Agric. Expl. Stn.*, No. 510

Inserra, R. N., Vovlas, N., Griffin, G. D. and Anderson, J. L. (1983). Development of the false root-knot nematode, *Nacobbus aberrans*, on sugarbeet. *J. Nematol*, **15**, 288–296

Jatala, P. and Jensen, H. J. (1976). Histopathology of *Beta vulgaris* to individual and concomitant infections by *Meloidogyne hapla* and *Heterodera schachtii. J. Nematol.*, **8**, 336–341

Jenkins, W. R. and Coursen, B. W. (1957). The effect of root-knot nematodes, *Meloidogyne incognita acrita* and *M. hapla* on *Fusarium* wilt of tomato. *Pl. Dis. Reptr.*, **41**, 182–186

Jepson, S. (1987). *Identification of Root-knot Nematodes.* C.A.B. International, Oxford

Jones, F. G. W. and Dunning, R. A. (1972). *Sugar Beet Pests.* MAFF Bulletin No. 162. HMSO, London

Jones, F. G. W. and Jones, M. G. (1984). *Pests of Field Crops*, 3rd edn. Edward Arnold, London

Jones, M. G. K. (1981). The development and function of plant cells modified by endoparasitic nematodes. In *Plant Parasitic Nematodes*, Vol. III (ed. B. M. Zuckerman and R. A. Rohde). Academic Press, New York, pp. 255–279

Kolodge, C., Radewald, J. D. and Shibuya, F. (1987). Revised host range and studies on the life cycle of *Longidorus africanus. J. Nematol.*, **19**, 77–81

Kontaxis, D. G., Lofgreen, G. P., Thomason, I. J. and McKinney, H. E. (1976). Survival of the sugar beet cyst nematode in the alimentary canal of cattle. *Calif. Agric.*, **30**(3), 15

Lamberti, F. and Roca, F. (1987). Present status of nematodes as vectors of plant viruses. In *Vistas on Nematology* (ed. J. A. Veech and D. W. Dickson). Society of Nematologists, Hyattsville, Maryland, pp. 321–328

Lamberti, F. and Taylor, C. E. (Eds.) (1979). *Root-knot Nematodes (Meloidogyne Species). Systematics, Biology and Control.* Academic Press, London

Lange, W., Jung, C. and Heijbroek, W. (1990). Transfer of beet cyst nematode resistance from *Beta* species of the section *Patellares* to cultivated beet. *Proc. 53rd Winter Congr. IIRB. Brussels 1990*, pp. 89–102

Lange, W. H. (1987). Insect pests of sugar beet. *Ann. Rev. Entomol.*, **32**, 341–360

Lehman, W. F., Ede, L., Marble, V. L., Nielson, M. W. and Radewald, J. D. (1983).

Registration of U.C. Cibola alfalfa. *Crop Sci.*, **23**, 1216

Lejealle, F. (1982). *Pests, Diseases and Disorders of Sugar Beet*. Deleplanque et Cie, Maison-Laffitte, France (English text by R. A. Dunning and W. J. Byford, distributed by Broom's Barn Experimental Station)

Libman, G., Leach, J. G. and Adams, R. E. (1964). Role of certain plant-parasitic nematodes in infection of tomatoes by *Pseudomonas solanacearum*. *Phytopathology*, **54**, 151–153

Maughan, G. L., Cooke, D. A. and Gnanasakthy, A. (1984). The effects of soil applied granular pesticides on the establishment and yield of sugar beet in commercial fields. *Crop Prot.*, **3**, 439–450

Miller, L. I. (1987). Economic importance of cyst nematodes in North America. In *Cyst Nematodes* (ed. F. Lamberti and C. E. Taylor). Plenum Press, New York, pp. 373–385

Minton, N. A., Bell, D. K. and Doupnik, B. (1969). Peanut pod invasion by *Aspergillus flavus* in the presence of *Meloidogyne hapla*. *J. Nematol.*, **1**, 318–320

Müller, J. (1979). Über die jährliche Generationszahl von *Heterodera schachtii* unter Feldbedringungen an Zuckerrüben. *NachrBl. dt. PflSchutzdienst, Braunsch.*, **31**, 92–95

Müller, J. (1985). Der Einfluss der Wirtspflanze auf die Geschlechts-determinierung bei *Heterodera schachtii*. *Mitt. biol. BundAnst Ld- u. Forstw.*, **226**, 46–63

Noling, J. W. and Ferris, H. (1985). Influence of *Meloidogyne hapla* on alfalfa yield and host population dynamics. *J. Nematol.*, **17**, 415–421

Olthof, T. H. A. and Potter, J. W. (1972). Relationship between population densities of *Meloidogyne hapla* and crop losses in summer-maturing vegetables in Ontario. *Phytopathology*, **62**, 981–986

Phipps, P. M. (1987). Control of northern root knot and ring nematode on peanut in Virginia, 1986. *Fungicide and Nematicide Tests*, **42**, Report 162, Am. Phytopath. Soc.

Pitcher, R. S. (1967). The host parasite relations and ecology of *Trichodorus viruliferus* on apple root, as observed from an underground laboratory. *Nematologica*, **13**, 547–557

Pitcher, R. S. and McNamara, D. G. (1970). The effect of nutrition and season of year on the reproduction of *Trichodorus viruliferus*. *Nematologica*, **16**, 99–106

Radewald, J. D., Osgood, J. W., Mayberry, K. S., Paulus, A. O. and Shibuya, F. (1969). *Longidorus africanus*, a pathogen of head lettuce in the Imperial Valley of Southern California. *Pl. Dis. Reptr.*, **53**, 381–384

Rohde, R. A. and Jenkins, W. R. (1957a). Effect of temperature on the life cycle of stubby-root nematodes. *Phytopathology*, **47**, 29

Rohde, R. A. and Jenkins, W. R. (1957b). Host range of a species of *Trichodorus* and its host–parasite relationships on tomato. *Phytopathology*, **47**, 295–298

Santo, G. S. and O'Bannon, J. H. (1982). Reaction of tomato cultivars to *Meloidogyne chitwoodi* and *M. hapla*. *Pl. Dis.*, **66**, 406–407

Santo, G. S. and Qualls, M. (1984). Control of *Meloidogyne* spp. on Russet Burbank potato by applying metham sodium through center pivot irrigation systems. *J. Nematol.*, **16**, 159–161

Sher, S. A. (1970). Revision of the genus *Nacobbus* Thorne and Allen, 1944 (Nematoda: Tylenchoidea). *J. Nematol.*, **2**, 228–235

Smith, H. G. (1986). Comparative studies of the sugar beet yellowing viruses: field incidence and effect on yield. *Asp. Appl. Biol.*, **13**, 107–113

Steele, A. E. (1965). The host range of the sugar beet nematode, *Heterodera schachtii* Schmidt. *J. Am. Soc. Sugar Beet Technol.*, **13**, 573–603

Stephan, Z. A. and Trudgill, D. L. (1983). Effect of time of application on the action of foliar sprays of oxamyl on *Meloidogyne hapla* in tomato. *J. Nematol.*, **15**, 96–101

Stephenson, J. W. (1960). The biology of *Brachydesmus superus* (Latz.) Diplopoda. *A. Mag. Nat. Hist.* (13th series), **3**, 311–319

Steudel, W., Schlang, J. and Müller, J. (1985). Untersuchungen zum Einfluss einiger Zwischenfruchte auf die Abundanzdynamik des Rübennematoden (*Heterodera schachtii* Schmidt) in verschiedenen Bodentiefen. *Mitt. biol. BundAnst. Ld- u. Forstw.*, **226**, 129–140

Stirling, G. R. and Wachtel, M. F. (1985). Root-knot nematode (*Meloidogyne hapla*) on potato in south-eastern South Australia. *Aust. J. Exp. Agric.*, **25**, 455–457

Stone, A. R. and Burrows, P. R. (1985). *Nacobbus aberrans. Commonwealth Institute of Helminthology Descriptions of Plant Parasitic Nematodes*. Set 8, No. 119. Commonwealth Agricultural Bureaux

Taylor, C. E. (1971). Nematodes as vectors of plant viruses. In *Plant Parasitic Nematodes*,

Vol. II (ed. B. M. Zuckerman, W. F. Mai and R. A. Rohde). Academic Press, New York, pp. 185–211

Taylor, C. E. (1978). Plant-parasitic Dorylaimida: biology and virus transmission. In *Plant Nematology* (ed. J. F. Southey). HMSO, London, pp. 232–243

Taylor, C. E. and Brown, D. J. F. (1981). Nematode–virus interactions. In *Plant Parasitic Nematodes*, Vol. III (ed. B. M. Zuckerman and R. A. Rohde). Academic Press, New York, pp. 281–301

Taylor, D. P. and Wyllie, T. D. (1959). Interrelationship of root knot nematodes and *Rhizoctonia solani* on soybean emergence. *Phytopathology*, **49**, 552

Theurer, J. C., Blickenstaff, C. C., Mahrt, G. G. and Doney, D. L. (1982). Breeding for resistance to the sugar-beet root maggot. *Crop. Sci.*, **22**, 641–645

Thomas, P. R. (1969). Population development of *Longidorus elongatus* in Scotland with observations on *Xiphinema diversicaudatum* on raspberry. *Nematologica*, **15**, 582–590

Thomason, I. J. and Fife, D. (1962). The effect of temperature on development and survival of *Heterodera schachtii* Schm. *Nematologica*, **7**, 139–145

Thomason, I. J. and McKenry, M. (1975). Chemical control of nematode vectors of plant viruses. In *Nematode Vectors of Plant Viruses* (ed. F. Lamberti, C. E. Taylor and J. W. Seinhorst). Plenum Press, London, pp. 423–439

Thornhill, W. A. (1983). Sugar beet seedling damage by *Atomaria linearis*, and its integrated control. In *Proc. 10th Int. Cong. Pl. Prot. Brighton*, Vol. 3, p. 1217

Towle, A. and Doncaster, C. C. (1978). Feeding of *Longidorus caespiticola* on rye-grass *Lolium perenne*. *Nematologica*, **24**, 277–285

van Hoof, H. A. (1970). Some observations on retention of tobacco rattle virus in nematodes. *Neth. J. Pl. Pathol.*, **76**, 329–330

Vrain, T. C. (1982). Relationship between *Meloidogyne hapla* density and damage to carrots in organic soils. *J. Nematol.*, **14**, 50–57

Weingartner, D. P., Smart, G. C. and Shumaker, J. R. (1980). Population dynamics of trichodorid nematodes in Florida Irish potato soils following soil fumigation. *J. Nematol.*, **12**, 241

Weischer, B. and Steudel, W. (1972). Nematode diseases of sugar beet. In *Economic Nematology* (ed. J. M. Webster). Academic Press, London, pp. 49–65

Whitehead, A. G., Fraser, J. E. and Greet, D. N. (1970). The effect of D-D, chloropicrin and previous crops on numbers of migratory root-parasitic nematodes and on the growth of sugar beet and barley. *Ann. Appl. Biol.*, **65**, 351–359

Whitehead, A. G. and Hooper, D. J. (1970). Needle nematodes (*Longidorus* spp.) and stubby-root nematodes (*Trichodorus* spp.) harmful to sugar beet and other field crops in England. *Ann. Appl. Biol.*, **65**, 339–350

Whitney, E. D. and Duffus, J. E. (1986). *Compendium of Beet Diseases and Insects*. The American Phytopathological Society, St. Paul, Minnesota

Williams, J. J. W., Savage, M. J. and Smallshire, D. (1981). *Longidorus vineacola* Sturhan & Weischer damaging spring barley and onions in England. *Pl. Pathol.*, **30**, 122

Winder, G. H. (1984). Soil pest control. *Br. Sugar Beet Rev.*, **52** (1), 29–30

Winder, G. H. and Dewar, A. M. (1985). Decreased severity of flea beetle damage to sugar-beet seedlings associated with experimental insecticide treatments incorporated in pelleted sugar beet seed. Tests of Agrochemicals and Cultivars 6, supplement to *Ann. Appl. Biol.*, **106**, 30–31

Winder, G. H. and Dewar, A. M. (1988). New insecticides for sugar beet seed. *Br. Sugar Beet Rev.*, **56**(4), 23–24

Winder, G. H. and Dunning, R. A. (1986). Effects of row application of insecticides at sowing on leaf miner (*Pegomya betae*) injury to sugar beet. *Crop Prot.*, **5**, 109–113

Winfield, A. L. and Cooke, D. A. (1975). The ecology of *Trichodorus*. In *Nematode Vectors of Plant Viruses* (ed. F. Lamberti, C. E. Taylor and J. W. Seinhorst). Plenum Press, London, pp. 309–341

Wyllie, T. D. and Taylor, D. P. (1960). *Phytophthora* root rot of soybeans as affected by soil temperature and *Meloidogyne hapla*. *Pl. Dis. Reptr.*, **44**, 543–545

Wyss, U. (1970). Zur Toleranz wanderner Wurzelnematoden gegenüber zunehmender Austrocknung des Bodens und hohen osmotischen Drüken. *Nematologica*, **16**, 63–73

Wyss, U. (1974). *Trichodorus similis*. Commonwealth Institute of Helminthology. Descriptions of plant-parasitic nematodes, Set 4, No. 59. Commonwealth Agricultural Bureaux

Wyss, U. (1975). Feeding of *Trichodorus*, *Longidorus* and *Xiphinema*. In *Nematode Vectors of Plant Viruses* (ed. F. Lamberti, C. E. Taylor and J. W. Seinhorst). Plenum Press, London, pp. 203–221

Wyss, U. (1981). Ectoparasitic root nematodes. Feeding behaviour and plant cell responses. In *Plant Parasitic Nematodes*, Vol. III (ed. B. M. Zuckerman and R. A. Rohde). Academic Press, New York, pp. 325–351

Wyss, U. and Zunke, U. (1986). Observations on the behaviour of second stage juveniles of *Heterodera schachtii* inside host roots. *Rev. Nematol.*, **9**, 153–165

Yarger, L. W. and Baker, L. R. (1981). Tolerance of carrot to *Meloidogyne hapla*. *Pl. Dis.*, **65**, 337–339

York, P. (1980). Relationship between cereal root-knot nematode *Meloidogyne naasi* and growth and grain yield of spring barley. *Nematologica*, **26**, 220–229

Yu, M. H. (1984). Resistance to *Heterodera schachtii* in Patellares section of the genus *Beta*. *Euphytica*, **33**, 633–640

Yun, Y. M. and Sullivan, E. F. (1980). Pest management systems for sugarbeets in the North American central Great Plains region. *J. Am. Soc. Sugar Beet Technol.*, **20**, 455–476

CHAPTER 3

Pests of Composite Crops: Lettuce

B. Emmett

3.1 INTRODUCTION

The lettuce (*Lactuca sativa* L.) has been developed from the wild lettuce (*Lactuca scariola* L.), a wasteland plant originating in the Eastern Mediterranean region. It is the basis of all salads and is usually eaten uncooked. There are several types of lettuce with varying horticultural characteristics. Recent developments in plant breeding have resulted in novel cultivars with serrated, dissected or crinkled leaves or with red leaf coloration. The lettuce is a frost-sensitive annual plant that grows rosettes of leaves forming a more or less compact heart. When allowed to grow naturally, a central raceme of composite flowers and seeds is quickly produced, after which the plant dies. There is little nutrient value but the white sap is rich in iron and contains some vitamins. Lettuce crops are grown either under glass or as field vegetables. They are sold fresh for immediate use or their shelf-life may be increased by cold-chain distribution of, e.g., crisphead cultivars through supermarket outlets. Demand varies with changing weather conditions, the popularity of salads increasing in warm weather.

Aphids are the most important pest of lettuce. In Northern Europe and the North American continent, the lettuce root aphid is particularly troublesome, because infested plants may be killed or fail to produce marketable heads. Several species of aphid infest the aerial parts and may render them unmarketable when present at harvest. Cast skins and other debris may also lead to rejection, even when all aphids have been killed. Leatherjackets, cutworms and other soil pests sometimes kill lettuce plants. They are usually of sporadic occurrence and attack crops infrequently. The short growing period of this crop and the fact that entire plants are eaten raw necessitate care in selection and use of insecticides to ensure that unacceptable residues are absent from harvested produce.

3.2 APHIDS

3.2.1 *Myzus ascalonicus* Doncaster: Shallot Aphid (Hemiptera: Aphididae)

The origins of the shallot aphid are a mystery. It was first collected on stored onions in Wyoming, USA, during 1940 and was found a year later on stored onions in Lincolnshire, UK (Blackman and Eastop, 1984). It is now distributed almost world-wide, including Europe, India, Japan, North and South America, Australia and New Zealand. In warm temperate and tropical areas it colonises plants in at least 20 families, but is most important as a pest of alliaceous, composite and cruciferous plants. The shallot aphid is polyphagous on a range of plants, including brassicas, lettuces, onions, potatoes, sugar beet and strawberries. It is frequently found infesting the leaves of lettuces, usually in small numbers and in association with other species of aphid. Shallot aphids also damage stored onions and shallots by feeding on the scale leaves. When very large numbers of aphids are present, onion and shallot bulbs may be unfit to plant out (Edwards and Heath, 1964). Shallot aphids are able to transmit virus diseases to stored potatoes.

Description

Adult winged shallot aphids are 1.3–2.4 mm long. The head is dark in colour, with a pair of slightly convergent prominences between the antennae. The thorax is also dark in colour, but the abdomen varies from a straw to a greenish-brown colour. The dorsal surface of the abdomen has a distinctive, central, dark-coloured patch. The distal halves of the siphunculi are swollen.

Adult, wingless, shallot aphids are 1.1–2.2 mm long and yellow to greenish-brown in colour (Figure 3.1). The abdomen is strongly convex and does not have the distinctive, central, dark-coloured patch of the winged aphid. The cauda in life is held up so that it is not easily seen from above. The apices of the tibiae and the tarsi are conspicuously dark in colour (ADAS, 1988b).

Life History

The life history of the shallot aphid is not well understood: males, sexual forms and eggs are not known. The following outline probably applies only to north temperate regions.

In the autumn winged female shallot aphids migrate to strawberry crops and onions or other vegetables in store. Winged aphids are also found in glasshouses, sometimes on potted plants. There is considerable mortality in cold winters. The aphids breed throughout winter and spring. The first winged forms appear in May or early June and fly to summer hosts, e.g.

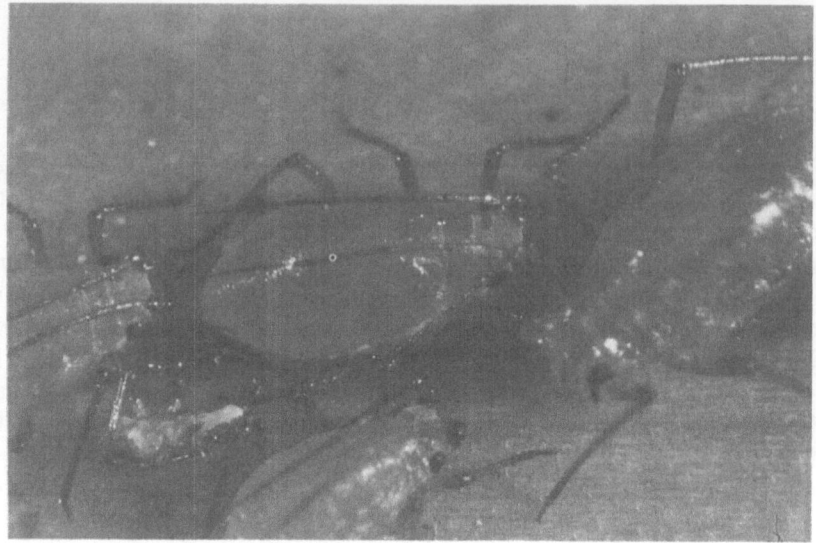

Figure 3.1   *Myzus ascalonicus* on leeks. © MAFF

lettuces, especially growing under protection. During this period colonies on strawberries decline and finally disappear (Alford, 1984). Towards the end of summer winged forms are produced. These aphids migrate and may then be found on strawberries and stored vegetable crops.

Damage

Damage to lettuces by shallot aphids is probably limited to fouling of leaves with honeydew. This sticky exudate is unacceptable to markets and can seriously reduce the value of crops (Anon., 1983). Shallot aphids may also contribute to the leaf curling and stunting generally associated with leaf-feeding aphids.

The shallot aphid is a known vector of about 20 plant viruses, including beet mosaic, beet yellows, onion yellow dwarf, potato leaf roll and, most importantly from the point of view of lettuce production, dandelion yellow mosaic (DYMV). DYMV can cause severe mosaic and necrosis of lettuce plants. Shallot aphids do not transmit lettuce mosaic virus, which is transmitted mainly by the peach-potato aphid (see p. 263).

Control

Specific recommendations on the control of lettuce leaf-feeding aphids such as the shallot aphid will be found on page 78 under 'Lettuce Aphid'.

3.2.2 *Nasonovia ribisnigri* (Mosley): Lettuce Aphid (Hemiptera: Aphididae)

Essentially an insect of north temperate regions and occurring mainly in Europe eastwards to the Ukraine, the lettuce aphid has also been introduced into Canada, north-eastern USA and South America (Blackman and Eastop, 1984). Wherever it is found, the lettuce aphid is probably the chief aphid pest of lettuces growing both under glass and outdoors. It occurs on chicory, composite weeds, especially hawkweeds (*Hieracium* L. spp.), and *Veronica* L. spp., *Nicotiana* L. spp. and *Petunia* Juss. spp. Winter hosts are most commonly gooseberries, occasionally red or black currants (Edwards and Heath, 1964).

Description

Adult, winged, lettuce aphids are 1.5–2.5 mm long and spindle-shaped. The dark-coloured head has a prominent, median tubercle between the antennae, giving a W-shaped appearance to the front of the head (Furk, 1988). The tips of the antennae and the thorax are black in colour. The abdomen is yellowish-green to dark green in colour and shiny with a dorsal pattern of small, black or dark-green patches. The siphunculi are moderately long, smooth and dark in colour at the apices, with well-developed flanges. The cauda is elongate (ADAS, 1988a).

Adult, wingless lettuce aphids are 1.3–2.7 mm long. The head, thorax and abdomen are pale yellow to apple green in colour. The abdomen is patterned with seven pairs of spots of the same colour as the body colour, only slightly darker. Otherwise, the antennae, front of head, siphunculi and cauda are all similar in shape and colour to the winged form.

Life History

In autumn, overwintering eggs of the lettuce aphid are laid on the wood, the foliage or near to the base of the buds of gooseberry or currant bushes. The eggs hatch usually in March or April into nymphs which infest the tips of the young shoots. Colonies are formed on the developing leaves, causing some leaf curling which becomes noticeable in late April or early May (Alford, 1984). In May or June, winged aphids migrate to lettuces and other composite plants, on which successive generations are produced until September or October. These generations feed more on the outer leaves of lettuces than in the plant centres. During October and November, winged aphids migrate to the winter hosts, where the eggs are laid. Although this life history occurs commonly in Europe, 'summer aphids' continue to breed on successive crops of outdoor lettuces throughout the winter (Anon., 1983).

Damage

Lettuce aphids colonise lettuces very rapidly and, when heavily infested, the leaves appear curled and blistered, often with a mottling of necrotic spots where aphids have been feeding. Plants may be so stunted that they fail to heart. Honeydew excreted by the aphids forms a sticky layer over the leaves to which dust and cast aphid skins adhere. Severely infested lettuces are unmarketable.

Lettuce aphids are able to transmit several viruses, including mosaic diseases of cauliflower and cucumber (Blackman and Eastop, 1984). Lettuce mosaic virus is, apparently, not usually transmitted by the lettuce aphid; more usually by the peach-potato aphid. In south-eastern Europe, especially Italy, lettuce necrotic yellows (LNYV) is a severe rhabdovirus disease of lettuces. The lettuce aphid was shown by Ragozzino *et al.* (1985) to transmit LNYV with acquisition-feeding periods of 24 h and transmission-feeding periods of 72 h.

Cultural and Chemical Control

Weeds and old lettuce crops are important sources of both leaf-feeding aphids and viruses. Where lettuces are grown intensively, losses from virus infection, particularly lettuce mosaic virus, may be severe unless strict attention is given to isolation of crops and to effective aphid control measures.

Infestation of lettuce seedlings with aphids can usually be avoided by raising them in weed-free isolation from other lettuce crops. They should be examined frequently for aphid infestation and, if found, promptly sprayed. One or two days before transplanting, lettuce seedlings should be treated with an aphicide to ensure that they are free from aphids. Where seedlings are raised in soil blocks, the same care is needed: they should be sprayed one or two days before setting out if any aphids are found. If treated in this way, winter-sown lettuces should remain free of aphids until harvest, because reinfestation by winged aphids does not take place on a large scale before May. However, the control of aphids on outdoor summer crops is more costly, requiring regular spraying, because winged, migrant aphids can reinfest the crop almost any time from May to September. If aphid colonies are found on summer lettuces at any time up to 3 weeks before cutting, they should be sprayed with one of the more persistent, systemic insecticides aimed at the hearts of the plants. Nevertheless, examination of the plants should continue to within a week of cutting. If a spray is necessary less than 3 weeks before cutting, short-persistence materials should be used. However, dead aphids may remain on sprayed crops, reducing their market value.

As aphids carrying lettuce mosaic virus can infect lettuce crops with the virus more quickly than insecticides applied after the arrival of the aphids can control them, applying insecticides after aphids have migrated to crops is not an effective way of controlling virus diseases. Aphid-borne virus diseases of lettuce crops are controlled only by constant and regular spray applications of insecticides, with the first application timed to coincide with the arrival of the first aphid migrant.

Investigations in the UK have identified lettuce cultivars with complete resistance to the lettuce aphid (Dunn and Kempton, 1980). The most promising cultivars were Iceberg and Ithaca, together with derivatives of Iceberg. However, without resistance to the peach-potato aphid, these cultivars have only limited value for the control of aphids and viruses on lettuces.

A range of insecticides is recommended in the UK for use on lettuces against leaf-feeding aphids. Demeton-*S*-methyl must not be applied within 21 days of harvest. Dimethoate and formothion each have a 14-day harvest interval, while malathion, pirimicarb and nicotine may be used 4, 3 and 2 days prior to harvest, respectively. There is a nil harvest interval for the pyrethroid insecticides cypermethrin and resmethrin.

If peach-potato aphids survive a properly applied treatment with an organophosphorus compound, it must be assumed that they are resistant. The next application of insecticide should be either a carbamate (e.g. pirimicarb) or a pyrethroid (e.g. cypermethrin or resmethrin).

3.2.3 *Pemphigus bursarius* (L.): Lettuce Root Aphid (Hemiptera: Aphididae)

The lettuce root aphid is a widespread pest of lettuces throughout western Europe, including Sweden, but has been little reported east of Germany. It is an extremely important pest in North America, occurring also in the Middle East, central Asia, north and south Africa and the southern states of Australia. *P. bursarius* is mainly a pest of lettuces, but other composite crops (e.g. chicory and endives) are also colonised. Considerable losses to lettuce root aphid colonies on curly endives are reported from Italy (Ciampolini, 1975). Among weed species colonised by the root aphid are dandelions (*Taraxicum officinale* Weber), sowthistles (*Sonchus oleraceus* L.), hawksbeards (*Crepis* L. spp.) and nipple-worts (*Lapsana* L. spp.).

Description

Winged lettuce root aphids are rarely seen on lettuce crops, because immigrants in early summer give rise quickly to wingless forms and emigrants, as soon as they have matured in autumn, fly to poplar trees.

Adult, winged lettuce root aphids are 2.2 mm long, each with a dark-coloured head bearing short antennae which have greatly reduced sixth joints. The thorax is dark brown or black in colour. The abdomen is round to pear-shaped, brownish-orange in colour, with a slight powdering of wax. The siphuncular apertures open not at the tips of characteristic stalks, but at the surface of the rear abdominal dorsum, appearing as dark-coloured rings. The legs are short and darker in colour than the body (ADAS, 1988c).

Adult, wingless lettuce root aphids are 2.0–2.2 mm long, each with a yellow-coloured head bearing green- to greyish-coloured antennae which are much shorter than the body. The body is oval-shaped, yellowish white in colour and often obscured by copious quantities of white or grey-coloured wax which is produced in tufts on the posterior part of the abdomen. Siphunculi are absent. The legs are short and dark in colour (Furk, 1988).

Life History

In late summer and autumn, winged, female lettuce root aphids migrate from lettuce crops to poplar trees, on which the aphids lay eggs. These eggs are laid in cracks in the bark of both Lombardy and black poplars. In March or April the eggs hatch into nymphs which feed on developing petioles. As a result of the feeding, the petioles enlarge to form yellow- or reddish-coloured, purse-shaped or flask-shaped galls, each with a lateral opening that allows access for the aphids. Within these galls the nymphs live and mature to produce a further generation of aphids. Winged aphids appear over a 4–5 week period in late June and July and migrate to lettuces (Ramert, 1977). These migrants already contain mature embryos, so that reproduction begins immediately after arrival on lettuce leaves. The wingless progeny feed first on the leaves and then migrate to the roots, where numbers increase with the production of several wingless genera-tions. The roots may become completely infested by aphids which charac-teristically are covered with a mealy, blue-grey-coloured, waxy secretion. When winged females appear in August, they move to the soil surface by walking up cracks or through channels made by roots. They remain around the collar of the host plant for a short period before flying to poplar trees. On the poplars, male and female, wingless, aphids are produced. These sexual forms do not feed. After mating, the females, called oviparae, lay eggs and die.

A small proportion of wingless lettuce root aphids may overwinter on lettuces or weed hosts only to colonise lettuces planted in adjacent soil during the following year (Anon., 1983). All stages of this aphid species survive winter conditions well. Alleyne and Morrison (1978) found that in Quebec, Canada, lettuce root aphid apterae could survive temperatures of 0°C for more than 40 weeks.

Damage

Large infestations of lettuce root aphids cause lettuces to wilt, dry up and turn yellow in colour. The symptoms are most severe in dry seasons, when plants quickly die (Figure 3.2). Unprotected crops sown or planted into infested soils or grown during periods of peak aphid invasion may suffer plant losses in excess of 60% (Ramert, 1977). Less severe attacks can reduce yield and marketability of heads. Where high-quality and uniformly sized lettuces are required, infestations of root aphids may cause symptoms leading to rejection of entire crops. The lettuce root aphid is apparently unable to transmit virus diseases.

Cultural Control

Weeds and old lettuce crops are important sources of lettuce root aphids. Crop residues should be removed and destroyed and the soil should be well ploughed to destroy as many overwintering aphids as possible. Crop rotation should be practised, avoiding, for instance, spring-sown crops being grown in soil that grew aphid-infested lettuces the previous year. The effects of lettuce root aphid attacks are worst when plants are under stress from a lack of soil moisture. Irrigation is, therefore, very beneficial and is especially recommended for crops showing signs of damage. The Lombardy poplar should not be planted as a windbreak and growers should discourage neighbours from planting poplars adjacent to areas where lettuces are grown. Destruction of existing poplars is not recommended, because they occur widely in plantations and gardens.

Figure 3.2 Lettuce crop badly attacked by lettuce root aphids, *Pemphigus bursarius* (Glazenbury, England, 1955). © MAFF

Lettuce cultivars with resistance to lettuce root aphids have been developed in the UK. The crisp lettuce cultivars Avoncrisp, Avondefiance and Lakeland are almost completely immune (Dunn, 1977) and cultivars Butterhead and Iceberg have some resistance. Resistant cultivars attract winged migrant lettuce root aphids but colonies do not subsequently develop. The mechanism of resistance is poorly understood but it may be linked to a gene for specific resistance to downy mildew (*Bremia lactucae* Regel) (Crute and Dunn, 1980). The resistance of cultivars Avoncrisp and Avondefiance probably originated from a Russian plant of Prickly Lettuce (*Lactuca serriola* L.).

Biological Control

Predators and parasitoids of the lettuce root aphid have been monitored in Quebec, Canada (Alleyne and Morrison, 1977). Hoverfly larvae, anthocorid bugs and parasitic wasps were common in galls formed by lettuce root aphids on poplars. Ladybirds and hoverflies were found preying on aphid colonies on lettuce roots. A neoaplectanid nematode was also present in large numbers of root aphid apterae. However, all of these natural enemies are unlikely to exert a significant influence on aphid populations.

Chemical Control

Established lettuce root aphid infestations cannot be effectively controlled by insecticidal sprays, although soil drenches of pirimicarb or formothion proved effective in Switzerland (Hauri *et al.*, 1973). Preventative treatments are necessary where the growing of resistant cultivars is unacceptable. These treatments are worth while only when trouble is expected—i.e. when lettuces are sown or transplanted between mid-April and end of June in areas where lettuce root aphids are usually common. Crops sown in July are usually not infested (Edwards and Heath, 1964). Suett and Padbury (1980) concluded that the use of phorate was unlikely to lead to large residues in field-sown lettuces, but was likely to lead to unacceptably high residues in lettuces raised in peat blocks. In the UK diazinon liquid, wettable powder or granules may be incorporated into soil before sowing or planting of field crops. Diazinon wettable powder may also be applied as a spray when the first winged lettuce root aphids are seen in crops. For peat blocks, diazinon wettable powder may be incorporated at the time of wetting prior to blocking and seeding. For lettuces grown in loose-filled cells, there are no recommendations.

The treatment of poplar trees with insecticides should not be considered as a means of controlling lettuce root aphid populations.

3.3 FLIES

3.3.1 *Tipula paludosa* Meigen: Crane Fly (Diptera: Tipulidae)

The crane fly is abundant throughout Europe, northern Asia, Canada and northern USA (Hill, 1987). The host crops, biology and habits of the crane fly are similar throughout its range. It is especially abundant where there are large areas of upland grass or permanent pastures. Almost all pasture land contains at least a few crane fly larvae, known because of their tough, leathery skins as leatherjackets (Edwards and Heath, 1964). They are important agricultural pests, especially of cereals. Any crop grown in newly ploughed grassland or in fields or stubbles infested with grass weeds during the previous autumn may be attacked. Bean, sugar beet, brassica, lettuce, sweet corn and other vegetable crops are frequently damaged by leatherjackets. Lettuce plant losses can be severe where large leatherjackets are present at the time of sowing or planting out.

Description

The adult crane fly is an elongate, slender insect with narrow wings and long, fragile legs. The body, which is 17–25 mm long, has a dull, brown or yellow-grey colour. The small head is projected forward and bears bristly, 14-segmented antennae. The thorax has faint longitudinal stripes and a dusty-grey-coloured appearance. The female abdomen tapers to a shiny, brown-coloured pointed apex which extends beyond the wings when folded. The male abdomen, which does not extend beyond the wings when folded, is truncated, with a club-like, slightly upturned apex.

The egg of the crane fly is shuttle-shaped, 0.9×0.4 mm, shiny and black in colour. The larva or leatherjacket has a dull, brownish-grey-coloured, wrinkled body which is 30–45 mm long at maturity. The leatherjacket's body, which is plump and slightly tapered anteriorly, with a soft, but tough, leathery skin, is capable of considerable contraction. The larval head is black in colour, small and indistinct (Alford, 1984). The anal segment of the leatherjacket has a single pair of elongate, ventral papillae which are located below two large, black-coloured spiracles. The pupa is 20–28 mm long, pale brown in colour and slender. The locations of head, wings and legs of the developing adult insect are visible at the anterior end of the pupa. A pair of prominent, respiratory horns projects from the pupa at the head of the adult insect. The adult abdominal segments are clearly marked on the pupa, each segment bearing several small, pointed papillae.

Life History

Adult crane flies begin to emerge from pupae during June and are abundant in late summer and early autumn. After mating, females fly to grass and lay up to 500 eggs each. Most eggs are laid from mid-August to the end of September. They are placed just below soil level in small batches. The eggs hatch about 2 weeks later into larvae or leatherjackets which feed on plant roots. By November the larvae are usually 10 mm long. They stop feeding entirely or feed only slowly during the winter. Feeding activity and growth resume in the spring. The larvae are fully grown by June, when they pupate in the soil. The pupal period lasts for about 2 weeks. The pupae move towards the soil surface to protrude beyond it shortly before adult emergence.

Damage

Leatherjackets usually feed just below the surface of the soil, destroying plant roots and underground stems. Seeds can also be damaged. On warm, damp nights leatherjackets feed on the soil surface, making ragged holes in leaves and cutting off plants at soil level rather like cutworms (see page 352). Vegetable seedlings such as lettuces may be severed from their roots within a day or two of planting. The leatherjackets can usually be found by digging away the soil from around damaged plants. In row crops such as brassicas, where several neighbouring plants have been killed, the leatherjacket responsible for the damage will frequently be found feeding at an adjacent healthy plant. In heavily infested fields bare patches may appear where many plants have been destroyed.

Cultural Control

To avoid leatherjacket attack, vegetable crops should not be sown or planted in soil which has been recently ploughed out of grass. If this sequence of cropping is unavoidable, a good soil tilth with firm consolidation are necessary to stimulate quick plant growth, so avoiding the most severe damage. Crops planted after mid-summer should be safe from attack. If the soil is then kept free of grass weeds, there should be no trouble in subsequent years (Anon., 1979).

Pest Monitoring

In the UK the advisory services forecast likely leatherjacket attacks to crops in spring and early summer. These forecasts are based on winter surveys of leatherjacket populations in permanent pasture. There are no recognised treatment threshold numbers in vegetable crops. For lettuces and other row-crop vegetables, few leatherjackets can severely damage crops. Populations of leatherjackets in unbroken grass intended for vegetable cultivation can be estimated by using a commercially available leatherjacket monitoring kit. It involves counting leatherjackets brought to the soil surface as a result of applying an aqueous salt solution to the soil.

Biological Control

Leatherjackets are attacked by viruses, parasitic fungi, parasitoids (e.g. tachinid flies of genus *Trichopareia* Brauer and Bergenstamm) and birds. Although *Tipula* iridescent virus is widespread, it does not usually affect a large proportion of the leatherjackets in a population. Starlings are the main bird species feeding on leatherjackets, considerably decreasing populations between late autumn and early spring. Cultivations help to decrease populations of leatherjackets by bringing them to the soil surface, where they die of exposure.

Abiotic Factors

The eggs and larvae of the crane fly are very sensitive to desiccation and probably many are killed, because they are commonly found close to the soil surface. To survive, the eggs require cool, moist soil conditions. Probably the most important environmental factors causing death of larvae are the severity and the duration of cold periods during winter, as well as the alternation of dry, frosty weather with rain during periods of thaw (Edwards and Heath, 1964).

Chemical Control

In the UK there are no approved recommendations for the chemical control of leatherjackets in vegetables, other than for lettuce crops. For lettuces, sprays of carbaryl may be applied when damage is first seen. Alternatively, gamma-HCH may be mixed with bran to form a bait which is applied, preferably in the evening, either by hand or by mechanical applicator. Baits are more effective than sprays when soil conditions are damp. Baiting efficiency depends on leatherjackets coming to the soil surface to feed at night.

REFERENCES

ADAS (1988a). *Lettuce Aphids*. Identification Cards of the Ministry of Agriculture, Fisheries and Food, London. No. 191
ADAS (1988b). *Lettuce Aphids*. Identification Cards of the Ministry of Agriculture, Fisheries and Food, London. No. 194
ADAS (1988c). *Lettuce Aphids*. Identification Cards of the Ministry of Agriculture, Fisheries and Food, London. No. 195

Anon. (1979). *Leaflet 179*. Ministry of Agriculture, Fisheries and Food, Alnwick

Anon. (1983). *Leaflet 392*. Ministry of Agriculture, Fisheries and Food, Alnwick

Blackman, R. L. and Eastop, V. F. (1984). *Aphids on the World's Crops*. Wiley-Interscience, Chichester

Ciampolini, M. (1975). *Informatore Agrario*, **31**, 21

Crute, I. R. and Dunn, J. A. (1980). *Euphytica*, **29**, 483–488

Dunn, J. A. (1977). *Bulletin SROP*, 59–60

Dunn, J. A. and Kempton, D. P. H. (1980). Tests of Agrochemicals and Cultivars (*Annals of Applied Biology*, **94**, supplement), No. 1, 58–59

Edwards, C. A. and Heath, G. W. (1964). *Principles of Agricultural Entomology*. Chapman and Hall, London

Furk, C. (1988). *Lettuce Aphids*. Identification key to accompany Identification Cards 190–195. Leaflet of the Ministry of Agriculture, Fisheries and Food, London, U-96

Hauri, P., Freuler, J., Bertuchoz, P. and Hugi, H. (1973). *Revue Suisse de Viticulture, Arboriculture et Horticulture*, **5**, 179–182

Hill, D. S. (1987). *Agricultural Pests of Temperate Regions and Their Control*. Cambridge University Press, Cambridge

Ragozzino, A., Iengo, C. and Camele, I. (1985). *Informatore Fitopatologico*, **35**, 43–46

Ramert, B. (1977). *Vaxtskyddsnotiser*, **41**, 83–87

Suett, D. L. and Padbury, C. E. (1980). *Pesticide Science*, **11**, 351–360

CHAPTER 4

Pests of Cruciferous Crops

S. Finch and A. R. Thompson

4.1 INTRODUCTION

Although members of the family Cruciferae are cosmopolitan, they are found chiefly in north temperate regions. Many are annual or ephemeral herbs which grow in dry open habitats and, generally, they are among the first plants to colonise disturbed soils. Their Latin name, Cruciferae (*crux* = a cross; *fero* = I bear) describes their flowers, which consist characteristically of four petals in the shape of a cross. The plants are also characterised by a wide range of secondary plant compounds, known as glucosinolates, which, with their breakdown products, give the plants characteristic tastes and often pungent odours. These chemicals are toxic to most insects but some other insects have overcome them, using the chemicals as essential cues in host location and selection. The insects that have adapted in this way are regarded as pests only when they attack Cruciferae cultivated by man. As many of the adapted insects can colonise successfully a wide range of the 220 genera of Cruciferae, many wild plants and garden flowers are reservoirs of pests of cultivated crops. The two genera of Cruciferae cultivated most extensively are *Brassica* L. and *Raphanus* L. Until about 15 years ago, cultivars of *Brassica oleracea* L. were the commonest crucifers grown on a field scale but, with the recent rapid expansion in the area of oilseed rape, the oil-bearing cultivars of *B. napus* L. and *B. campestris* L. are now commonest. Not only are these grown at a much higher density (80–120 plants/m^2) than are the *B. oleracea* cultivars (2–20/m^2), but also they occupy a much larger area than do other cruciferous crops and, hence, are the commonest host plants that pest insects are likely to encounter. This situation is exacerbated by seed shed before or during harvest giving rise to large numbers of volunteer plants in non-cruciferous crops in subsequent years.

In temperate regions cultivated Cruciferae can be attacked by 50–60 insect species, of which about 20 are major pests of crop plants, damaging all the different stages of growth of the crop and all parts of the plants. The relative importance of individual insect species as pests varies from crop to crop and, although large numbers of some species may be present, the species are considered as pests only when the large numbers coincide with vulnerable stages of particular crops.

4.2 APHIDS

4.2.1 *Brevicoryne brassicae* (L.): Cabbage Aphid (Hemiptera: Aphididae)

Geographical Distribution

B. brassicae is abundant on cruciferous crops throughout all temperate regions of the world.

Description and Life Cycle

The wingless aphids (apterae) are 1.6–2.6 mm long, greyish-green or dull mid-green in colour, with a dark head and characteristic black transverse bars on the dorsal surface of the thorax and abdomen (Figure 4.1). The body is covered with a greyish-white-coloured mealy wax which is also secreted onto the surface of host plants. The winged aphids (alatae) are slightly longer (1.6–2.8 mm) than the apterae and have a dark-coloured head and thorax. The characteristic, black in colour, transverse bars are visible only on the abdomen. The veins and the wings appear brown in colour, and all winged individuals are females.

Unlike other aphid pests of vegetables, *B. brassicae* is a 'one-host aphid' (cf. *Myzus persicae* on page 263). It does not spend the winter on woody plants but remains on herbaceous Cruciferae throughout its life cycle. In most northern areas of its distribution, *B. brassicae* overwinters as small, elongate, shiny black-coloured eggs laid particularly around leaf scars of stems of plants that remain in the field throughout the winter. In the UK the eggs hatch in late February and March and the resulting aphids colonise the flowering stems of seed crops (Figure 4.1) or harvested vegetable crops that have not been ploughed in. In mid-May alatae disperse to colonise new host plants on which further batches of apterae and, later, alatae are produced parthenogenetically. In warm years, when conditions are favourable for the rapid development of large aphid colonies, winged forms are produced periodically throughout the summer. In the more northerly temperate regions there are usually two peaks of aphids each year, one in late July and the other in early October. Falling temperatures and decreasing day length in the autumn induce aphids to produce sexual forms which mate and lay eggs from September to December, peak oviposition occurring generally in October in the UK. Even in northern Europe, some of the parthenogenetic asexual aphids manage to overwinter, especially in mild winters, in countries with a maritime climate.

Plant Damage

B. brassicae is a serious pest on most cruciferous crops and is potentially troublesome on crops of cabbage, cauliflower and Brussels sprouts. It also colonises radish, swede, mustard and, to a lesser extent, kale and rape

Figure 4.1 Close up of cabbage aphid colony on flower stem of brassica seed crop. © HRI

crops, but turnips are much less susceptible. Colonies are found also on a wide range of cruciferous weeds, including charlock (*Sinapis arvensis* L.), shepherd's purse (*Capsella bursa-pastoris* (L.) Medic.) and wild radish (*Raphanus raphanistrum* L.), but, even when cruciferous weeds are abundant, the aphids tend to be restricted to a few well-spaced plants. In contrast, colonies of the aphid can be found on most plants in cultivated crops, particularly plants belonging to the genus *Brassica*.

The first signs of attack are small bleached areas on the leaves of infested plants. The leaves then turn yellowish and become excessively crumpled, the aphid colonies being protected within the crumpled leaves. Seedling plants are most affected. They can be stunted and may die in unfavourable weather. Even when young plants are infested only lightly, the leaves of the plants when they are mature continue to show signs of the original attack. Infestations on larger plants reduce yield and also spoil the plants by contaminating them with wax, cast skins and honeydew. The latter is particularly troublesome, as it is an ideal culture medium for sooty moulds, which, as their name suggests, produce black marks on the surface of the produce and thereby lower its quality. When infestations are large, the aphids sometimes penetrate the hearts of Brussels sprout buttons, cabbage plants and cauliflower curds. It is essential to prevent this happening, as the mere presence of aphids, alive or dead, reduces the market value of the crop.

In addition to the direct crop damage it causes, *B. brassicae* also transmits cauliflower mosaic virus and turnip mosaic virus to cruciferous crops (see Table 8.1 on pp. 268–9 for description of viruses). Good crop hygiene, rather than trying to kill viruliferous aphids, is the only way of reducing the impact of these viruses, as the time taken by viruliferous aphids to infect new crops with these non-persistent viruses is often less than 1 min, too short a period for aphids to receive a toxic dose of insecticide.

Natural Control

The weather is the major natural agent restricting the build-up of *B. brassicae* infestations in cold temperate regions. In dry, warm seasons the aphids go largely unchecked and often produce extremely large infestations, whereas in wet, cool years aphid populations remain small. In wet years outbreaks of entomogenous fungi often coincide with periods of high humidity following rain, and these fungi can spread rapidly to reduce aphid populations even further. In contrast to the fungi, predators, mainly coccinellid beetles and syrphid flies, and parasitoids, mainly *Diaeretiella rapae* MacIntosh (Hymenoptera: Aphidiidae), have only a minor effect on populations of *B. brassicae*.

Cultural Control

Cultivated cruciferous plants that remain in fields throughout the winter are largely responsible for ensuring that large numbers of eggs and/or adults overwinter. Therefore, the most effective control measure is to eliminate as many of these sources of infestation as possible before the aphids spread to new crops. This is normally achieved by rotavating, or discing, overwintering brassica crops as soon after harvest as possible and then ploughing the fields to bury the plant debris. As crops grown for seed cannot be destroyed early enough, it is imperative that they be treated with appropriate insecticides in the spring to prevent them becoming refuges for overwintering aphids. As these treatments are not economic on oilseed rape crops, new cruciferous vegetable crops should be sited as far away as possible from overwintering oilseed rape crops. Mulches of transparent or blue plastic or straw placed on the soil between crop rows also reduce the numbers of *B. brassicae* attracted to *Brassica* crops.

Host Plant Resistance

It is now generally accepted that there is little chance of producing a brassica cultivar resistant to all pests, and, even when resistance is found, it is not always durable. An additional disconcerting problem in trying to breed brassicas resistant to *B. brassicae* is the number of aphid biotypes present in the field. These can have devastating effects on research aimed at producing plants resistant to the aphid. For example, of eight Brussels sprout clones selected over a number of years for their resistance to the Wellesbourne biotype of *B. brassicae*, only one was not completely susceptible to biotypes of the aphid collected from other localities in England. More recently, light-green lines from an Australian cauliflower bred for resistance to pests in New York State, USA, have proved resistant to the aphid in the Netherlands and offer further opportunity for breeding aphid-resistant cultivars.

Chemical Control

The practical problem of controlling *B. brassicae* with insecticides differs in detail from grower to grower, from crop to crop and from season to season. Granular formulations of insecticides incorporated into the soil remain effective longer than do foliar sprays but only a slow-release formulation of disulfoton, an organophosphorus compound, lasts sufficiently long to offer season-long control. When aphid colonies begin to establish on a crop, a clear indication that soil-applied granules have begun to lose their effectiveness, it is usual to start applying foliar sprays of appropriate insecticides. The insecticide selected is governed largely by the interval available between application and harvest, although a choice between carbamate and organophosphorus insecticides, and between insecticides of different toxicity to natural predators, may be available. Early in the growth of a crop, insecticides that remain effective for more than a week (e.g. demeton-*S*-methyl) can be used, but such chemicals are not suitable when aphids are present on crops near to harvest. When this happens, chemicals that require a harvest interval of a few days (e.g. dimethoate, pirimicarb), 1 day (e.g. heptenophos) or less than 1 day before harvest (e.g. cyfluthrin, cypermethrin, fenvalerate, permethrin, resmethrin) must be used. As many of the insecticides with more persistent effects against *B. brassicae* have a systemic action, they often fail to provide adequate protection under drought conditions when the plants stop growing and the insecticide is not taken up by the foliage or the roots for translocation into the foliage. However, soil-applied systemic insecticide 'bound' in dry soil does become available for translocation when sufficient soil moisture is restored. If irrigation is not available to provide adequate soil moisture, foliar sprays may have to be applied, using large volumes of water (at least 1500 l/ha for a Brussels sprout crop in full leaf) to make them effective. When applied in this way, insecticides that act usually by systemic action are used as contact insecticides and the spray must then actually 'hit' the aphids to be effective.

Integrated Control

The numbers of insecticidal sprays applied to crops can often be reduced by using a system of supervised control, in which crops are inspected or 'scouted' to determine when sufficient aphids are present to merit treatment, instead of applying sprays routinely. The level of aphid population at which insecticide is required is often referred to as the 'control threshold'. However, this threshold is better described as the 'tolerance level', as it is based on the tolerance of the crop to infestations of *B. brassicae*. Although a single tolerance level can be used throughout the life of a crop, less insecticide is required if the tolerance level is adjusted to match the different stages of plant growth. The 'tolerance levels' suggested by Theunissen (1984) for Brussels sprouts indicate that insecticidal treatments should be applied when 10–15% of the plants are infested with aphids 2 weeks after transplanting and that the threshold should be increased to 20–40% infested plants 4–8 weeks later. However, once the sprout buttons begin to form, the 'tolerance levels' are reduced to 10%, and during the 4 weeks prior to harvest the tolerance level is zero. Whether supervised control systems with adequate scouting can generate sufficient savings to make them attractive economically depends upon the amount of insecticide currently applied.

4.3 WHITEFLIES

4.3.1 *Aleyrodes proletella* (L.): Cabbage Whitefly (Hemiptera: Aleyrodidae)

Geographical Distribution

A. proletella is common throughout Europe and New Zealand but is not found in the USA, Canada or Australia. The cabbage whitefly is polyphagous but individuals appear to have a strong preference for members of the Compositae and Cruciferae. Other polyphagous species of *Aleyrodes* Latreille, such as *A. lonicera* Wlk. and *A. spiraeoides* Quaint, are pests of non-cruciferous crops in the USA and Canada but none has extended its host range to include cruciferous vegetable crops.

Description and Life Cycle

Adult *A. proletella* are about 1.5 mm long and have the general appearance of small white-coloured moths (Figure 4.2). The wings have much reduced venation and a span of about 3 mm. The head and thorax of the adults are dark in colour and the abdomen is yellow. Because the insects are covered normally in powdery wax, the adults appear to be completely white in colour: hence their common name.

A. proletella overwinter as adults on the undersides of leaves of brassica crops. Most of the overwintering insects are mated females. The eggs are prevented from developing by an ovarian diapause which, under field

Figure 4.2   Adults and scales of cabbage whiteflies. © HRI

conditions, requires a period of chilling for its completion and is normally terminated during late February or March. Post-diapause development then begins and, as soon as temperatures exceed 15°C, the adults start to lay their oval, stalked eggs in small groups on the undersides of leaves. The eggs hatch in 10–14 days and the six-legged nymphs that emerge from the eggs wander around on the leaf surface before they select a suitable feeding site. The nymph pierces the plant tissues with the stylets of its proboscis and then sucks the plant's juices. During the next few days the insect changes completely: its legs degenerate and it secretes a waxy covering and turns into the typical ovoid, low-conical shape commonly referred to as a 'scale' (Figure 4.2). The scales continue to feed, and moult three times before forming fourth instars, which are sometimes called pupae. The fourth instar secretes a much thicker and more opaque waxy coating than do the earlier instars. It also has two areas of dark-coloured cells, pigmented by symbiotic fungi, on either side of the mid-line at the proximal end of the abdomen. These are known as mycetomes. The adult *A. proletella* emerges from the pupal cuticle through a T-shaped dorsal rupture. In the summer *A. proletella* requires about 1 month to complete a generation, but when temperatures are cooler in the spring and autumn, development is less rapid. In most countries where this pest occurs, there are 4–5 generations each year with conditions favourable for whitefly reproduction occurring for about 7–8 months.

The scales and adults possess a vasiform orifice on the dorsal surface of the last segment. This orifice, a characteristic of whiteflies, also contains a small tongue-like organ called a lingula. Honeydew secreted into the vasiform orifice is flicked as far as 2 cm away by movements of the lingula. Therefore, honeydew does not accumulate to the same extent on leaves attacked by whiteflies as on leaves attacked by aphids.

Adult whiteflies feed mainly on the undersurfaces of leaves. With large populations of *A. proletella*, the adults rise in dense clouds when host plants are disturbed. During the summer months *A. proletella* disperses to find new host plants during the day, provided that the light intensity is sufficiently high, and ceases to fly towards evening. In the autumn the behaviour of cabbage whiteflies changes and they disperse under any light conditions as long as temperatures are favourable for flight.

Plant Damage

Crops infested most frequently with *A. proletella* are cabbage, Brussels sprout, cauliflower, broccoli and kale. Other crops infested to a lesser degree include swede, turnip, mustard, a wide range of wild cruciferous plants and several members of the Compositae. Although feeding by *A. proletella* often produces white or yellow patches on infested leaves, it rarely reduces crop yields. *A. proletella* is a problem because its immature stages (scales) and waste products (mainly honeydew) contaminate and thus reduce the quality of plant produce.

Biological Control

Coccinellid beetles and drosophilid flies have been observed feeding on *A. proletella* (Mound and Halsey, 1978). These authors also list ten species of chalcid wasps which have been reared from whitefly pupae. Most attempts to control this pest by biological means have involved species of *Encarsia* Foerster (Hymenoptera: Eulophidae).

Chemical Control

Insecticides are applied largely for 'cosmetic' reasons to obtain produce completely free of any signs of insect activity. Some pyrethroid insecticides (e.g. permethrin) are capable of killing all stages of whitefly, but most other insecticides are effective only against the adults. To control the pest adequately with organophosphorus insecticides (e.g. pirimiphos-methyl), several sprays have to be applied to infested plants at 3–5 day intervals to kill newly emerged adults before they lay eggs and start a new generation. As *A. proletella* adults are found mainly on the lower surfaces of leaves, treatments are usually effective only when the insecticide can be sprayed at sufficiently high volume and with appropriate equipment to wet both leaf surfaces thoroughly. Low-volume and ultra-low-volume sprays, although often easier to apply than high-volume sprays, are generally not effective against *A. proletella*.

Integrated Control

Control thresholds have been developed in Germany: sprays are applied when crop inspections indicate an average degree of infestation of more than 20 adults or 50 larvae/plant, rather than routinely every 2 weeks. Using this system, the numbers of sprays required have been reduced by 1 on savoy cabbages and by 3 on red cabbages.

4.4 BEETLES

4.4.1 *Meligethes* Stephens spp.: Blossom Beetles (Coleoptera: Nitidulidae)

Geographical Distribution

Meligethes spp., commonly referred to as 'blossom' beetles, are common throughout North America, Canada, the UK and the whole of western Europe. Of the many species of *Meligethes*, the two most damaging to cruciferous crops are *M. aeneus* Fabricius and *M. viridescens* Fabricius.

Description and Life Cycle

The adults are metallic, greenish-black-coloured beetles, 1.5–3.0 mm long, with short legs and distinct clubbed antennae (Figure 4.3). The eggs are

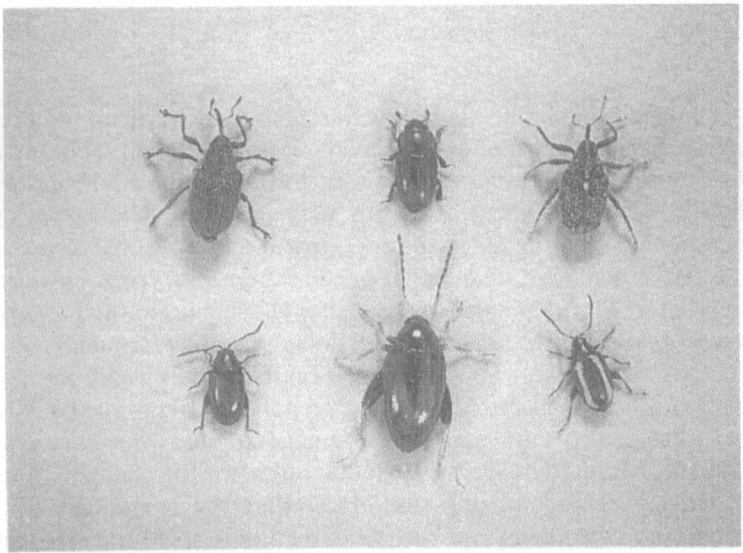

Figure 4.3 (Clockwise from top left) Adults of cabbage seed weevil; blossom beetle; cabbage stem weevil; flea beetle (*Phyllotreta nemorum*); cabbage stem flea beetle; and flea beetle (*P. cruciferae*). © HRI

white, about 0.7–0.9 mm long and 0.25–0.35 mm wide. The larvae are creamy-white in colour, with a distinct head and three pairs of legs. Final-instar larvae are more than 4 mm long and yellowish-white in colour, with each body segment bearing two dark flecks of pigment. The pupae are ovate, yellow to dirty-white in colour, and about 2 mm long.

The adults overwinter in the soil in sheltered, well-drained sites alongside hedgerows and thickets. Overwintering adults become active when mean daily temperatures rise above 9°C, usually towards the end of March or in early April in the UK. However, they do not fly to brassica crops until about a month later, when mean daily temperatures have risen above 15°C. On arrival at a host crop, the beetles aggregate alongside shelter barriers, but within days they move into the crop and feed on the buds and flowers. During feeding, females chew holes in the bases of unopened flower buds and lay in each hole 1–3 eggs which, depending on the temperature, hatch in 4–10 days. The larvae feed on pollen grains in the anthers. As the larvae grow, they move to buds in which no eggs have been laid, and, as buds without eggs become fewer, some larvae move on to the flowers. After feeding for 25–30 days, the fully grown, third-instar larvae descend to the soil to pupate in earthen cells. The young beetles emerge 2–3 weeks later. At first, they are light yellowish-brown in colour but within 1–2 days they darken to the characteristic metallic, greenish-black colour. Although some beetles feed for a short time on newly formed seed pods, most leave to feed on other plants. Beetles move to their overwintering sites when the photoperiod begins to shorten, towards the end of July or the beginning of August in the UK.

Plant Damage

The feeding of blossom beetles in the florets of some horticultural brassica crops (e.g. calabrese, cauliflower) may reduce marketability, but the main damage caused by *Meligethes* spp. is associated with the adults destroying the flowers of seed crops and thereby reducing yield. The extent of the damage depends on the stage of crop growth and on the severity of the infestation. When large numbers of beetles enter seed crops in which the buds are large but unopened, some beetles chew holes in the bases of the flowers to gain access to pollen and, in doing this, they frequently damage the stigmas and ovaries. Least damage occurs in fields where the flowers open before the insects attack. In this case pollen is freely available to the beetles, which also eat petals, and damage to the ovaries is reduced considerably. Buds damaged by beetles often wither and fall from the plant, leaving behind 'blind' stalks. Recording the percentage of blind stalks on each raceme in crops such as oilseed rape is of little assistance in estimating the effects of beetle damage, as many flower heads abort for reasons other than beetle damage and, even when beetle damage occurs, the racemes frequently compensate by producing additional flowers. Only unusually do large numbers of larvae cause crop loss. Larvae infesting less than 25% of flowers may have an overall beneficial effect as pollinators, as they move from flower to flower to find new feeding sites. Overwintering seed crops suffer less damage than do spring-sown crops. Autumn-sown plants are about 1 m tall, have many well-developed racemes and are often in flower when the beetles arrive in early summer. In contrast, spring-sown crops are often less than 10 cm tall and frequently have only a small primary flowering shoot. Early-flowering, large plants of autumn-sown crops are better able, therefore, to tolerate beetle attacks than are later-flowering, small plants of spring-sown crops.

Cultural Control

Effects of routine crop cultivations on the overwintering survival of three hymenopterous parasitoids (*Phradis interstitialis* Thomson, *P. morionellus* Holmgren and *Tersilochus heterocerus* Thomson—all Ichneumonidae) of *Meligethes* have been studied in southern Sweden, where, on average, four times as many parasitoids emerged from wheat crops direct-drilled into oilseed rape stubble than from wheat crops drilled into stubble plots which had first been disc-harrowed or ploughed. Direct-drilling of crops favoured survival of the parasitoids but no comparable data were provided on the numbers of blossom beetles. In Finland damage to spring-sown oilseed rape crops was reduced by growing autumn-sown trap crops alongside and to cauliflower crops in the summer by surrounding them with flowering crops of Chinese cabbage, calabrese, sunflower, marigold or oilseed rape. This strategy reduced annual damage by *Meligethes* spp. from 30% to between 2% and 5%. However, to be effective, the trap crops had to cover a larger area than the area growing the 'protected' crop.

Chemical Control

Because of the large difference in size between plants in autumn- and spring-sown oilseed rape crops, the spray thresholds which should be used are, respectively, 15–20 beetles and 3 beetles per plant at the green bud stage. As autumn-sown crops flower relatively early in the year, one spray application of, for example, azinphos-methyl+demeton-*S*-methyl sulphone, deltamethrin, endosulfan, fenvalerate, gamma-HCH, malathion or phosalone is usually sufficient to protect crops in fields where the threshold infestation is exceeded. However, on spring-sown crops a second treatment may be necessary at the yellow-bud stage if the crops become reinfested with beetles. Once a crop starts to flower, most damage has been done and the optimal time to spray has passed. For this reason sprays should not be applied to flowering crops. Apart from the possible detrimental effect of insecticides in killing bees and contaminating their products, insecticides applied at flowering also kill large numbers of other pollinators and may inadvertently decrease, rather than increase, final seed set.

A computer-based advisory system for farmers, designed specifically to prevent excessive use of insecticides by forecasting the levels of pest infestation and calculating the economics of control, has been developed for the control of pests (mainly *Meligethes* spp., *Dasineura brassicae* Winnertz (see page 125) and *Ceutorhynchus assimilis* Paykull (see page 103)) of oilseed rape crops in the UK.

4.4.2 *Phyllotreta* Foudras spp.: Flea Beetles (Coleoptera: Chrysomelidae)

Geographical Distribution

Flea beetles are characterised by their enlarged hind femora (Figure 4.3), with which they make long 'flea-like' leaps. Most flea beetles that damage cruciferous crops belong to the genus *Phyllotreta* and are abundant throughout North America, Europe and south-east Australia. The more important species are *P. aerea* Allard (Europe and Canada), *P. albionica* Le Conte (Canada), *P. armoricae* Koch (Canada and USA), *P. atra* Fabricius (Europe), *P. consobrina* Curtis (UK), *P. cruciferae* Goeze (Europe, Canada and USA), *P. nemorum* L. (Europe and south-east Australia), *P. nigripes* Fabricius (Europe), *P. pusilla* Horn (USA), *P. ramosa* Crotch (USA), *P. robusta* Le Conte (Canada), *P. striolata* Fabricius (Canada and USA) and *P. undulata* Kutschera (Europe). Most temperate countries have a large number of *Phyllotreta* species, the vast majority being pests of agricultural crops.

Description and Life Cycle

The adult beetles can be separated by colour into two major groups, one resembling *P. nemorum*, in which the elytra are black with two longitudinal yellow bands, and the other resembling *P. cruciferae*, in which the elytra are of one colour (Figure 4.3). Adults of all species have a metallic lustre and are about 1.5–3.0 mm long.

The adults overwinter in sheltered sites alongside hedgerows and copses or in other places where leaf litter and plant debris abound. They move out as soon as temperatures rise in the spring, at first staying close to shelter barriers and feeding mainly on weed seedlings. Adult movement remains limited until mid-day temperatures rise above 20°C, when most beetles become highly active and disperse, frequently in large numbers, on prevailing winds. When the dispersing adults locate a suitable host crop, they settle and start to feed, often on seedling tissues below ground. Towards the end of May the beetles mate and begin to lay their eggs, either singly (*P. nigripes*) or in groups (*P. undulata*), in the soil. The eggs are pale-yellow to yellow in colour, about 0.3 mm long and 0.15 mm wide. The larvae of all species are generally white or pale-yellow in colour and have very short legs. The larval head is always sclerotised and dark, and several species also have one or two other characteristically dark thoracic and abdominal plates. The larvae of most species move to plant roots to feed but the larvae of *P. nemorum* feed as leaf miners in the cotyledons. After feeding for about 4–5 weeks, the fully grown larvae are about 5–6 mm long and resemble small wireworms (see page 33). They pupate in earthen cells in the soil. The 2.5-mm-long pupa is white to yellow in colour at first, but later turns darker. In the UK the pupal stage lasts about 4 weeks and a new generation of adults emerges towards the end of July or in early August, when cruciferous plants are large and there is plenty of food for the beetles to build up their reserves for overwintering. On days suitable for flight in the autumn, the beetles seek shelter and fly into hedgerows and thickets, where they become inactive at the lower temperatures. The beetles do not tend to overwinter in the relatively exposed soil under cruciferous crops. In general, the beetles become active in the spring in reverse order to the order in which they entered shelter in the autumn. *P. nigripes* is usually the last species to enter overwintering sites in the autumn and the first to leave them in the spring. Although *Phyllotreta* spp. are generally considered together, damage occurring at different times of the year at individual sites is usually caused by different species.

Plant Damage

Damage by *Phyllotreta* spp. is most evident on seedling cruciferous crops in the spring, generally from mid-April and throughout May. Severe damage caused by beetles feeding on seedlings below the soil surface sometimes results in patchy crops. With emerged seedlings, the beetles chew holes, particularly in the cotyledons, giving them a characteristic shot-hole appearance. Apart from the direct damage they cause, flea beetles can also transmit turnip yellow and turnip mosaic viruses (see Table 8.1 on pp. 268–9 for descriptions of plant viruses). Crops are again at risk at the end of July and in early August, when the new generation of *Phyllotreta* emerges. Although most plants are by then large, damage may be severe if large numbers of beetles enter crops, as when local seed crops are harvested and large populations move from these crops into vegetative crops.

Natural Control

In fine warm springs the soil surface warms up faster than the layer 1–2 cm deep in which cruciferous seedlings germinate. These conditions favour flea beetles and feeding activity is then invariably high. However, dry conditions subject seedlings to water stress and the additional loss of water from tissues caused by beetles' feeding may be sufficient to kill the seedlings. In contrast, alternating periods of hot and cold weather inter-spersed with showers are not favourable to the survival of flea beetles. In showery weather, when the surface soil is more or less permanently moist, seedlings establish much quicker and are soon large enough to tolerate most infestations of *Phyllotreta* spp. Hence, favourable growing conditions shorten the sensitive growth stage of the plant and reduce damage.

Cultural Control

Attempts to reduce the impact of this pest by cultural methods have concentrated on growing crops in living mulches or with a weed cover so that the crop plants are not silhouetted against a bare soil background and are therefore not so obvious to colonising flea beetles. Although this practice reduced the numbers of beetles attacking undersown crops, the plants were significantly smaller than monocropped ones. Similarly, cruci-ferous plants growing in alternate rows with other (intercropped) crop plants were less damaged by flea beetles, but the intercropped plants were considerably smaller than plants growing in weed-free, monocropped situations. Unless these cultural practices can overcome reduced crop yields, their usefulness will remain minimal as an alternative to insecticides for controlling *Phyllotreta* spp. A better approach may be to fertilise crops with organic manures rather than with inorganic chemicals, as there is evidence that crops grown 'organically' support fewer flea beetles.

Biological Control

Single applications before or after crop-plant colonisation by flea beetles of water suspensions of the nematode *Neoaplectana feltiae* Filipjev (Rhabditi-da: Steinernematidae), using approximately 2.5×10^6 and 10^5 infective juveniles per m^2 of crop, were not effective against adult *P. cruciferae* or their progeny. Few nematodes survived in the field for more than 6 days, an unexpectedly high mortality, which could not be explained solely by the high concentration of herbicide applied to the test plots.

Chemical Control

Considerable damage can occur before cruciferous seedlings emerge from the soil. Thus, it is often prudent to apply insecticide (e.g. gamma-HCH) to the seed or to the soil at drilling (e.g. aldicarb, carbofuran, carbosulfan), to ensure adequate establishment of cruciferous crops at a time when activity by *Phyllotreta* spp. can be severe. If attacks are heavy, additional applications of insecticidal dusts or sprays containing, for example, carb-aryl, gamma-HCH or cyfluthrin can be made when the seedlings have emerged from the soil.

4.4.3 *Psylliodes chrysocephala* (L.): Cabbage Stem Flea Beetle (Coleoptera: Chrysomelidae)

Geographical Distribution

P. chrysocephala occurs throughout Europe, particularly in coastal regions and, recently, has been found in Newfoundland, Canada. In Canada the closely related 'hop flea beetle' (*P. punctulata* Melsheimer) feeds on a wide range of plant species, including cultivated cruciferous crops, but it is less damaging than *P. chrysocephala*, as its larvae feed only on plant roots.

Description and Life Cycle

The adult beetles are about 5 mm long, much larger than the other flea beetles associated with cruciferous crops. Although they are usually shiny greenish- or bluish-black in colour (Figure 4.3), bronze forms occur commonly in some localities. Apart from size, beetles of the genus *Psylliodes* Berthold differ from other flea beetles by having the hind tarsus inserted at some distance from, rather than at, the apex of the tibia. In addition, the antennae are 10- rather than 11-jointed.

As the adult beetles live for about a year and a half, generations overlap considerably, and at some times of the year all stages of the life cycle may be present in the same crop. Adult beetles overwinter in plant debris in similar sites to *Meligethes* and *Phyllotreta* spp. but their larvae overwinter within the stems of host plants. In late winter, generally at some time during February, these larvae move into the soil and pupate in earthen cells. Adults from these pupae and from overwintering sites in hedgerows emerge in the spring and, after a short bout of feeding, begin to lay single, yellowish-coloured eggs, each about 0.5 mm long, in the soil alongside cruciferous plants or on plant tissues. Each female probably lays up to 1000 eggs but, particularly during the spring period of oviposition, many are destroyed by other soil arthropods during the 20 days or so that they require to hatch. Shortly after they emerge, the small larvae chew into suitable host plants, usually at the base of petioles, and feed in the leaf veins of large plants or in the main stems of smaller plants. The larvae, which eventually grow to between 6 mm and 8 mm in length, are white in colour and have numerous small dark spots on their dorsal surface. In common with other beetle larvae, they have three pairs of walking legs and the chitin of the head segment is heavily sclerotised, appearing black in colour. In this species the dorsal plates behind the head and on the last segment of the body are also black. The larvae grow rapidly and, depending on the situation, leave their host plants to pupate in the soil, either towards the end of April or throughout May. A new generation of adults emerges from the soil in June and July which, with those adult beetles surviving the previous winter, is induced into a summer resting phase (aestivation) by genetically fixed adaptation rather than by a response to changes in ecological factors such as temperature, photoperiod, humidity or food. They become active again and start laying eggs towards the end of August and this phase of oviposition continues into the winter. Cabbage stem flea beetle larvae are adapted to develop at cold temperatures, ceasing development only when the ambient temperature falls below 3°C. The beetles lay eggs singly, but a beetle may lay several eggs close to one plant and several beetles may lay on the same plant. Hence, an infested plant may contain 30 or more larvae in early winter. Larvae present in such numbers soon destroy the growing points of the primary and many of the secondary racemes, and this damage reduces subsequent seed production. Severe infestations may kill the plants. Most of the older beetles die after laying eggs but, following oviposition, the younger beetles move to sheltered sites, where about 30% overwinter successfully.

Plant Damage

In the UK larvae of *P. chrysocephala* are considered to be major pests of brassica seed crops. In the autumn and winter, the larvae often feed within the stems of spring cabbages, kale and, particularly, autumn-sown oilseed rape. In the past this pest was prominent in nursery beds of swede seedlings and brassica transplants, as the females appear to prefer to lay their eggs in dense plant stands. When large numbers of larvae are present, the plant becomes flabby, loses its larger leaves and sprouts many new small leaves from the side of the growing point to replace leaves already lost. If the insects continue to feed after plants have reached this stage, the plants are usually killed and crop stands become patchy. With crops, such as oilseed rape, that are grown at a high plant density, damage caused by cabbage stem flea beetle larvae has less effect on yield than in spaced crops such as cabbage and kale, as surviving oilseed rape plants, unlike surviving cabbage and kale plants, are able to grow larger and compensate to some extent for yield loss. However, compensatory seed yield is usually reduced and, furthermore, may not be ripe at harvest.

Damage caused by plant pathogens such as canker, *Leptosphaeria maculans* (Desm.) Ces. and de Not. (Dothideales: Pleosporaceae), to the crown of cruciferous plants is often related to the amount of beetle damage, and nearly all canker on the upper stems of plants is associated with insect damage.

Chemical Control

In those cropping areas where this insect predictably reaches pest status annually, granular formulations of insecticides (e.g. carbofuran, fonofos, phorate) should be incorporated into the soil at the time of drilling the seed, to provide effective protection. Alternatively, crops can be sprayed in the early autumn to kill egg-laying adults and later (in October) to kill surviving larvae. In the UK pyrethroids (e.g. bifenthrin, cyfluthrin, cypermethrin, deltamethrin, fenvalerate) have proved exceptionally effective when applied as sprays. Another option is not to apply insecticide against the adults but to inspect the crop in early October and, if more than five larvae are found to a plant, to apply an insecticidal treatment which is probably justified economically.

4.4.4 *Ceutorhynchus assimilis* (Paykull): Cabbage Seed Weevil (Coleoptera: Curculionidae)

Geographical Distribution

C. assimilis is a pest of seeding cruciferous crops throughout Europe, Canada and the USA but is not found in the southern hemisphere. A wide range of cruciferous weeds provides alternative hosts.

Description and Life Cycle

Different *Ceutorhynchus* Germar species are difficult to identify, as the adults resemble each other closely and there are many species. For example, there are more than 50 species in the UK and the vast majority feed on Cruciferae. Adult *C. assimilis* overwinter in sheltered sites in hedgerows and leaf litter. The adult beetles are 2–3 mm long and black in colour, the colouring being muted slightly by a covering of greyish-white scales (Figure 4.3). These scales and the absence of a small tooth on the ventral side of the middle and hind femora are the characteristics that separate this species from *C. pleurostigma* Marsham, the turnip gall weevil (see p. 107). The adult weevils emerge from overwintering during May and June and feed on the foliage of many cruciferous plants, though only rarely doing much damage. After this initial feeding phase, males and females mate, with the females moving subsequently onto newly forming pods within seeding crops. The time of this migration can be forecast from air temperatures. The female bores a round hole in the pod with her rostrum and lays an egg alongside or inside a seed. Several eggs may be laid in the same seed pod, although there is evidence that, to deter further oviposition in the same pod, the female marks it with a pheromone. The females oviposit until about mid-June. The eggs, which hatch in 8–10 days, are creamy white in colour and are 0.4–0.5 mm long and 0.2–0.3 mm wide. The larvae feed within the pods, totally destroying some of the developing seeds and damaging many others. The legless larvae are creamy-white in colour, with a distinctive yellow-brown, lightly sclerotised head. Each larva destroys 3–5 seeds. After feeding for 4–5 weeks, the larvae are fully grown and 4–6 mm long. At this stage each larva chews a characteristic, round emergence hole, about 0.5 mm in diameter, through the wall of the seed pod and drops to the soil to pupate in an earthen cell. The pupal stage lasts about 12 days. Adult weevils emerge from the pupae throughout July and August and, although many may be seen on flowers of wild and cultivated plants, they feed little. When the first cold weather occurs, the weevils move to sheltered sites associated with hedgerows and similar barriers.

Plant Damage

C. assimilis is considered to be a serious pest of seed crops in the UK, particularly as a result of the recent large increase in the area planted with oilseed rape. In untreated crops infestations of up to 70% of pods have been recorded. Mustard crops are generally damaged more than are oilseed rape crops, largely because mustard is drilled to a smaller plant stand. This pest is doubly troublesome because the holes made by the adults feeding on the seed pods are used by *Dasineura brassicae* Winnertz (see page 125), the brassica pod midge, to gain access to the pods.

Biological Control

Although the larval instars of the cabbage seed weevil can be parasitised by more than ten other insect species, several of which have been introduced into Canada as biological control agents, none has managed to control weevil infestations on a field scale.

Chemical Control

In the UK, oilseed rape crops are partially protected against *C. assimilis* adults by applying insecticidal sprays before flowering, but, as the weevils drop to the ground at the slightest disturbance, foliar sprays are generally not completely successful. Sprays containing, for example, azinphos-methyl + demeton-*S*-methyl sulphone, endosulfan, gamma-HCH, malathion, phosalone should not be applied when the crop is in full flower, as insecticides will then kill too many pollinators and other beneficial insects. After flowering, when the crop is predominantly green, *C. assimilis* larvae hatch from the eggs. At this stage the seed pods are small and appropriate insecticides can pass through the thin pod walls and kill larvae that have entered the pods. Spraying insecticides to kill larvae is considered to be economically justified only in crops where more than one adult weevil is recorded per plant during the flowering stage. The large size and intertwining nature of individual plants in seeding brassica crops, together with the weevils' habit of falling to the ground when disturbed, make the threshold difficult to assess accurately. However, control of *C. assimilis* is essential if subsequent damage by *D. brassicae*, the pod midge (see page 125), is to be maintained at a low incidence in areas where the midge is likely to cause damage.

4.4.5 *Ceutorhynchus quadridens* (Panzer): Cabbage Stem Weevil (Coleoptera: Curculionidae)

Geographical Distribution

C. quadridens occurs as a pest throughout Europe and the northern states of America but is not found in the southern hemisphere.

Description and Life Cycle

C. quadridens is distinguished from *C. assimilis*, the cabbage seed weevil, by usually being slightly larger (3–4 mm long), by having yellowish-white-coloured body scales (which make the adults appear to have a rough, granular, surface texture), by being grey-brown rather than grey-black in appearance and, most characteristically, by having a distinct patch of white-coloured scales behind the scutellum (Figure 4.3). These scales and the reddish-coloured antennae and legs can be seen easily with a ×5–10 magnification hand lens. *C. picitarsis* Gyllenhal, the rape winter stem weevil (see p. 106), is more difficult to distinguish from *C. quadridens*, but *C. picitarsis* does not have scales on the elytra and has only a few small scales behind the scutellum.

Adult *C. quadridens* emerge from their overwintering sites about mid-April in the south of the UK, at about the same time as *Meligethes* spp. and *C. assimilis*. However, few eggs are laid until mid-May, when mated females chew holes in the main veins and stalks of the lower leaves and stems of host plants (frequently species of *Brassica* L.) and lay 2–6 eggs in each hole. The translucent eggs are oval-shaped and measure about 0.7×0.5 mm. The area surrounding the site where the eggs are laid turns white in colour and forms characteristic, slightly swollen, blisters as the parenchymatous pith cells degenerate and die as a direct result of chemicals released during egg-laying. The head capsules of newly hatched larvae are black in colour, but later turn yellow-brown. The larvae tunnel initially through the leaf veins and stalks, which become brittle and can be broken easily, and then move to the main stem, where communal feeding, often by 20–30 individuals, may destroy the plant. Fully grown larvae, 4–6 mm long, chew their way out of the stem, often leaving by holes in lower leaf scars which mark the entry points of the larvae when they moved from the leaves into the stem, and drop to the soil surface to pupate in earthen cells about 5 mm long. The new-generation adults begin to emerge from the pupae during July onwards and generally feed little before entering their overwintering sites.

Plant Damage

Although all cultivated Cruciferae can be attacked, the numbers of weevils are often small and infestations tend to be sporadic. Plants attacked in nursery seed beds become flabby and spongy to the touch and should be discarded, as they usually die within weeks of transplanting. *C. quadridens* became numerous in spring-sown oilseed rape crops growing in England in the early 1980s. However, cabbage stem weevil damage was spread evenly across crops with few plants being killed and infestations remained largely unnoticed by growers. Probably as a result of the major switch from spring- to autumn-sown oilseed rape crops on which it establishes less successfully, *C. quadridens* is again on the decline. However, it can cause cosmetic problems in high-value crops such as cauliflower if systemic insecticides are not applied to control *Delia radicum* L., the cabbage root fly (see page 129). Tunnels where weevil larvae have entered cauliflower curds are clearly visible in the cut ends of stems and many curds may be rejected. .

Chemical Control

In the UK carbofuran and carbosulfan applied to control the cabbage root fly (see page 133) and gamma-HCH and malathion applied to control blossom beetles (see page 97) and/or seed weevil infestations (see page 104) also protect plants from damage by *C. quadridens*. Pyrethroid insecticides, including deltamethrin, will also reduce infestations of this weevil.

4.4.6 *Ceutorhynchus picitarsis* Gyllenhal: Rape Winter Stem Weevil (Coleoptera: Curculionidae)

Geographical Distribution

In the 1930s *C. picitarsis* was a rare species restricted mainly to the wild cruciferous plant *Sisymbrium officinale* Scop. ('hedge mustard'). With the expansion of overwintering oilseed rape in many countries in Europe, this winter-feeding weevil was provided with an abundance of an appropriate alternative cultivated host plant. It established successfully on winter oilseed rape and, as a result, was given its present common name. Since 1982 *C. picitarsis* has been of minor pest status in some parts of the UK and Europe. It is not found in North America.

Description and Life Cycle

The adult weevils can be separated morphologically from *C. quadridens* as described on page 105. They also differ behaviourally: *C. picitarsis* is active in the autumn, whereas *C. quadridens* is active mainly in the spring. Adult

C. picitarsis emerge in the summer from pupae in the soil beneath the previous year's rape crops and move to hedgerows and other shelter, where they enter a resting phase until the prevailing temperatures become lower in the autumn. They then become active and, after mating, the females move into young rape crops. Eggs are laid in crevices in stem axils or in holes chewed by the female weevils in the stalks of the leaves. The eggs hatch throughout the autumn and winter, and the legless larvae tunnel into the leaf stalks and the crowns of infested plants. The larvae are white in colour and 3–4 mm long when fully grown and, like the larvae of other weevil pests, have a distinct light-brown-coloured head. They complete their development in April or May and then leave the plants, to pupate in earthen cells in the soil.

Plant Damage

Although some oilseed rape plants attacked at the seedling stage in the autumn may be killed, most plants survive *C. picitarsis* infestations but the terminal growing points may be destroyed by the feeding larvae. This damage results in the development of many side-shoots, delayed flowering and consequent loss of yield.

Chemical Control

C. picitarsis is usually controlled by insecticides applied to control infestations of *P. chrysocephala*, the cabbage stem flea beetle (see page 102). In areas where the weevil is prevalent it may be necessary to apply treatments targeted specifically towards the control of *C. picitarsis* when infestations of *P. chrysocephala* are too light to merit treatments.

4.4.7 *Ceutorhynchus pleurostigma* (Marsham): Turnip Gall Weevil (Coleoptera: Curculionidae)

Geographical Distribution

Although *C. pleurostigma* is found on brassica crops throughout the UK and Europe, it is not considered a pest of major economic importance. Small numbers occur on the roots of most brassica crops and the weed *Sinapis arvensis* L. (charlock), but they are not found on the roots of most other cruciferous weeds in cultivated soils.

Description and Life Cycle

Adult weevils are about 3 mm long and appear to be black in colour on the dorsal surface and to be grey on the ventral surface. Although appearing black, the elytra are sparsely covered with scales white to grey in colour.

The adults emerge from pupae in the soil in late spring and throughout the summer but, like *C. picitarsis*, they remain quiescent until the cooler conditions of August and September, when they become sexually mature. They feed on the leaves, flowers and young seed pods of a wide range of wild and cultivated Cruciferae. The adults copulate repeatedly, during which the males may be carried by the females for up to 12 h. The eggs are almost transparent, soft, oval-shaped and measure 0.35 mm × 0.25 mm. They are laid singly, just below the surface of host plant roots, in small holes made by the females with their mouthparts. Although laid individually, several eggs, which hatch in 5–7 days, are often laid in the same root. On hatching, the larvae feed on the root tissues, and chemicals released by the larvae during feeding induce the plant cells to produce galled tissue which surrounds each of the 'invading' larvae. The larvae are legless, have a white-coloured fleshy body and a light-brown head typical of the curculionids, and pass through three instars to reach between 4 mm and 5 mm long when fully grown. Being somewhat transparent, the larvae that feed on swedes appear reddish-yellow because of the colour of the food within each of their guts. As the larvae feed, the galls continue to grow and may reach 10–15 mm in diameter. Where several larvae feed close together on a root, the galls often coalesce to form a large outgrowth with a larva feeding at the centre of each gall. The larvae feed on the surrounding tissues for about 10 weeks and then, when fully grown, they chew their way out of the galls to pupate in the soil in earthen cells in which the surrounding soil particles are glued together by a sticky material secreted by the larvae. Although many larvae enter these cocoons in the early spring, they do not pupate until as much as 30 days later, which helps to synchronise the later emergence of adults. Some later-developing larvae may continue to feed well into the summer. The pupal stage lasts about 35 days in the summer and the adults can force their way out of the pupal cells only when the soil is moist. In dry soil adults are able to survive within the cells for at least 40 days without food. When newly eclosed adults reach the surface of the soil, they are not immediately active but remain under soil clods and similar objects for about a day until their integument hardens fully.

Plant Damage

The most common galls found on roots of brassica plants are formed as a result of attack by the club root fungus, *Plasmodiophora brassicae* Voronin (Plasmodiophoromycetes: Plasmodiophorales). Some other galls occur as a result of 'hybridisation nodules' which arise when certain cultivars of swedes initiate numerous new growing points producing gall-like swellings on the part of the 'root' below soil level. These two types of gall can be differentiated from the galls caused by the weevil, by cutting them open. In contrast to club root and hybridisation nodules, weevil galls have larvae feeding in them or else each gall has a larval feeding chamber with a hole through which the larva left to pupate (Figure 4.4). Hybridisation nodules and weevil galls are usually round but hybridisation nodules are usually smaller and always solid. Similarly, club root galls are solid but they tend to be elongate rather than round, particularly when the fungus attacks small roots. In addition, club root galls are translucent when cut and, because of their high water content and the thin walls of their cells, they tend to rot quickly. In contrast to club root galls, galls of the turnip gall weevil rot slowly after the weevils have left and hybridisation nodules rot extremely slowly.

Cultural Control

In regions where the turnip gall weevil is particularly troublesome, roots from infested crops should be lifted as soon as possible in the spring and heaped to dry, as the larvae are unable to escape from galls in which the surface has dried out. Galled seedlings should be rejected when crops are transplanted in the autumn and crop rotation should be practised whenever feasible.

Biological Control

Although as many as 10% of weevil larvae may be parasitised by *Diospilus oleraceus* Haliday (Hymenoptera: Braconidae), they are not killed until they pupate and therefore the parasitoid has no effect on the damage caused by feeding weevil larvae. Late larvae and pupae may also be killed

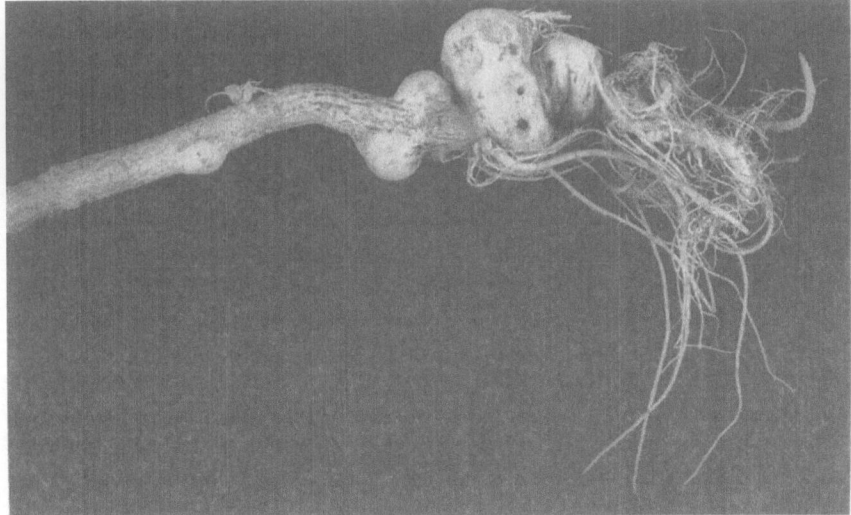

Figure 4.4 Turnip gall weevil damage showing holes in galls through which larvae left to pupate in soil. © HRI

by species of the fungus *Actinomucor* Schostakowitsch (Mucorales: Mucoraceae), but, again, generally after the pest has caused most of its damage.

Chemical Control

Specific insecticidal treatments are rarely considered to be economically justified and, consequently, approved treatments are likely to be available only locally.

4.5 BUTTERFLIES

4.5.1 *Pieris brassicae* (L.): Large Cabbage White Butterfly (Lepidoptera: Pieridae)

Geographical Distribution

P. brassicae is common throughout Europe and Asia but is not found in North America or Australasia. It is a major pest of cruciferous crops when large numbers overwinter successfully or migrate *en masse* between adjacent countries—for example, from France across the English Channel to England.

Description and Life Cycle

The adults which emerge in the spring from overwintering chrysalids are most active from mid-May to mid-June in the UK. The wingspans of the males and females average 63 mm and 70 mm, respectively. The scales covering the tips of the forewings of both sexes are black in colour Figure 4.5) and the female has, in addition, two large conspicuous black spots on the upper surface of each forewing. The females lay their flask-shaped eggs in batches of 20–100 on Cruciferae, usually on the undersides of the leaves. The eggs hatch in about 14 days and the young caterpillars first eat their eggshells and then feed communally on adjacent leaf tissues. After the third moult, the caterpillars disperse, many moving onto new leaves. Because most of the caterpillars feed on only one plant, they are extremely conspicuous. At first they are pale green in colour, but they rapidly become mottled blue-green. When fully grown, each caterpillar is 25–40 mm long and possesses three yellow-coloured longitudinal stripes along the body. Each is covered also with black markings and has groups of short, stiff, white hairs that arise from fleshy protuberances along the body. The caterpillar has a well-developed spinneret on the head, prolegs on abdominal segments 3–6 and a pair of claspers on the last segment. After feeding for about 30 days, they leave the host plant and

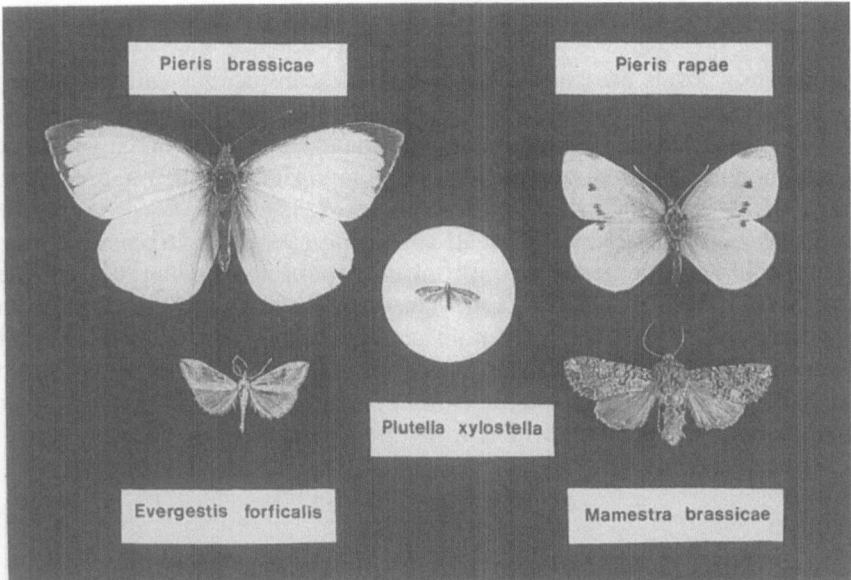

Figure 4.5 Adults of: (top left) large cabbage white butterfly, male; (top right) small cabbage white butterfly, female; (middle) diamond-back moth; (bottom left) garden pebble moth; and (bottom right) cabbage moth. © HRI

search for sheltered pupation sites on vertical or overhanging surfaces of trees, walls, fences or other similar objects. Each larva spins a silken pad to which it attaches itself in an upright position. It then secretes a silken girdle around its waist to hold it in position and pupates within the chrysalis. The chrysalis is grey-green in colour, with yellow and black marks and prominent abdominal spines. After 10–20 days in the summer, the chrysalids split open and the adults emerge. Their wings are crumpled at first, but within hours they expand and dry. A second generation occurs in the UK during July and August and the progeny of this generation overwinter in the pupal stage. In warmer parts of Europe there may be 3–4 generations each year.

Plant Damage

P. brassicae does relatively little damage to farm crops, as insecticidal treatments are usually very effective. Most damage occurs at field margins, where the butterflies often aggregate to take advantage of the food and shelter provided by hedgerow plants, and in gardens and allotments, particularly where insecticides are not applied. The adults lay their eggs on many cruciferous species and the larvae, if unchecked, often skeletonise leaves. Damage occurs most frequently on cabbage and Brussels sprout crops and on garden nasturtiums (*Tropaeolum majus* L.). The second generation is larger than the first and usually causes more damage.

Natural Control

Population levels are stabilised because the conspicuous caterpillars are eaten by some birds and predatory carabid beetles. A baculovirus and the parasitoid *Apanteles glomeratus* L. (Hymenoptera: Braconidae) achieve exceptionally high mortality of the caterpillars, aided by their aggregation. The small, bright-yellow-coloured cocoons of *A. glomeratus* can often be seen clustered alongside the dead and dying caterpillars from which the parasitoid has emerged. Natural control by parasitoids generally acts late in the life cycle of the pest and, hence, rarely reduces pest damage on existing crops, but high parasitoid activity does ensure that fewer pest insects survive to damage subsequent crops. The major effect of parasitoid activity is therefore to regulate insect pest populations so that their numbers do not fluctuate widely from one generation to the next.

Biological Control

In the Netherlands 60 strains and species of the egg-parasitoid *Trichogramma* Westwood (Hymenoptera: Trichogrammatidae) have been considered for inundative release against mixed populations of lepidopterous pests of cruciferous crops. However, in field experiments in which 180000 individuals of *T. evanescens* Westwood were released every 2 weeks from June until September, caterpillar densities frequently continued to exceed the threshold for economic damage. A major constraint to this strategy of biological control is that strains of *Trichogramma* often parasitise only one of the caterpillar species present in the field.

The potential for using different isolates of the fungal pathogen *Entomophthora* Fresenius (Entomophthorales: Entomophthoraceae) as a mycological insecticide is now being studied, although the use of this pathogen is likely to be restricted, because fungal pathogens tend to be costly to produce, susceptible to climatic factors and frequently of low pathogenicity. Furthermore, hosts often show great variation in their susceptibility to fungal pathogens. One pathogen that has been successfully used against *P. brassicae* is the bacterium *Bacillus thuringiensis* Berliner (Eubacteriales: Bacillaceae). The commercial preparations of *B. thuringiensis* spores and toxin now available for the treatment of crops are used like biological insecticides. Sprays containing the *B. thuringiensis* spores and toxin are highly effective against young caterpillars but need to be applied at 14 day intervals if they are to provide effective protection throughout the season. In effect, therefore, the only practical difference between mycological and chemical insecticides is that, with the mycological products, the toxicant is obtained from a biological source, rather than being synthesised chemically.

Cultural Control

Hand-picking of egg masses and young larvae is recommended on a garden scale for reducing the damage done by *P. brassicae*.

Chemical Control

Although *P. brassicae* is one of the commonest butterflies associated with agricultural land, it is much less common than formerly and insecticidal treatments of field crops are rarely considered to be economically justified. However, when local populations are boosted by large numbers of adults immigrating from other countries, insecticidal treatments may be needed. Only one spray is usually required to kill the caterpillars, especially if a pyrethroid insecticide (e.g. cyfluthrin, cypermethrin, deltamethrin, fenvalerate, permethrin, resmethrin) is used, although it is advisable to add a wetting agent to the spray to ensure adequate retention of the insecticide by the waxy brassica foliage.

4.5.2 *Pieris rapae* (L.): Small Cabbage White Butterfly (Lepidoptera: Pieridae)

Geographical Distribution

P. rapae is widespread throughout the northern hemisphere between the Arctic Circle (latitude 67° N) and the Tropic of Cancer (latitude 23.5°N) and is now well established in Australia and New Zealand. Its common name in the USA and Canada is 'the imported cabbage worm' and in Australia 'the cabbage white butterfly', as the larger *P. brassicae* (see page 110) does not occur on the Australian continent.

Description and Life Cycle

The adults emerge from overwintering chrysalids in the spring. The upper surfaces of the wings are a gleaming white colour and the forewings have conspicuous black tips. As with *P. brassicae*, the females have two black spots in the middle of each forewing (Figure 4.5). These spots often appear grey and less pronounced in females that emerge from the overwintering generation. The male has one black spot in the middle of each forewing, although occasionally the wings may be pure white without dark scales at the tips. The undersides of the wings of both sexes are yellow in colour, dusted with grey and without a pattern, a characteristic feature that distinguishes *P. rapae* from *P. napi* L. (green-veined white) and females of *Anthocharis cardamines* L. (orange tip), as all three species have an average wingspan of about 50 mm. *P. rapae* becomes active slightly earlier in the year than does *P. brassicae*. Mated females lay eggs on wild and cultivated cruciferous plants. In crops, they do not restrict themselves to the sheltered areas around the hedgerows, and eggs are often laid on many plants within the crops. Females alight on the upper surface of a chosen leaf and stretch their abdomen and ovipositor around the edge of the leaf to lay eggs on the lower surface. Eggs are laid individually and the females fly between laying successive eggs often on different plants. The pale-yellow-coloured, bottle-shaped eggs hatch in 10–15 days and the caterpillars feed solitarily, usually in the heart or around the growing point of the plant. Later instars rest openly along the mid-ribs of the leaves. Although dark green in appearance and well-camouflaged against the background, the larvae are relatively easy to find, as their presence is revealed by damaged leaves with accompanying frass. The caterpillars have a green velvety appearance, a characteristic dorsal yellow line and elongate, yellow spiracular patches along their sides. When fully grown, they are about 25 mm long. Caterpillars of *P. napi* occur mainly on wild cruciferous plants, although a few are found on cultivated crops. They can be distinguished from the caterpillars of *P. rapae* by the absence of the yellow dorsal line. *P. rapae* usually pupates on the plants, although sometimes the chrysalids are found on sheds and walls and under window-sills in similar positions to the chrysalids of *P. brassicae*. The chrysalis, although smaller, is similar to the *P. brassicae* chrysalis and rests vertically on a pad of silk to which it is attached by a silken girdle. The chrysalids of *P. rapae* usually colour-match their surroundings to a remarkable degree, so that chrysalids found on actively growing plants during the summer are invariably green, whereas chrysalids found on dead and decaying plant material during the winter are generally dull grey and speckled to match the colour patterns characteristic of diseased and senescing foliage. In the UK the second generation emerges over an extended period from July to September. *P. rapae* is univoltine in localities close to the Arctic Circle but can pass through eight generations annually in localities close to the Tropic of Cancer.

Plant Damage

In general, damage caused by *P. rapae* to foliage is slight, although it can be pronounced in seasons when infestations of caterpillars are severe. This species is considered a pest, not because of the amount of crop it destroys, but because of the amount it makes unmarketable by frass contamination. Frass is particularly troublesome in cabbage crops near to harvest, as the pungent-smelling frass lodges frequently between the wrapper leaves and the cabbage heads.

Cultural Control

Results from attempts to reduce infestations of *P. rapae* by intercropping, companion planting or using various types of mulches have been inconsistent. Although some results may indicate that one of these particular approaches has a beneficial controlling effect, there are other results indicating that the remaining treatments either have no effect or are detrimental because they result in increased pest damage.

Biological Control

Many *P. rapae* are killed by the hymenopterous parasitoids *Apanteles* Foerster (Braconidae) spp., *Pteromalus puparum* L. (Pteromalidae) and *Meteorus versicolor* Wesmael (Braconidae). In many seasons in Virginia, USA, between 60% and 70% of overwintering pupae of *P. rapae* may be parasitised by *P. puparum*. A method for using species of *Heterorhabditis Poina.* (Rhabditida: Heterorhabditidae) and *Neoaplectana* Steiner (Rhabditida: Steinernematidae) nematodes to control *P. rapae* under field conditions in Canada has been suggested, although it has not yet been tested. Attempts to control caterpillars of white butterflies with naturally occurring viral and bacterial diseases produced promising results under laboratory and field conditions in Maryland, USA, and in the UK. In Maryland a granulosis virus of *P. brassicae* was as effective as a commercial formulation of *B. thuringiensis* at controlling larvae of *P. rapae* and a non-commercial formulation of *B. thuringiensis* gave better control than did commercial formulations. However, under the maritime conditions of the UK only 7–33% of a purified preparation of *P. brassicae* granulosis virus remained infective 1 day after application, demonstrating the need for frequent applications of microbial preparations (or improved formulations), particularly where the insects live in a relatively damp climate and emerge over a protracted period.

Chemical Control

The caterpillars are difficult to contact with insecticide once they have entered the hearts of brassica plants, and sprays are therefore best applied to coincide, as closely as possible, with egg hatch. One correctly timed spray is usually sufficient, although, if the adults emerge over a protracted period, a second spray may be needed. Attacks by first-generation caterpillars do not often affect the yield of Brussels sprout plants, which may be safely left untreated. Attacks nearer to harvest, however, usually have to be controlled. Selection of the insecticide is governed largely by the time interval remaining between attack and crop harvest, only chemicals with a 7 day harvest interval (e.g. carbaryl, iodofenphos, quinalphos), a 3 day interval (e.g. pirimiphos-methyl), a 2 day interval (e.g. trichlorfon) or a 0 day interval (e.g. the pyrethroids: cyfluthrin; cypermethrin; deltamethrin; fenvalerate; permethrin; and resmethrin) being available when damage becomes apparent shortly before harvest. Although treatments with foliar sprays, using especially pyrethroid insecticides, are favoured by most commercial growers, dusts are often preferred by private gardeners, as these formulations can be applied more readily and without specialised application equipment.

4.6 MOTHS

4.6.1 *Mamestra brassicae* (L.): Cabbage Moth (Lepidoptera: Noctuidae)

Geographical Distribution

M. brassicae is abundant throughout Europe and temperate Asia. It is not present in Canada and the USA, where its agroecological niche is filled by *M. configurata* Walker, the Bertha Armyworm. Although *M. brassicae* is commonly called the 'cabbage moth', its caterpillars, like those of many other noctuid moths, are extremely polyphagous, feeding on vegetables such as beets, lettuces, onions, peas and potatoes, on many types of garden flowers and on a wide range of tree species.

Description and Life Cycle

The adult moths (Figure 4.5), similar in appearance to many other sombrely coloured members of the Noctuidae, emerge from pupae in the soil during May and June. The forewings are mottled and may appear grey, green, brown or black. The hindwings are generally lighter in colour than the forewings. Adults have a characteristic curved dorsal spur on each tibia of the forelegs. Other features that distinguish *M. brassicae* from similar Noctuidae are the white-bordered, kidney-shaped mark towards the centre of each forewing and the irregular white transverse line near the margin. The wingspan of the moths varies from 30 mm to 50 mm, the size being governed largely by the nutritive value of the caterpillars' host plants. Shortly after emergence from the soil, the moths mate and the females start to lay hemispherical, ribbed eggs which, although white in colour when newly laid, become pink later. The eggs are laid singly but sufficiently close to one another to form neat groups, each containing up to 50 eggs. In the UK most eggs are laid in May and June on the undersides of leaves and hatch in 6–10 days. Initially, the young caterpillars are coloured green and feed gregariously on leaves. They continue to feed together during the first three instars, but during the fourth instar they disperse over the host plant, at the same time changing colour, with the dorsal region darkening to a brownish hue and the ventral region turning yellowish green. Some individuals may turn almost totally black in colour. At this stage the caterpillars also have a dusky dorsal stripe, speckled with white, and a yellowish, light-green or dusky-brown stripe low down on the sides. Even slight infestations of fourth-instar caterpillars can be troublesome, particularly in crops such as heading cabbage, as the caterpillars frequently bore into the hearts of plants, rendering them unmarketable. Larval development normally takes 4–5 weeks, by which time the larvae are fully grown, are 40–50 mm long and have a characteristically small, dorsal hump on abdominal segment 8. Mature larvae leave plants to pupate into glossy brown-coloured chrysalids within flimsy cocoons in the soil. Although many pupae overwinter in the UK, some produce a second generation. The larvae of this generation enter the soil during October, either to pupate then or to overwinter as caterpillars with pupation occurring in the spring. As a result, generations may run into each other in subsequent years.

Plant Damage

Severe infestations of caterpillars rapidly skeletonise the outer leaves of large plants and can sometimes destroy small plants. Plant death does not occur frequently, because, when the moths emerge, most crops are well-established and the females lay their eggs on large rather than small plants. Most crop loss caused by *M. brassicae* caterpillars occurs as a result of plants becoming contaminated with frass rather than from the amount of plant material consumed.

Biological Control

Cabbage moth populations can be reduced, though not eradicated, by the egg parasitoid *T. evanescens* (see page 112), by the larval parasitoids *A. glomeratus* (see page 112) and *Hyposoter ebenius* Gravenhorst (Hymenoptera: Ichneumonidae) and by the pupal parasitoids *P. puparum* (see page 115) and *Pimpla instigator* Fabricius (Hymenoptera: Ichneumonidae). Apart from these hymenopterous parasitoids, *M. brassicae* caterpillars can be parasitised by the dipterous parasitoids *Phryxe vulgaris* Fallén and *Compsilura concinnata* Meigen (both Larvaevoridae). None of these parasitoids is specific to *M. brassicae*. They all attack a wide range of caterpillars. Mixed populations of *T. evanescens*, *T. confusum* Viggiani and *T. dendrolimi* Matsumura have been released recently in the field in Germany to control *M. brassicae*. In the first year of release, 1981, the summer was cold, the caterpillar populations were small and the degree of parasitism was increased only slightly from 32% in untreated plots to 43% in treated plots. However, in the following year, when the weather was warm and the pest infestation was high, release of a local strain of *T. evanescens* achieved 93% parasitism of caterpillars compared with only 15% parasitism in the untreated crop.

Laboratory studies in France have indicated that *Paecilomyces fumoso-*

roseus Wize (Deuteromycetes: Moniliaceae), an entomopathogenic fungus, is potentially a valuable control agent: five out of ten fungal isolates tested were extremely active against *M. brassicae*. A baculovirus is also produced commercially in France for the control of *M. brassicae* and some other noctuid species. *M. brassicae* is not highly susceptible to the existing commercial strains of *B. thuringiensis*.

Host Plant Resistance

An Australian cauliflower, line PI234599, bred for its resistance to cruciferous pests in New York State, USA, was also found to be resistant to infestations of *M. brassicae* in the Netherlands.

Integrated Control

In Germany a mixture of a nuclear polyhedrosis virus of *M. brassicae*, a formulation of *B. thuringiensis* subsp. *kurstaki* Kurstak and the relatively selective carbamate insecticide pirimicarb was compared with some non-selective insecticides used extensively to control infestations of lepidopterous pests in brassica crops. Using 1×10^{13} viral polyhedra/ha, insect control achieved by the virus and the most effective non-selective insecticide was similar. The effect of the virus was not influenced adversely by the bacterium and the selective insecticide, and the mixture seems appropriate to integrated control programmes. The stability of the virus was enhanced by the use of protectants against UV light and the virulence of several isolates of the virus was high against all strains of *M. brassicae* tested.

Chemical Control

A single insecticidal spray treatment, using, for example, one of the compounds recommended for the protection of brassicas against *Pieris* spp. (see pages 113 and 115), is usually sufficient to control this pest in the UK. Treatments to control *M. brassicae* are most effective when applied before the caterpillars reach the fourth-instar stage of development.

4.6.2 *Plutella xylostella* (L.): Diamond-back Moth (Lepidoptera: Plutellidae)

Geographical Distribution

P. xylostella, a cosmopolitan species, is the most widespread of lepidopterous pests of vegetable brassica crops. It is a major pest of cruciferous crops in America, South-east Asia, Australia, Canada, Europe and New Zealand.

Description and Life Cycle

Adult moths are about 6 mm long, with a wingspan averaging 15 mm (Figure 4.5). They are brownish in colour and have three light-brown to white, triangular marks on the trailing edge of each forewing. Usually, the females are lighter-coloured than the males. Both sexes have grey hind-wings fringed with hairs, the fringe along each trailing edge being about as broad as the wing. The wings of adults at rest are closed above the body, so that, when viewed from above, the triangular marks on the forewings meet to form diamond shapes from which the moth gets its common name.

Diamond-back moths overwinter as adults if the temperatures do not fall too low. As most moths die during the winter in the more northern parts of their distribution, infestations start in the following year only when immigrating moths from warmer areas reinfest the area. The moths are relatively poor fliers but they are often transported long distances on the wind. In the cool maritime climates of the UK and northern Europe, there are usually one or two generations, albeit ill-defined, each year. The moths do not survive the winter in Canada, but 2–3 generations occur annually in the coastal regions of British Columbia and as many as 6 generations per year in Ontario. Each year, new infestations are started by moths entering Canada from the USA. The numbers of moths caught in pheromone traps in early summer in the USA provide a useful method for predicting the severity of infestations in Canada later in the year. In hot southern areas of Europe and North America, *P. xylostella* may complete its life cycle in 12–15 days and there may then be as many as 15 generations in one year. The moths are active mainly towards dusk and mate shortly after emerging from overwintering. The small, yellow-coloured eggs are laid singly, or in groups of 2–3, mainly alongside the mid-rib, or the smaller leaf veins of leaves of Cruciferae. Each female lays between 50 and 150 eggs. The exact positions in which eggs are laid are governed largely by the genera of the host plants: in *Brassica* spp. most eggs are laid on the upper leaf surfaces; in *Barbarea* Beckmann spp. (water-cress, etc.) most are laid on the lower leaf surfaces; and in *Arabis* L. spp. (rock-cress) similar numbers are laid on both leaf surfaces. The eggs hatch in 2–3 days and the small caterpillars then crawl to the lower leaf surfaces, where they chew their way into the leaves and feed as leaf miners on the spongy mesophyll cells. After the first moult the caterpillars emerge from the leaves and feed from the lower surfaces. As the caterpillars eat practically all of the leaf material except the upper epidermis, leaves are often covered by numerous windows (fenestrations). When infestations are severe, feeding is more intensive and leaves can be destroyed entirely. The caterpillar is coloured light green, with a tapering body that is widest in the middle. The first segment of the thorax has two circles of black spots. Each of the remaining two thoracic segments and all segments of the abdomen have up to 14 white patches and 1–2 dark spots. In the first two of the four larval instars, the

head capsule is black in colour but it becomes pale yellow as the larva
matures. Caterpillars disturbed while feeding wriggle violently and often
drop for a short distance from the leaf surface, remaining suspended from
it by a fine silken thread. Fully grown caterpillars construct a flimsy cocoon
on the leaf surface and the insects then remain quiescent in a prepupal
stage for one or two days before developing into pupae about 9 mm long.
The prepupal and pupal stages occupy about one-third of the complete
generation time.

Plant Damage

P. xylostella caterpillars destroy the foliage of most types of brassica crops
and, in some parts of the world (e.g. south-east Asia), are considered to be
the most important factors limiting the successful production of cruciferous
vegetables. Severe infestations of *P. xylostella* are particularly destructive,
as, unlike many lepidopterous pests, the caterpillars attack and skeletonise
plants at the seedling or newly transplanted stage. Destruction of the apical
buds of seedlings by feeding larvae results in headless plants or plants with
multiple undersized heads. Early crop damage reduces the photosynthetic
tissue and lowers crop yield. Although late damage may have little effect
on yield, perforated wrapper leaves down-grade the quality and thus the
value of the harvested crop.

Biological Control

Although many parasites of *P. xylostella* have been recorded, only a few
are effective in the field. The effective parasitoids belong to the hymenopterous genera *Diadegma* Foerster (Ichneumonidae), *Apanteles* (see
page 112) and *Microplitis* Foerster (Braconidae), and, of these, only
Diadegma eucerophaga Horstmann and possibly *D. fenestralis* Holmgren
alone seem able to control *P. xylostella*. Unless local key species of
parasitoids can be identified and their populations supplemented by
laboratory-reared individuals, control of *P. xylostella* where it is a major
pest will continue to rely on routine insecticidal treatments, often applied
at very frequent intervals. The parasitoids responsible for stabilising *P.
xylostella* infestations in the northern parts of this pest's distribution are
described in detail by Harcourt (1986).

The most common pathogens of *P. xylostella* are the fungi *Entomophthora blunckii* Lakon and *E. radicans* Brefeld (see page 112), although *P.
xylostella* caterpillars can be killed by other entomophthoraceous fungi, a
granulosis virus, a nuclear polyhedrosis virus and *B. thuringiensis* var.
kurstaki (see page 112). Apart from preparations of *B. thuringiensis*, none
of the other pathogens tested on a commercial scale has protected field
crops adequately against *P. xylostella*, although most have shown promise
in the laboratory. A more comprehensive discussion of the current

situation concerning biological control of this pest can be found in Talekar *et al.* (1985) and Talekar (1986).

Host Plant Resistance

Although *P. xylostella* caterpillars completed their development on 71 different lines and cultivars of *Brassica* crops in the field in New York State, USA, all lines were not equally suitable as hosts: the cabbage line G8329, the cauliflower line PI234599, mature Chinese cabbage and mature Pak choi were damaged only slightly. Laboratory tests indicated that the 'resistance' expressed by cabbage line G8329 was caused by the apparent difficulty experienced by first-instar larvae, the leaf-mining stage, in establishing themselves in the leaves.

Integrated Control

Using action thresholds based on the percentages of plants infested rather than on the numbers of larvae on plants, the number of insecticidal sprays applied to cabbage crops in Ohio, USA, was reduced from 6 to 3 on spring plantings and from 11 to between 4 and 7 on summer plantings. Crop monitoring—in this case, to determine action thresholds—is an economical proposition to a farmer if he/she can do it, but, if staff have to be hired specifically for the monitoring, routine weekly sprays will usually be cheaper.

Chemical Control

In northern countries one well-timed, well-directed spray containing, for example, only one of the insecticides recommended against *Pieris* spp. (see pages 113 and 115) is usually required on each crop to prevent yield losses from damage by *P. xylostella*. However, a second spray may be needed when immigration occurs over an extended period. In the warmest parts of the temperate region (e.g. Italy, Spain and southern states of North America) insecticidal treatments often have to be applied weekly or even more frequently to counter the intense pest pressure produced by up to 15 generations of the insect each year. Under these regimens of insecticidal usage, *P. xylostella* has developed resistance to most, and in some cases all, conventional insecticides at present available. Insect growth regulators, e.g. diflubenzuron, and rapidly acting biological control agents (e.g. *B. thuringiensis*) appear to offer the most promise for the future protection of cruciferous crops in areas such as southern Italy and the southern states of North America.

4.6.3 *Evergestis forficalis* (L.): **Garden Pebble Moth (Lepidoptera: Pyralidae)**

Geographical Distribution

E. forficalis seems to be Euro-Asiatic in origin. It was recorded in North America only in about 1930. Although the main pest species in Europe is the garden pebble moth, *E. pallidata* Huf. (the purple-backed cabbage-worm) and *E. rimosalis* Guen. (the cross-striped cabbageworm) are the species of primary importance in Canada and the USA.

Description and Life Cycle

The moths appear first in the spring, generally in May and June in the UK. Unless disturbed from crop plants, they are seen rarely, as they fly only at dusk and during the night. Adults have a wingspan of 25–30 mm. The forewings are yellowish white in colour, with brown veins, and they are covered with a series of oblique brown lines and shaded areas (Figure 4.5). The discal spot is elongate and coloured dark-reddish brown, sometimes with a white centre. The hindwings are pale ochreous white in colour, streaked with grey.

The females lay batches of about 20 eggs on the undersides of the leaves of cruciferous plants. The eggs are shiny, oval and flattened and initially translucent but later becoming a yellow colour. Depending on the weather, the eggs hatch in 4–10 days. The caterpillars pass through four instars and are coloured yellowish-green at first, but later become glossy pale green, with yellowish mid-dorsal and lateral stripes. A fully grown larva is 18–20 mm long, with a row of black spots along each side. The caterpillars feed on the undersides of leaves, frequently as a group within leaf folds and beneath protective silk webbing. Fully grown caterpillars crawl into the soil and spin tough silken cocoons which, in organic soils, become covered with loose material and are extremely difficult to distinguish from small loose clods of soil. The insects form yellowish-brown-coloured pupae. A second generation of adults emerges in August and September and fully grown caterpillars of this generation overwinter in cocoons in the soil, pupating only when development resumes with rising soil temperatures in the following spring.

Plant Damage

Formerly regarded as a minor pest in the UK, *E. forficalis* now often causes significant damage, not only in the UK, but also throughout Europe, and in some years is as damaging as the cabbage white butterflies. The caterpillars feed on the leaves of older plants and sometimes mine into the hearts. Crops are fouled by frass and the caterpillars' webbing, which also retains frass on the plants.

Cultural Control

Damage by *E. forficalis* caterpillars to Brussels sprout plants can be reduced by intercropping with *Spergula arvensis* L. (spurry). However, intercropping is likely to be less effective in commercial crops, because crop yields tend to be reduced when the *Spergula* spreads enough to reduce *E. forficalis* populations.

Biological Control

A range of parasitoids—in particular, species of the hymenopterous egg parasitoid *Trichogramma* (see page 112)—assists in checking populations of *E. forficalis*. Even under the most favourable conditions, *Trichogramma* spp. rarely parasitise more than 30% of the eggs laid by *E. forficalis*. The nematode *Heterorhabditis heliothidis* has shown promise in laboratory experiments for controlling the garden pebble moth, although all large insects such as *E. forficalis* are likely to be susceptible to *Heterorhabditis* spp. in the laboratory.

Chemical Control

Crops are protected against *E. forficalis* by the insecticidal spray treatments recommended against the range of Lepidoptera which infests summer cruciferous crops (see pages 110–123). It is imperative that the insecticide reach the undersides of the leaves.

4.7 SAWFLIES

4.7.1 *Athalia rosae* (L.): Turnip Sawfly (Hymenoptera: Tenthredinidae)

Geographical Distribution

A. rosae is common throughout mainland Europe. Although it was considered to be the most important insect pest of cruciferous crops in the UK in the early part of this century, it seems to have been eradicated.

Description and Life Cycle

Adult *A. rosae* have a black-coloured head and thorax and a bright orange-yellow abdomen, the females being noticeably yellower than the males. There are many other yellow sawflies of the genus *Athalia* Leach, which are frequently mistaken for *A. rosae*. When disturbed, both sexes feign death. The adults emerge from the soil in May and are 6–8 mm long, with a wingspan of 14–16 mm. They live for 12–14 days. Each female lays

250–350 eggs along the edges of leaves of cruciferous plants, showing an apparent preference for *B. rapa* L. (turnip). The eggs hatch in 5–12 days and the newly emerged, 2-mm-long, black larvae feed on leaf tissue. First-instar larvae fall off the leaf when disturbed, but remain attached to it by a silken thread emitted from their mouthparts. Later instars fall to the soil and curl up when disturbed. The larvae feed for about 3 weeks and frequently skeletonise host plants. They moult three times before the fully grown, 20–25-mm-long larvae move into the soil to pupate in silken cocoons. During the summer months, pupation lasts about 3 weeks. In Bulgaria, Romania and Germany, where *A. rosae* is a pest, there are usually 3–5 generations each year.

Cultural Control

In the early 1900s populations of *A. rosae* often built up to high numbers on cruciferous weeds alongside cultivated areas. The introduction of routine crop rotation and effective herbicides reduced populations of this pest to very low numbers.

Natural Control

The hymenopterous parasitoid *Labrossyta scotoptera* Gravenhorst (Ichneumonidae) has been reared from *A. rosae* pupae in Romania but the extent of the contribution of this parasitoid towards suppressing populations of *A. rosae* has not been determined.

Plant Damage

In some years *A. rosae* causes serious damage to Chinese cabbage crops in Germany. The turnip sawfly is considered to be of pest status in most eastern European countries, particularly in crops of oilseed rape. The larvae eat the leaves of host plants, often leaving only the mid-ribs.

Physical Control

A novel approach using light to reduce the numbers of overwintering *A. rosae* has been proposed in Hungary. The method consists of illuminating crops in the autumn, when the insects are ready to be induced into diapause, for 30 min periods in the third and eighth hours after dusk, to prevent the insects receiving enough continuous darkness to induce diapause. When this method was used, only 38% of the insects entered diapause. The remaining 62% did not enter diapause and were killed subsequently by low winter temperatures.

Chemical Control

As sawflies are generally extremely susceptible to insecticides, infestations of *A. rosae* are often eradicated by treatments applied to protect crops against caterpillars and aphids rather than specifically against the sawflies. In countries such as Bulgaria, where specific treatments against *A. rosae* may be needed, the numbers of adults caught in glass-pane interception traps provide a good forecast of the size of larval infestations likely to develop 3 weeks later. In practice, insecticides are usually applied when more than 8% of the leaf area of cabbage crops or 11% of the leaf area of cauliflower crops has been damaged by any foliar-feeding insect species (butterflies, moths, sawflies).

4.8 FLIES

4.8.1 *Dasineura brassicae* (Winnertz): Brassica Pod Midge (Diptera: Cecidomyiidae)

Geographical Distribution

D. brassicae is restricted to seeding brassica crops growing in Europe.

Description and Life Cycle

The adults are delicate midges which, because of their small size, are impossible to identify accurately in the field. Adults emerge from the soil about mid-April in southern regions of Europe and about mid-May in northern regions, and live for between 1 and 3 days. Eggs are laid in large clusters in seed pods which have been previously damaged either by the feeding and egg laying of *C. assimilis* (cabbage seed weevil; see page 103) or by cultural practices. Females also lay eggs in pods damaged by fungal pathogens. The eggs are transparent and spindle-shaped and have a reddish-coloured central spot which can only be seen under magnification. On hatching, the larvae are transparent but later become white in colour and clearly visible. They feed for about 4 weeks and, when fully grown, are about 2 mm long. In each infested pod there may be 50 or more larvae, reproducing paedogenetically and feeding on the seeds and the pod walls (Figure 4.6). Most fully grown midge larvae crawl out of the pods and fall down to the soil surface, although some may be released prematurely by pod dehiscence. The larvae wriggle into the soil to pupate in cocoons, and adults of the next generation emerge about 2 weeks later. *D. brassicae* usually has about 4 generations each year. Progeny from the last generation overwinter in cocoons in the soil.

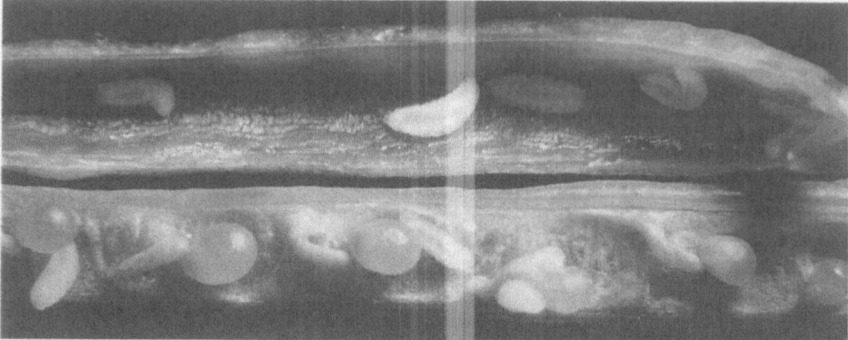

Figure 4.6 Larvae of brassica pod midge within brassica seed pod. © HRI

Plant Damage

The midge attacks a wide range of seeding brassica crops, including *B. juncea* Coss. (leaf mustard). It is not found on *Sinapis alba* L. (white mustard), although *S. arvensis* L. (charlock), a weed species, is considered to be an extremely important reservoir of *D. brassicae*. Where the midge occurs, damage to brassica seed crops is extremely variable, generally being more severe along the headlands than in the open fields. Apart from the direct damage done by the larvae destroying the seeds, larval feeding also induces attacked pods to swell, a condition referred to commonly as 'bladder pod'. Swollen pods split prematurely and discharge damaged and undamaged seeds on to the soil surface. This loss of seed yield from 'bladder pod' increases the economic importance of *D. brassicae*.

Cultural Control

As the adult midges are weak fliers, infestations in new seed crops can be minimised by drilling crops at least 0.5 km away from the previous year's crops.

Host Plant Resistance

Not all seeding crops are equally attractive to the pod midge. Lesser numbers of eggs are laid on *B. nigra* Koch (black mustard), *B. juncea* (leaf mustard) and *B. carinata* A.Br. (Abyssinian mustard), which are also less suitable for larval development, than on *B. campestris* (turnip rape) and *B. napus* (swede rape). Female midges land less and stay for shorter periods on 'resistant' than on susceptible cultivars. Although similar numbers of eggs were laid on low- and high-glucosinolate cultivars of swede rape, breakdown products from the glucosinolates which were present in greater quantities in resistant than in susceptible host plants were toxic to pod midge larvae. *B. juncea* and *B. carinata* would appear to have inherited some as yet unknown parental characteristic from *B. nigra* that makes them partially resistant to *D. brassicae* larvae.

Integrated Control

Computer-based farm advisory services have been developed in the UK for the protection of oilseed rape crops against insect pests, including *D. brassicae*. A less complex forecasting system that has enabled the numbers of insecticidal sprays applied against *D. brassicae* to be reduced from 5 to 3 is now being used in Poland.

Chemical Control

Some of the post-flowering insecticidal spray treatments available to control infestations of *C. assimilis* (see page 104)—for example, endosulfan and phosalone—are also effective against brassica pod midge larvae. Pyrethroid insecticides (e.g. fenvalerate) are generally most effective against *D. brassicae*. The severity of *D. brassicae* damage is related directly to *C. assimilis* damage and, in crops where numbers of *C. assimilis* are not large enough to warrant treatment, *D. brassicae* damage is generally minimal. Thresholds have not yet been produced to indicate the degree of *D. brassicae* infestation which would justify applying an insecticide specifically against the midge.

4.8.2 *Delia floralis* (Fallén): Turnip Root Fly (Diptera: Anthomyiidae)

Geographical Distribution

D. floralis is restricted to the most northern areas of the cold temperate region. It is common in Canada, northern America and northern Europe, where it is found frequently feeding with larvae of *D. radicum* L. (see page 129) on the roots of many cultivated cruciferous crops. In the most northerly regions *D. floralis* tends to displace *D. radicum*.

Description and Life Cycle

Generally from the middle of June until the end of July adult turnip root flies emerge from pupae in the soil. In parts of Denmark a small proportion of the population emerges, however, during early June, which permits the development of a partial second generation. Outside Denmark the pest is otherwise restricted to one generation each year. The adults are 7–9 mm long, about 2 mm longer than *D. radicum*. Unlike male *D. radicum*, male *D. floralis* lack a tuft of hairs at the base of the hindleg. Female *D. floralis* have a much lighter grey colour than *D. radicum* and do not have the median stripe and shimmering spots on the abdomen that characterise female *D. radicum*. In addition, female *D. radicum* have a strong anteroventral bristle near the base of the middle femur, whereas this bristle is usually missing in *D. floralis* or is, at best, weak. The adults live for about

35 days and begin to lay eggs in the soil around host plants about 12 days after emergence. The eggs are white in colour, 1.0–1.2 mm long and 0.28–0.38 mm wide. They hatch in 8–9 days and the larvae feed on the roots of host plants for 5–8 weeks, moulting twice. The fully grown third-instar larva moves a short way into the soil and pupates, the final larval skin forming the dark-brown-coloured puparium in which the pupa overwinters. The puparia of *D. floralis* are 6.5–7.5 mm long and 3–4 mm wide, and are distinguished from the puparia of *D. radicum* by being generally larger. More specifically, the arrangement of the sclerotised remains of the larval tubercles at the distal end of the puparium differs between the two species. Although both species have characteristic bifid tubercles, the bifid process is formed by the fusion of the fifth (outer apical) and sixth (inner apical) tubercles in *D. floralis* and of the sixth and seventh (ventro-apical) tubercles in *D. radicum*. Viewed in plan, there are no tubercles between the bifid ones in *D. radicum*, but one pair of ventro-apical tubercles is present in *D. floralis*. Most pupae remain within the puparia for between 8 and 10 months.

Plant Damage

D. floralis larvae feed primarily on the roots of cruciferous plants, generally about 2–6 cm below the soil surface, occupying a similar agroecological niche to *D. radicum*. In localities where both species are present, the larvae often feed together on the same plant. Larvae of both species destroy plant roots and mine the 'roots' of swedes and turnips (Figure 4.7). Mining of swedes and turnips that is restricted to the surface can often be trimmed to make plants marketable but excessive trimming is undesirable, as it is labour-intensive and costly.

Natural Control

Like the eggs of *D. radicum*, the eggs of *D. floralis* are eaten by a wide range of predatory ground beetles (Carabidae) and rove beetles (Staphylinidae). In Norway the most effective carabid predators of the eggs of *D. floralis* are *Bembidion lampros* Herbst, *Calathus melanocephalus* L. and *Pterostichus melanarius* Illiger, and the most effective staphylinid predators are *Aleochara bilineata* Gyllenhal and *Philonthus ochropus* Gravenhorst. The fungus *Strongwellsea castrans* Batko and Weiser (Entomophthorales: Entomophthoraceae) helps to lower *D. floralis* populations by sterilising females but, as it often infests flies only after some eggs have been laid, the fungus exerts only partial control. In addition, the effect of the fungus depends on the weather and is sustained more in cool, wet years, when conditions are unfavourable for fly development, than in hot, dry years, when conditions favour development of the fly.

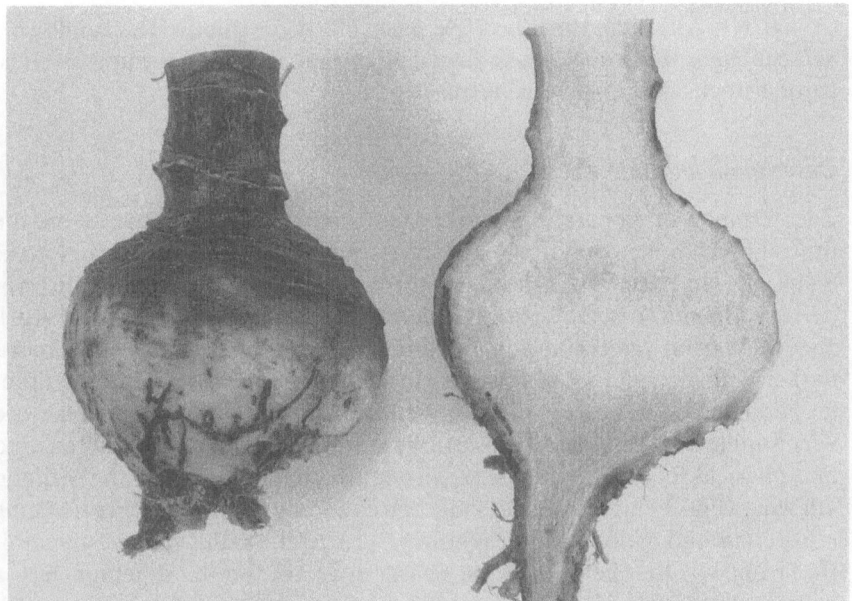

Figure 4.7 External and internal damage to swede roots by turnip root fly larvae. © HRI

Host Plant Resistance

Field experiments have now shown that the resistance of some cultivars of swede, such as Angus and Melfort, to attack by *D. floralis* is not correlated with the dry matter of roots, as was suggested previously. The resistance of plants to this pest appears instead to be governed largely by chemicals present on the surfaces of the leaves, deterring the females during host plant selection. The chemicals involved are not thought to be any of the glucosinolates or their derivatives that give cruciferous plants their characteristic odours and tastes.

Chemical Control

Although many of the numerous methods developed for controlling *D. radicum* with insecticides (see page 133) have been adopted for this pest whenever control measures are required, few recommendations specific to *D. floralis* exist. One current recommendation in the UK is for a mid-season application of the carbamate insecticide carbofuran, which is also recommended against *D. radicum*.

4.8.3 *Delia radicum* (L.): Cabbage Root Fly (Diptera: Anthomyiidae)

Geographical Distribution

D. radicum is restricted to the temperate zone of the holarctic region

(35–60°N). Although the larvae are particularly destructive to cauliflowers and cabbages, they also attack Brussels sprouts, radishes, turnips, swedes, garden stocks and many cruciferous weeds.

Description and Life Cycle

The number of generations of *D. radicum* in a year depends on the prevailing climatic conditions, and varies from 1 in northern Europe to as many as 5 in parts of Canada and the USA. The active, greyish-brown-coloured adults (resembling small, slender houseflies) emerge in late April and May from overwintering puparia. The male is smaller (5–6 mm), darker-coloured and more bristly than the female (6–7 mm) and the tip of its abdomen is more broadly rounded. Several other Anthomyiidae are very similar to *D. radicum* and can be distinguished only by reference to fine physical features: adult *D. radicum* are characterised by the 3rd and 4th wing veins being parallel at their apices and by the pre-alar bristle being more than half as long as the following bristle. In males the eyes meet at the front of the head and the basal half of the hind femur has a characteristic tuft of hairs. Females are distinguished by having well-separated eyes, a strong anteroventral bristle near the base of the middle femur and two posterior bristles on each front tibia. Male and female flies mate when 4–7 days old and females generally begin to lay eggs on the day following mating. The white-coloured eggs, about 0.9–1.0 mm long with one side convex and the other slightly concave, are laid mainly in the soil around the stems of cruciferous plants, but under some conditions, they are laid on the heads of cauliflower and Brussels sprout plants. Each female lays 50–70 eggs. At emergence, the larva forces its way through a thin area of the chorion between the two ridges that extend along the concave side of the egg. The larva is a more or less transparent, tapering, legless maggot with the head reduced and deeply retracted into the broad prothorax. The three larval instars are distinguishable by the size of their mouthparts. Measurements of the distance between the tip of the mandible and the posterior end of the ventral process of the pharyngeal sclerite in the first-, second- and third-instar larvae fall into three distinct groups with means of 0.29 mm, 0.53 mm and 0.88 mm, respectively. Under field conditions in the UK, the first instar usually feeds for about 4 days before moulting, the second for about 6 days, and the third for 10–20 days. Fully grown larvae, 5–8 mm long, crawl out of the plant into the soil near the roots to pupate. The dark-brown-coloured puparia are sub-elliptical in shape, with smoothly rounded sides. They can be identified by the hardened, posterior protruberances which are the remains of the larval tubercles. Most puparia are to be found 8–12 cm below the soil surface. The pupal stage of the early generations lasts about 12–18 days, but pupae of the later generations, which overwinter in the soil, may last 5–10 months. Temperatures higher than 22°C in the summer induce *D. radicum* to enter a resting phase called aestivation. Cool temperatures and shortening photoperiods in the autumn induce *D. radicum* to enter diapause, the overwintering resting phase.

Plant Damage

D. radicum larvae feed primarily on the roots of cruciferous plants, although the maggots sometimes attack aerial parts, including the buttons of Brussels sprout plants. Plants attacked as seedlings or soon after transplanting may be damaged so severely that they are killed (Figure 4.8). In some seasons in North America, 90% of the plants in some crops may be killed. Losses in the UK are occasionally as high as 60% but average estimates are nearer to 24%. Vigorously growing plants can support large populations of larvae without showing signs of attack, but when the larvae attack the part of the plant used for human consumption (e.g. swede and radish roots and Brussels sprout buttons), even small amounts of damage lower quality and thus crop value.

Figure 4.8 Cauliflower plants undamaged (left) and damaged (right) by cabbage root fly larvae. © HRI

Natural Control

Although anthomyiid and cordilurid (both Diptera) predators of *D. radicum* adults have been reported, predation of the immature stages has been studied more extensively. Eggs are preyed on by trombidiid mites, ants and carabid and staphylinid beetles, and larvae are the prey of ants, beetles and other anthomyiid larvae. Of the beetle predators associated with *D. radicum* in the UK, staphylinids are about twice as numerous, though not necessarily twice as effective, as carabids. The carabids *B. lampros* (see page 128) and *Harpalus affinis* Fabricius are important predators during April and May, whereas *P. melanarius* (see page 128), *Harpalus rufipes* Degeer and *Trechus quadristriatus* Schrank are more important by July.

Although the larvae of *D. radicum* are attacked by many hymenopterous parasitoids, they are killed only after they have pupated. Five braconid, three cynipid and four ichneumonid species have been reared from *D. radicum* puparia, but the cynipid *Trybliographa rapae* Westwood is the only hymenopterous parasitoid of major importance. It has been recorded from more than 60% of *D. radicum* pupae in some samples. Parasitism of pupae by staphylinid beetles of the genus *Aleochara* Gravenhorst is also common, *A. bilineata* (see page 128) usually being more common than *A. bipustulata* L., although the two species have similar life cycles. The two *Aleochara* species frequently parasitise 10–20% of pupae and occasionally as many as 60%.

The fungus *Entomophthora muscae* Cohn (see page 112) kills *D. radicum* adults, whereas *Strongwellsea castrans* (see page 128) only sterilises them. During the summer months both fungi become noticeable: flies killed by *E. muscae* can readily be seen clinging to the foliage of many hedgerow plants; and many flies, particularly females, can be observed with the sunken, white, circular body between the abdominal sclerites indicative of infection by *S. castrans*. It is not easy to estimate the effect of these fungi on populations of *D. radicum*, as adult insects tend to become infected late in life, when females have laid most of their eggs.

Cultural Control

Attempts to reduce *D. radicum* populations have included crop rotation, destruction of infested plants, avoidance of autumn-sown crops (which encourage large overwintering populations) and growing of seed crops away from the main areas of vegetable brassica production. Infestations are also reduced by sowing or transplanting susceptible crops out of phase with the main period of egg laying by *D. radicum* females.

Physical Control

Egg laying on plants can be prevented by covering seed beds with cheese cloth or by placing tarred felt discs, or carpet underlay collars, on the soil around the base of each plant immediately after transplanting. The modern practice of covering crops with plastic mulch to encourage earlier maturation helps to reduce damage caused by *D. radicum* when the crop remains covered for all or part of a generation of the fly.

Host Plant Resistance

Attempts to select radishes resistant to *D. radicum* have been unsuccessful, largely because the attractiveness of the different radish lines changes relative to each other as the plants age. Similarly, attempts to produce lines of cauliflower resistant to *D. radicum* have so far been unsuccessful.

Biological Control

Attempts are being made in Belgium and Denmark to rear sufficient numbers of *A. bilineata* (see page 128), to try to control *D. radicum* by inundative releases. *A. bilineata* is a predator and a parasitoid of the immature stages of *D. radicum*. Experiments have been done to try to control infestations by releasing sterile males and by trying to sterilise natural field populations of the fly with chemosterilant baits. Biological control has also been attempted by spraying crop foliage with suspensions of entomopathogenic fungi and by drenching the soil around plant stems with suspensions containing high inocula of entomophilic nematodes. Although brassica plants receiving some of these treatments were less damaged by *D. radicum* than were untreated plants, the degree of damage exceeded acceptable limits.

Chemical Control

The range of cruciferous crops that can be attacked at many stages of plant growth and by several generations of *D. radicum* in any one year is wide and, as a result, the range of available insecticidal treatments is wide. Resistance of *D. radicum* to organochlorine insecticides is extensive and protection of crops now relies heavily on a relatively limited number of organophosphorus and carbamate compounds, including aldicarb, carbofuran, carbosulfan, chlorfenvinphos, chlorpyrifos, diazinon, dimethoate, fonofos and triazophos. Plants grown in nursery seed beds for transplanting require protection against *D. radicum*, usually with a soil-incorporated insecticide, if the seedlings are expected to be in the seed bed when the first generation of flies starts egg laying in the spring. Retreatment of the plants is necessary at transplanting. As effective curative measures for killing larvae established on plant roots are not available, it is important with field-sown and transplanted crops to ensure that prophylactic treatments are applied to the soil with precision and accuracy. These treatments include liquid and granular formulations applied along crop rows in the field and also preplanting treatments with, for example, chlorpyrifos applied to modules used to raise transplants. Although considerable effort is currently being expended to investigate the potential of film-coating technology to seed treatments in order to reduce phytotoxicity hazards associated with many present-day insecticides, effective seed treatments are currently not available. The dual-component granular products containing chlorpyrifos, fonofos or quinalphos to control cabbage root fly and disulfoton to protect treated crops against aphids offer an opportunity to UK growers for a sophisticated crop protection strategy. Although treatments applied at drilling or transplanting are usually adequate for short-season cruciferous crops exposed to only one well-defined generation of *D. radicum*, longer-season crops are likely to require supplementary treatments. The timing and precision of application of these supplementary treatments is of critical importance. The phenomenon of accelerated degradation of insecticidal residues in soil demonstrated largely, but not exclusively, with carbamate compounds, emphasises the need for treatments against *D. radicum* to be based not on one insecticide or group of insecticides, but on a rotation of insecticidal groups, not only for successive crops, but even for sequential treatments of long-season crops.

REFERENCES AND BIBLIOGRAPHY

Adams, A. J. (1985). The critical field photoperiod inducing ovarian diapause in the cabbage whitefly, *Aleyrodes proletella* (Homoptera: Aleyrodidae). *Physiological Entomology*, **10**, 243–249

Ahman, I. (1986). Resistensegenskaper hos *Brassica* mot skidgallmyggan (*Dasineura brassicae* Winn.). *Vaxyskyddsrapporter, Jorbruk*, **39**, 45–50

Alborn, H., Karlsson, H., Lundgren, L., Ruuth, P. and Stenhagen, G. (1985). Resistance in crop species of the genus *Brassica* to oviposition by the turnip root fly, *Hylemya floralis*. *Oikos*, **44**, 61–69

Altieri, M. A., Wilson, R. C. and Schmidt, D. L. (1985). The effects of living mulches and weed cover on the dynamics of foliage- and soil-arthropod communities in three crop systems. *Crop Protection*, **4**, 201–213

Anderson, A., Hansen, A. G., Rydland, N. and Oyre, G. (1983). Carabidae and Staphylinidae (Col.) as predators of eggs of the turnip root fly *Delia floralis* Fallén (Diptera, Anthomyiidae) in cage experiments. *Zeitschrift für Angewandte Entomologie*, **95**, 499–506

Atak, U. and Atak, E. D. (1983). Marmara bolgesinde lahana gal bocegi (*Ceutorhynchus pleurostigma* Marsh.) nin biyo-okologisi ve mucadelesi uzerinde arastirmalar. (Investigations of the bio-ecology and control of the cabbage gall weevil (*Ceutorhynchus pleurostigma* Marsh.) in the Marmara district). *Bitki Koruma Bulteni*, **23**, 92–113

Baker, G. J. (1984). Damage assessment studies and the derivation of action thresholds for larvae of the cabbage moth, *Plutella xylostella* (L.) on cabbages. In *Proceedings of the Fourth Australian Applied Entomological Research Conference, Adelaide, 1984. Pest Control: Recent Advances and Future Prospects* (ed. P. Bailey and D. J. Swincer). Department of Agriculture, South Australian Government, Adelaide, pp. 175–180

Balachowsky, A. S. and Mesnil, L. (1935). *Les Insectes Nuisibles aux Plantes Cultivées*, 2 vols. Min. Agric., Paris

Balagurunathan, R. and Lebrun, P. (1984). Comparative efficacy of different frequencies of Bactospeine application in the control of cabbage caterpillars. *Mededelingen van de Faculteit Landbouwwetenschappen, Rijksuniversiteit Gent*, **49**, 901–908

Blackman, R. L. and Eastop, V. F. (1984). *Aphids on the World's Crops. An Identification Guide*. John Wiley, Chichester

Bracken, G. K. and Bucher, G. E. (1986). Yield losses in canola caused by adult and larval flea beetles, *Phyllotreta cruciferae* (Coleoptera: Chrysomelidae). *Canadian Entomologist*, **118**, 319–324

Buchi, R. (1986). Biologie und Bekampfung des Schwarzen Triebrusslers, *Ceutorhynchus picitarsis* Gyll. (Col., Curculionidae). (Biology and control of the black winter stem weevil, *Ceutorhynchus picitarsis* Gyll. (Col., Curculionidae)). *Anzeiger für Schadlingskunde, Pflanzenschutz, Umweltschutz*, **59**, 51–56

Cantelo, W. W. and Sandford, L. L. (1984). Insect population response to mixed and uniform plantings of resistant and susceptible plant material. *Environmental Entomology*, **13**, 1443–1445

Carter, D. J. (1984). *Pest Lepidoptera of Europe with Special Reference to the British Isles*. Series Entomologia, Vol. 31. Junk, Dordrecht

Charpentier, R. (1985). Host plant selection by the pollen beetle *Meligethes aeneus*. *Entomologia Experimentalis et Applicata*, **38**, 277–285

Coaker, T. H. and Williams, D. A. (1963). The importance of some Carabidae and Staphylinidae as predators of the cabbage root fly, *Erioischia brassicae* (Bouché). *Entomologia Experimentalis et Applicata*, **6**, 156–164

Culliney, T. W. and Pimentel, D. (1986). Ecological effects of organic agricultural practices on insect populations. *Agricultural Ecosystems and Environment*, **15**, 253–266

Davies, C. R. and Gilbert, N. (1985). A comparative study of the egg-laying behaviour and larval development of *Pieris rapae* L. and *P. brassicae* L. on the same host plants. *Oecologia*, **67**, 278–281

Dmoch, J. and Starzynski, A. W. (1983). Zwalczanie szkodnikow luszczynowych na nasiennikach kapusty glowiastej (*Brassica oleracea* L. var. *capitata*). *Roczniki Nauk Rolniczych*, **10**, 59–74

Dunn, J. A. and Kempton, D. P. H. (1972). Resistance to attack by *Brevicoryne brassicae* among plants of Brussels sprouts. *Annals of Applied Biology*, **72**, 1–11

Edwards, C. A. and Heath, G. W. (1964). *Principles of Agricultural Entomology*. Chapman and Hall, London

Ellis, P. R. and Hardman, J. A. (1988). Investigations of the resistance of cabbage cultivars and breeders lines to insect pests at Wellesbourne. *Report of the Joint Experts Meeting. Integrated Plant Protection in Field Vegetables, Rennes, France, 20–22 November 1985*, pp. 99–105

Feltwell, J. (1982). *Large White Butterfly: The Biology, Biochemistry and Physiology of* Pieris brassicae *(Linnaeus)*. Series Entomologica, Vol. 18. Kluwer, Holland

Finch, S. (1987). Horticultural crops. In *Integrated Pest Management* (ed. A. J. Burn, T. H. Coaker and P. C. Jepson). Academic Press, London, pp. 257–293

Finch, S. (1988). Entomology of crucifers and agriculture. Diversification of the agro-ecosystem in relation to pest damage to cruciferous crops. In *The Entomology of Indigenous and Naturalized Systems in Agriculture* (ed. M. K. Harris and C. E. Rogers). Westview Press, Boulder, Colorado, pp. 39–71

Finch, S. and Collier, R. H. (1986). Forecasting the times of attack of the cabbage root fly (*Delia radicum*) on vegetable brassica crops. *Aspects of Applied Biology*, **12**, 37–46

Finch, S. and Skinner, G. (1976). The effect of plant density on populations of the cabbage root fly (*Erioischia brassicae*) (Bch.) and the cabbage stem weevil (*Ceutorhynchus quadridens* (Panz.)) on cauliflowers. *Bulletin of Entomological Research*, **66**, 113–123

Finney, J. R. (1983). The feasibility of using nematodes (*Heterorhabditis* sp. and *Neoaplectana* sp.) for control of *Brassicae*, apple and blueberry pests. In *Proceedings of the 30th Annual Meeting Canadian Pest Management Society*, **30**, 39–41

Hansen, K. E. (1984). Forsog med bekaempelse of glimmerbosser (*Meligethes aeneus*), skulpesnudebiller (*Ceutorhynchus assimilis*) og skulpegalmyg (*Dasyneura brassicae*) i vinter- og varraps. (Trials on the control of blossom beetles (*Meligethes aeneus*), brassica seed weevils (*Ceutorhynchus assimilis*) and brassica pod midges (*Dasyneura brassicae*) in winter- and spring-rape). *Tidsskrift for Planteavl*, **88**, 91–100

Harcourt, D. G. (1986). Population dynamics of the diamondback moth in southern Ontario. In *Diamondback Moth Management. Proceedings of the First International Workshop, Tainan, Taiwan, 11–15 March 1985*. Asian Vegetable Research and Development Center (1986), Shanhua, Taiwan, pp. 3–15

Hassan, S. A. and Rost, W. M. (1985). Mass production and utilization of *Trichogramma*: 6. Studies towards the use against lepidopterous pests. *Mededelingen van de Faculteit Landbouwwetenschappen, Rijksuniversiteit Gent*, **50**, 389–398

Heiningen, T. G. van, Pak, G. A., Hassan, S. A. and Lenteren, J. C. van (1985). Four years' results of experimental releases of *Trichogramma* egg parasites against Lepidoptera pests in cabbage. *Mededelingen van de Faculteit Landbouwwetenschappen, Rijksuniversiteit Gent*, **50**, 379–388

Helle, W. (1987). *World Crop Pests*, Vol. 2A, *Aphids. Their Biology, Natural Enemies and Control* (ed. A. K. Minks and P. Harrewijn). Elsevier, Amsterdam

Hertveldt, L., Van Keymeulen, M. and Pelerents, C. (1984). Large scale rearing of the entomophagous rove beetle, *Aleochara bilineata* (Coleoptera: Staphylinidae). *Mitteilungen, Biologische Bundesanstalt für Land- und Forstwirtschaft*, **218**, 70–75

Hill, D. S. (1987). *Agricultural Insect Pests of Temperate Regions and Their Control*. Cambridge University Press, Cambridge

Hommes, M. (1983). Untersuchungen zur Populationsdynamik und integrierten Bekampfung von Kohlschadlingen. *Mitteilungen, Biologische Bundesanstalt für Land- und Forstwirtschaft*, **213**

Horsakova, M. (1985). Vliv teploty ovzdusi a vyvojove faze rostlin repkyozime na nalet skudcu do porostu. (The effect of air temperature and plant development stage on the infestation of winter rape by pests). *Rostlinna Vyroba*, **31**, 747–754

Hughes, R. D. and Mitchell, B. (1960). The natural mortality of *Erioischia brassicae* (Bouché) (Dipt., Anthomyiidae): life tables and their interpretation. *Journal of Animal Ecology*, **28**, 343–357

Hulden, L. (1986). The whiteflies (Homoptera, Aleyrodidae) and their parasites in Finland. *Notulae Entomologicae*, **66**, 1–40

Husberg, G. B., Granlund, H. and Hokkanen, H. (1985). [Control of rape beetles with the aid of trap crops.] *Vaxtskyddsnotiser*, **49**, 98–101

Isaac, P. V. (1923). I. The turnip gall weevil, *Ceutorhynchus pleurostigma*, Marsh. (Coleoptera, Curculionidae). Part I. Life-history and bionomics. *Annals of Biology*, **X**, 151–170

Isaac, P. V. (1923). The turnip gall weevil, *Ceutorhynchus pleurostigma*, Marsh. Part II. Larval anatomy. *Annals of Biology*, **X**, 171–193

John, M. E., Vaughan, J. and Evans, E. J. (1984). *Control of Pests and Diseases of Oilseed Rape*. Booklet No. 2387, Ministry of Agriculture, Fisheries and Food, UK

Jones, F. G. W. and Jones, M. G. (1984). *Pests of Field Crops*, 3rd edn. Edward Arnold, London

Kearney, P. C. and Kellogg, S. T. (1985). Microbial adaptation to pesticides. *Pure and Applied Chemistry*, **57**, 389–403

Kozlowski, M. W., Lux, S. and Dmoch, J. (1983). Oviposition behaviour and pod marking in the cabbage seed weevil, *Ceutorhynchus assimilis*. *Entomologia Experimentalis et Applicata*, **34**, 277–282

Krieg, A. (1983). In vivo-Production von Baculovirus-Praparaten zur Bekampfung von Schradinsetten im Rahmen eines Forschungsprogrammes für Biotechnologie. *Mitteilungen Deutschen Geselloschaft für Allegemeine und Angewandte Entomologie*, **4**, 39–43

Lagenbruch, G. A., Hommes, M. and Groner, A. (1986). Feldversuche mit dem Kernpolyedervirus der Koheule (*Mamestra brassicae*). *Zeitschrift für pflanzenkrankheiten und pflanzenschutz*, **93**, 72–86

Lamb, D. J. and Foster, G. N. (1986). Some observations on *Strongwellsea castrans* (Zygomycetes: Entomophthorales), a parasite of root flies (*Delia* spp.), in the south of Scotland. *Entomophaga*, **31**, 91–97

Lasota, J. A. and Kok, L. T. (1986). Parasitism and utilization of imported cabbage worm pupae by *Pteromalus puparum* (Hymenoptera: Pteromalidae). *Environmental Entomology*, **15**, 994–998

Latheef, M. A. and Ortiz, J. H. (1983). Influence of companion plants on oviposition of imported cabbage worm, *Pieris rapae* (Lepidoptera: Pieridae), and cabbage looper, *Trichoplusia ni* (Lepidoptera: Noctuidae), on collard plants. *Canadian Entomologist*, **115**, 1529–1531

Latheef, M. A., Ortiz, J. H. and Sheikh, A. Q. (1984). Influence of intercropping on *Phyllotreta cruciferae* (Coleoptera: Chrysomelidae) populations on collard plants. *Journal of Economic Entomology*, **77**, 1180–1184

Lin, J., Eckenrode, C. J. and Dickson, M. H. (1983). Variation in *Brassica oleracea* resistance to diamondback moth (Lepidoptera: Plutellidae). *Journal of Economic Entomology*, **76**, 1423–1427

Maniania, N. K. and Fargues, J. (1984). Specificité des hyphomycetes entomopathogenes pour les larves de lepidoptères Noctuidae. *Entomophaga*, **29**, 451–464

Mann, B., Wratten, S. D. and Jepson, P. (1986). Pest advisory packages and their development for a viewdata system. In *Proceedings of the British Crop Protection Conference—Pests and Diseases*, **3**, 1027–1034

Meier, W. and Jossi, W. (1983). Einsatz von Pyrethroid-Insektiziden in Feldkulturen. (Use of pyrethroid insecticides in field crops). *Mitteilungen-Schweizerische Landwirtschaft*, **31**, 29–42

Mietkiewski, R., Geest, L. P. S. van der and Balasy, S. (1986). Preliminary notes on the pathogenicity of some *Erynia* strains (Mycophyta, Entomophthoraceae) towards larvae of *Pieris brassicae*. *Journal of Applied Entomology*, **105**, 499–504

Morris, O. N. (1987). Evaluation of the nematode *Steinernema feltiae* Filipjev, for the control of the crucifer flea beetle, *Phyllotreta cruciferae* (Goeze) (Coleoptera: Chrysomelidae). *Canadian Entomologist*, **119**, 95–101

Mound, L. A. and Halsey, S. H. (1978). *Whiteflies of the World. A Systematic Catalogue of the Aleyrodidae (Homoptera) with Host Plant and Natural Enemy Data*. British Museum (Natural History) and John Wiley, Chichester

Nilsson, C. (1985). Impact of ploughing on emergence of pollen beetle parasitoids after hibernation. *Zeitschrift für Angewandte Entomologie*, **100**, 302–308

Northwood, P. J. and Verrier, C. (1986). Control of major pests of oilseed rape in west Europe. In *Proceedings of the 1986 British Crop Protection Conference—Pests and Diseases*, **2**, 745–752

Pak, G. A. and Lenteren, J. C. van (1984). Selection of a candidate *Trichogramma* sp. strain for inundative releases against lepidopterous pests of cabbage in the Netherlands. *Mededelingen van de Faculteit Landbouwwetenschappen, Rijksuniversiteit Gent*, **49**, 827–837

Ponti, O. M. B. de (1984). Chancen der Resistenzzuchtung gegen kohlschadlinge. *Mitteilungen, Biologische Bundesanstalt für Land- und Forstwirtschaft*, **218**, 25–28

Pouzet, A. and Ballanger, Y. (1984). Les piégèage des ravageurs du Colza a l'automne. *Phytoma*, **362**, 21–23

Purvis, G. (1986). The influence of cabbage stem flea beetle (*Psylliodes chrysocephala* (L.)) on yields of oilseed rape. In *Proceedings of the 1986 British Crop Protection Conference—Pests and Diseases*, **2**, 753–759

Raworth, D. A. (1984). Population dynamics of the cabbage aphid, *Brevicoryne brassicae* (Homoptera: Aphididae) at Vancouver, British Columbia. V. A simulation model. *Canadian Entomologist*, **116**, 895–911

Rost, W. M. and Hassan, S. A. (1983). Naturlich auftretende Eiparasiten der Gattung *Trichogramma* (Trichogrammatidae, Chalcidoidea, Hymenoptera) und deren Bedeutung als Gegenspieler von Kohlschadlingen. *Nachrichtenblatt des Deutschen Pflanzenschutzdienstes*, **35**, 184–188

Rygg, T. and Kjos, O. (1982). Kjemisk bekjempelse av kalfluer *Delia floralis* Fallén og *D. brassicae* (Wiedeman) i kalrot. *Forskning og Forsok i Landbruket*, **33**, 111–118

Saringer, G. (1983). Illumination for half an hour at a time in autumn, in the scotophase of the photoperiod, as a possible ecological method of controlling the turnip sawfly *Athalia rosae* L. (Hym., Tenthredinidae). *Zeitschrift für Angewandte Entomologie*, **96**, 287–291

Saringer, G. (1984). Summer diapause of cabbage stem flea beetle, *Psylliodes chrysocephala* L. (Col., Chrysomelidae). *Zeitschrift für Angewandte Entomologie*, **98**, 50–54

Scarisbrick, D. H. and Daniels, R. W. (eds.) (1986). *Oilseed Rape*. Collins, London

Schaaf, D. C. van der, Kaskens, J. W. M., Kole, M., Noldus, L. P. J. J. and Pak, G. A. (1984). Experimental releases of two strains of *Trichogramma* spp. against lepidopteran pests in a Brussels sprouts field crop in the Netherlands. *Mededelingen van de Faculteit Landbouwwetenschappen, Rijksuniversiteit Gent*, **49**, 803–813

Schultz, R. R. and Daebeler, F. (1984). Zum Schaden durch den Rapserdfloh (*Psylliodes chrysocephala* L.) insbesondere seiner Imagines. *Nachrintentblatt für den Pflanzenschutz in der DDR*, **38**, 113–115

Schutte, F., Bartels, G. and Niemann, P. (1984). Zur Methodik biozonotischer Untersuchungen bei der flugtuchtigen Rubsenblattwespe *Athalia rosae* L. *Mitteilungen, Biologische Bundesanstalt für Land- und Forstwirtschaft*, **221**, 63–68

Smith, D. B. and Sears, M. K. (1983). Evidence for dispersal of diamondback moth, *Plutella xylostella* (Lepidoptera: Plutellidae), into southern Ontario. In *Proceedings of the Entomological Society of Ontario*, **113**, 21–27

Smith, D. M. and Hewson, R. T. (1984). Control of cabbage stem flea beetle and rape winter stem weevil on oilseed rape with deltamethrin. In *Proceedings of the British Crop*

Protection Conference—Pests and Diseases, **2**, 755–760

Suett, D. L. and Jukes, A. A. (1988). Evidence and implications of accelerated degradation of organophosphorus insecticides in soil. *Journal of Toxicological and Environmental Chemistry*, **18**, 37–49

Suett, D. L. and Thompson, A. R. (1985). The development of localised insecticide placement methods in soil. In *Application and Biology*. BCPC Monograph No. 28, pp. 65–74

Talekar, N. S. (Ed.) (1986). *Diamondback Moth Management. Proceedings of the First International Workshop, Tainan, Taiwan, 11–15 March 1985*. Asian Vegetable Research and Development Center (1986), Shanhua, Tainan, Taiwan

Talekar, N. S., Yong, H. C., Lee, S. T., Chen, B. S. and Sun, L. Y. (1985). *Annotated Bibliography of Diamondback Moth*. Asian Vegetable Research and Development Center Publication, Shanhua, Tainan, Taiwan, pp. 85–229

Tatchell, G. M. and Payne, C. C. (1984). Field evaluation of a granulosis virus for control of *Pieris rapae* (Lep.: Pieridae) in the United Kingdom. *Entomophaga*, **29**, 133–144

Theunissen, J. (1984). Supervised pest control in cabbage crops: theory and practice. *Mitteilungen, Biologische Bundesanstalt für Land- und Forstwirtschaft*, **218**, 76–84

Theunissen, J. and Ouden, H. den (1980). Effects of intercropping with *Spergula arvensis* on pests of Brussels sprouts. *Entomologia Experimentalis et Applicata*, **27**, 260–268

Thioulouse, J., Debouizie, D. and Ballanger, Y. (1984). Structures spatiales et temporelles des populations d'un ravageur du Colza (*Psylliodes chrysocephala* L. (Col., Chrysomelidae)) dans plusiers parcelles de culture. *Acta Oecologia, Oecologia Applicata*, **5**, 335–353

Thompson, A. R. and Goodwin, M. C. (1983). Effects of some insecticide and insect growth regulator treatments on the immature stages of the cabbage whitefly (*Aleyrodes proletella* L.). *Mededelingen van de Faculteit Landbouwwetenschappen, Rijksuniversiteit Gent*, **48**, 309–315

Tompkins, G. J., Linduska, J. J., Young, J. M. and Dougherty, E. M. (1986). Effectiveness of microbial and chemical insecticides for controlling cabbage looper (Lepidoptera: Noctuidae) and imported cabbage worm (Lepidoptera: Pieridae) on collards in Maryland. *Journal of Economic Entomology*, **79**, 497–501

Ward, J. T., Basford, W. D., Hawkins, J. H. and Holliday, J. M. (1985). *Oilseed Rape*. Farming Press, Ipswich, Suffolk, England

Wheatley, G. A. and Finch, S. (1984). Effects of oilseed rape on the status of insect pests of vegetable brassicas. In *Proceedings of the British Crop Protection Conference—Pests and Diseases*, **2**, 807–814

Wilson, L. T., Pickel, C., Mount, R. C. and Zalom, F. G. (1983). Presence-absence sequential sampling for cabbage aphid and green peach aphid (Homoptera: Aphididae) on Brussels sprouts. *Journal of Economic Entomology*, **76**, 476–479

CHAPTER 5

Pests of Cucurbit Crops: Marrow, Pumpkin, Squash, Melon and Cucumber

A. York

5.1 INTRODUCTION

The Cucurbitaceae comprises about 90 genera and more than 700 species of plants. They are among the most widely grown and important vegetable crops in the world. Originally spreading from three semitropical or tropical centres, they are currently grown almost world-wide, either outdoors or under protection (glasshouses). According to Whitaker and Davis (1962), the genus *Cucurbita* L. (squashes, pumpkins, gourds), including *C. pepo* L. (field pumpkin), *C. mixta* Pang. (striped cushaw), *C. moschata* Duchesne (cushaw and winter crookneck squashes) and *C. maxima* Duchesne (autumn and winter squashes), has a ten-thousand year history of association with Man, as indicated by its presence at archaeological sites in Mexico and Central America. The cucumber (*Cucumis sativus* L.), the muskmelon (*C. melo* L.) and the luffa gourd (*Luffa cylindrica* Roem.) are probably all indigenous to India. The watermelon (*Citrullus vulgaris* Schrad.) probably developed in Africa, although some authorities have made a strong case for India as its area of origin. The origin of the white-flowered or 'bottle' gourd (*Lagenaria siceraria* Standl.) has caused considerable debate. Archaeological evidence from 4000 BC places it in South America, yet it has been identified from an Egyptian tomb of the fifth dynasty (3500–3300 BC). Many authorities think that it developed in Africa and reached the Americas by natural means—i.e. ocean currents. Regardless of whether the 'bottle' gourd originated in South America or in Africa, its use throughout the world has been extremely important as food containers, water bottles, dippers, scoops, dishes, etc.

Grown in temperate as well as tropical areas, the cucurbits tend to be long-season crops, 55–120 days, favoured by warm, dry climates. As a consequence, they are exposed to a wide variety of insect pests. Having been grown for centuries in some areas of the world, farming strategies have evolved with the cucurbit crops. In many areas (e.g. south-east Asia and Africa) virus diseases and minor leaf feeding are taken for granted by local growers. In other areas (e.g. the USA), because cucurbits are high-value crops which are managed very intensively, local growers try to prevent the occurrence of virus diseases and leaf feeding.

5.2 APHIDS

5.2.1 *Aphis gossypii* Glover: Melon Aphid (Hemiptera: Aphididae)

The melon aphid (or cotton aphid; Figure 5.1) occurs nearly world-wide but in northern temperate regions is often confined to glasshouses. It is particularly abundant and damaging in tropical regions. It is a small to medium-sized aphid (0.9–1.8 mm) with antennae about half the length of the body. Apterae are variable in colour, from yellow through green to dark green, nearly black. Under crowded conditions, however, apterous melon aphids may be pale yellow to white in colour. Commonly, they are coloured light green, are mottled in appearance and have uniformly dark siphunculi and pale or dusky caudae. The head and antennae of alate melon aphids are coloured black, the eyes are dark red to black, the abdomen is green to dark green with darker patches on either side, the siphunculi are almost black and the cauda is dark green to black.

Melon aphids overwinter as eggs in temperate regions, but in semi-tropical and tropical regions, they are continuously parthenogenetic and viviparous. Females mature in 4–20 days, depending on temperature, and produce from 20 to 140 offspring each at a rate of 2–9 per day. Melon aphid numbers build up rapidly in dry weather.

The melon aphid is reported to feed on 220 crop plants in 46 families, including 30 crops in the Solanaceae (e.g. eggplant, okra, pepper, tomato) and 19 crops in the Cucurbitaceae (e.g. muskmelon, watermelon, squash, pumpkin) (Roy and Behura, 1983). In addition to melons and cotton, it attacks many other crops, including coffee, cocoa and many Cruciferae and Compositae. Many ornamental plants are also attacked. Feeding near the plant base or on the undersides of young leaves results in cupping of the leaves as they expand. When plants are heavily infested, leaf distortion and stunting are common. Fruit set may be reduced. Upper leaf surfaces may be covered with honeydew and patches of sooty mould fungus. In addition to feeding damage, the melon aphid is known to transmit over 50 plant viruses both non-persistent and persistent. Of particular importance are cucumber mosaic virus, watermelon mosaic virus 2 and zucchini yellow mosaic virus. Virus infection in severe cases may be as high as 50% of the plants in a field. Many of the virus diseases have distinctive symptoms. Muskmelon plant resistance may play an important role in preventing acquisition of watermelon mosaic virus 2 by melon aphids (Romanow *et*

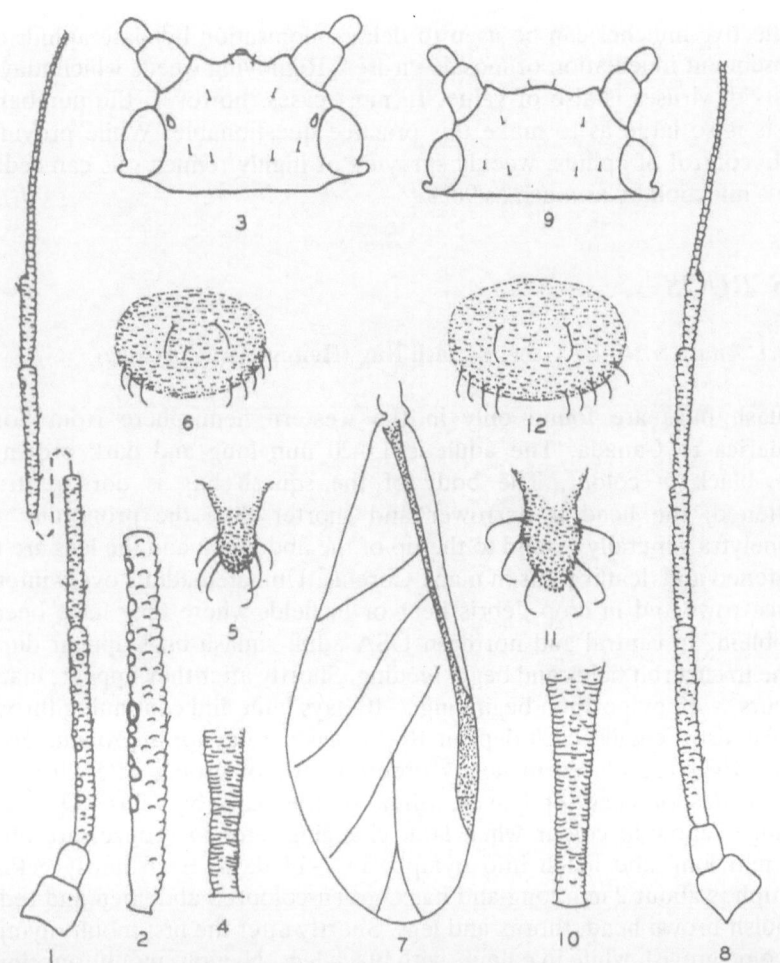

Figure 5.1 *Aphis gossypii*. Alate viviparous female: 1, antenna (×100); 2, segment 3 of antenna (×200); 3, head; 4, cornicle (×100); 5, cauda (×100); 6, genital plate (×100); 7, forewing. Apterous viviparous female: 8, antenna (×100); 9, head; 10, cornicle (×100); 11, cauda (×100); 12, genital plate (×100). Reproduced with permission from Cottier (1953)

al., 1986). In addition, muskmelon varieties AR Hale's Best Jumbo, AR5 and AR Topmark are known to be resistant to the aphids.

The role of natural enemies in controlling melon aphids is unknown. Many natural enemies can be found wherever the aphid occurs. Syrphid flies have been used in glasshouses with mixed success (Chambers, 1986). Natural populations of melon aphids are occasionally wiped out by fungal outbreaks of *Entomophthora* spp. (Entomophthorales: Entomophthoraccae) and fungal pathogens have been used in glasshouses with some success. Over-fertilisation with nitrogen can increase aphid populations.

Reflective mulches can be used to delay colonisation by alate aphids and subsequent inoculation of mosaic viruses. Removing weeds which may be hosts of viruses is also of value. In most cases, however, the number of hosts is so large as to make this practice questionable. While providing little control of aphids, weekly spraying of highly refined oils can reduce virus infection by as much as 90%.

5.3 BUGS

5.3.1 *Anasa tristis* De Geer: Squash Bug (Hemiptera: Coreidae)

Squash bugs are found only in the western hemisphere from South America to Canada. The adult is 15–20 mm long and dark brown to grey-black in colour. The body of the squash bug is dorsoventrally flattened: the head is narrower and shorter than the pronotum; the hemelytra generally extend to the tip of the abdomen; and the legs are not flattened and 'leaflike', as in many Coreids. Unmated adults overwinter in fence rows and in crop debris near or in fields where they have been a problem. In central and northern USA adult squash bugs appear during June in cucurbit fields and begin feeding. Shortly after they appear, mating occurs, with oviposition beginning 7–10 days later and continuing through to August. Females each deposit 10–20 eggs per day for approximately 45 days. Eggs are placed in neatly ordered rows or groups of 5–20 on the undersides of cucurbit leaves, often in the axils of veins. The eggs, orange-yellow in colour when laid, changing later to bronze, are about 1.5 mm long and hatch into nymphs in 7–14 days. Each newly eclosed nymph is about 2 mm long and has a green-coloured abdomen and red to reddish-brown head, thorax and legs. Shortly after the first moult, nymphs become greyish white in colour, with black legs. Nymphs moult four times, requiring 5–6 weeks to reach maturity in the first generation. Second-generation nymphs require somewhat longer to mature—7 weeks—probably owing to plant nutrient deficiencies. The nymphs, which are whitish in colour, with black legs, are often present in large numbers on a single cucurbit vine. They have a reclusive habit and appear to prefer feeding on the undersides of fruit and leaves. In central USA only 1–1.5 generations occur each year, but all squash bug life stages can be found in the field almost any time during the growing season. First-generation females which emerge before late July do not enter diapause but begin immediately to oviposit. Females appearing thereafter increasingly enter diapause, until 100% of females appearing in early September enter diapause (Nechols, 1987). Diapausing females overwinter unmated. In cool springs when first-generation nymphs develop slowly, more first-generation adults enter diapause, resulting in significantly lower populations later in the season. When squash bugs are numerous, they emit a disagreeable odour. This odour, emitted also when the insect is squeezed, arises from two scent glands located on the dorsal abdomen.

Summer squash (zucchini, crookneck), Hubbard squash, pumpkin and marrow squash seem to be the crops most severely attacked by squash bugs, but other cultivars will also be attacked (Howe and Rhodes, 1976). As adults and nymphs feed on the undersides of leaves, the leaves yellow in colour and appear to senesce. Severe infestations cause plants to wilt rapidly, a condition termed 'Anasa wilt'. Usually if squash bugs are controlled, affected plants will regain their turgour, in contrast to bacterial wilt, which causes permanent wilting.

Control of squash bugs is extremely difficult. Pyrethroid insecticides have been reasonably successful in controlling high populations of squash bugs on canning pumpkins. Little or no control was obtained from the use of insecticides previously registered for control of squash bugs, although trichlorfon is moderately effective against immature stages. Both in home gardens and in the field, crop residues after harvest should be destroyed, either by being buried deeply or by being removed. These practices remove much of the overwintering protection for the pest. Natural enemies of squash bugs are numerous, but only a parasitic wasp (*Ooencyrtus* Ashmead spp. (Hymenoptera: Encyrtidae), up to 26% parasitism of eggs) and a parasitic tachinid fly seem to be important.

5.3.2 *Leptoglossus australis* (Fabricius): Leaf-footed Bug (Hemiptera: Coreidae)

The leaf-footed bug (Figure 5.2) is a persistent, moderate to severe pest of cucumbers, gourds and similar crops in Africa, India, south-east Asia, China, Pacific islands and Australia (Jadhav *et al.*, 1979; Visalakshi *et al.*, 1980). It is a minor pest of passion fruit (Murray, 1976). The dark-grey–brown-coloured female is 20–25 mm long, with a pale-orange-coloured stripe across the anterior edge of the mesonotum. The antennae have alternating dark- and pale-orange-coloured zones (Hill, 1983). Females live for about 40 days, laying eggs in rows along stems and twigs, averaging 50 eggs per female. Eggs hatch in 5–9 days and five nymphal instars develop subsequently over a period of about 30 days (Jadhav *et al.*, 1980). Adults and nymphs feed by puncturing stems, flowers and tender fruit, leaving dark spots. Terminal shoots with heavy feeding damage may die. Immature fruits may abscise after heavy feeding.

Since preventing population build-up is important in preventing damage, timely monitoring of crops (scouting) is very important in the management of leaf-footed bug infestations. Sanitation by removing crop debris from harvested fields at the end of the growing season to prevent harbourage and use of the male pheromone in a baiting or scouting programme are also of value. Currently, control relies mostly on repeated applications of insecticides whenever the pest seems to be too numerous.

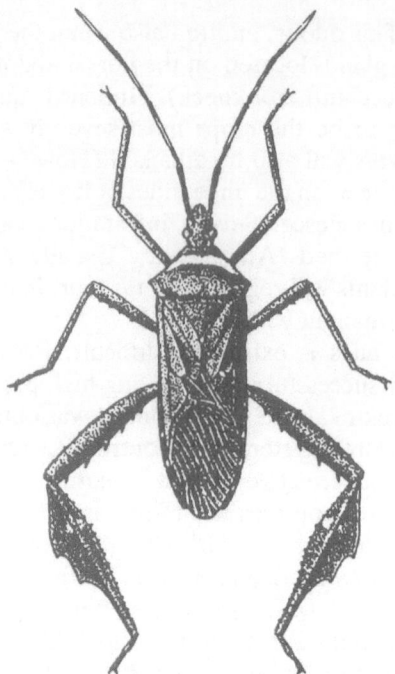

Figure 5.2 Adult leaf-footed bug, *Leptoglossus australis* (×2). Reproduced with permission from Kalshoven (1981)

5.4 BEETLES

5.4.1 *Acalymma vittatum* (Fabricius): Striped Cucumber Beetle (Coleoptera: Chrysomelidae)

The striped cucumber beetle (Figure 5.3) is restricted in distribution to the western hemisphere from South America to Canada. The adult beetle measures 6–8 mm long and is distinctively coloured, with a black head, yellow-orange thorax and yellow and black stripes running to the tips of the wing covers. Only unmated adults overwinter, protected by fallen leaves, trash and debris in hedgerows, forests or other areas of shelter. They emerge from overwintering during early spring in southern and central USA, feeding upon wild cucurbits, as well as blossoms of several foliage (ornamental) and fruit trees. Bright-yellow-coloured eggs are laid in cracks in the soil at the base of plants. A female striped cucumber beetle may oviposit up to 1300 eggs in 2 months at a rate of 10–20 per day. Eggs hatch in 5–8 days into larvae, which feed for 10–20 days on the underground parts of plants, often destroying the root systems of seedlings. Pupation occurs in the soil and lasts for about 7 days. Adult beetles emerge from pupae in late

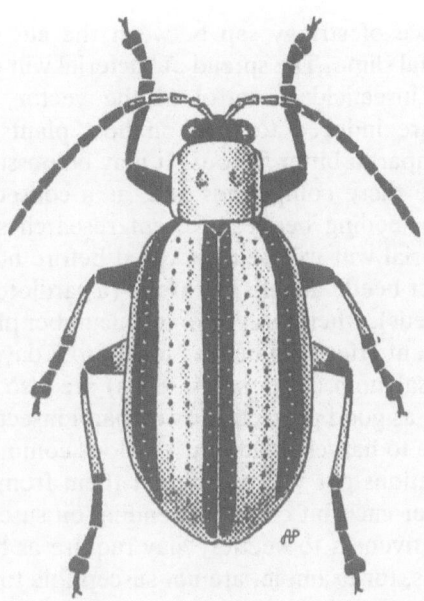

Figure 5.3 Adult striped cucumber beetle, *Acalymma vittatum*. © Purdue University

June or early July and are active until autumn. In southern USA there may be several generations per year.

The crops most seriously attacked by the striped cucumber beetle are cucumber, muskmelon (honeydew, cassaba melon), Hubbard squash, pumpkin, summer squash and watermelon. Adult feeding on the stems of seedlings is very serious and often kills the young plants outright (Brewer *et al.*, 1987). Serious stem damage to seedlings under 10 cm tall can be caused by a beetle population density of less than 1 per plant (Burnside and Barry, 1976). Cucumber plants at the cotyledon stage are often killed within 1 day of emerging from the soil. In addition to the damage caused by the foliar-feeding adults, larval feeding on roots can significantly delay early growth and subsequent maturity of plants (Latin and Reed, 1984). While the direct damage done to cucurbit plants by striped cucumber beetles is serious, the beetles' involvement in the transmission of a bacterium (*Erwinia tracheiphila* (Smith) Bergey, Harrison, Breed, Hammer and Huntoon) which causes a disease known as bacterial wilt is even more serious. Cucumbers, cantaloupes and some related melons, such as cassaba and honeydew, are severely affected by bacterial wilt. The bacterium overwinters in the body of the beetle, to be transmitted in the spring on the mouthparts or in the faeces to young plants. After transmission, the disease may not appear for several weeks, but when it does, it rapidly kills plants. Bacterial wilt can often be diagnosed quite simply in a plant by cutting through the stem with a knife and then allowing the two cut surfaces to come into contact with each other before slowly pulling them apart. The presence of stringy sap between the cut surfaces is a good indicator of bacterial slime. The spread of bacterial wilt disease in cucurbits is suppressed by insecticidal control of the vector, striped cucumber beetles. Beetles are induced to feed on host plants by cucurbitacins, compounds that impart a bitter flavour. It may be possible to breed plants which do not have these compounds and, as a consequence, are not as attractive hosts to feeding beetles. Recent research suggests that plant resistance to bacterial wilt will be discovered before host-plant resistance to striped cucumber beetle attack. Neriifolin (a cardiotonic glycoside from yellow oleander seeds), when sprayed on to cucumber plants, exhibits toxic and feeding deterrent effects on beetles for up to 7 days after application. Azadirachtin and salannin (from neem seeds) are also feeding deterrents but do not provide as good protection as carbaryl insecticide.

From cold frame to harvest field, muskmelons commonly require 10–13 insecticidal applications per year to protect them from striped cucumber beetle attack. Other cucurbit crops, depending on susceptibility to bacterial wilt and attractiveness to beetles, may require as few as 2–3 applications. Watermelons, for example, are not susceptible to bacterial wilt and, as a consequence, young plants need to be protected for a 2–3 week period only after planting. Insecticides commonly used as sprays to control striped cucumber beetles on cucurbits are carbaryl, methoxychlor (low hazard to honeybees) and endosulfan. As an alternative to foliar-applied insecticides, carbofuran granules applied to the soil before planting or transplanting provide excellent control for 5–8 weeks of both adult beetles feeding on stems and leaves and larvae feeding on roots.

5.4.2 *Diabrotica undecimpunctata howardi* Barber: Spotted Cucumber Beetle (Coleoptera: Chrysomelidae)

Like the striped cucumber beetle (see page 144; Figure 5.4), the spotted cucumber beetle is restricted in distribution to the western hemisphere from South America to Canada, although it has close relatives in Africa and Asia. Where it is found, the spotted cucumber beetle is annually a serious pest of cucurbits. The yellow- or yellowish-green-coloured adult beetle is about 7–9 mm long, with a black head and 12 black spots on the wing covers. It overwinters in crop debris near to the soil surface and becomes active when daytime temperatures reach 18–22°C, often not appearing for 2–4 weeks after the emergence of striped cucumber beetles. Eggs (120 per female) are laid in the soil around the base of plants. The larvae which hatch from the eggs feed on plant roots for a 3–4 week period. Pupation requires 10–14 days. Under artificial rearing conditions, there is a preoviposition period of 7–10 days and a generation is completed in 60 days (Rose and McCabe, 1973). A single generation only, perhaps with a partial second generation, occurs in central and northern USA.

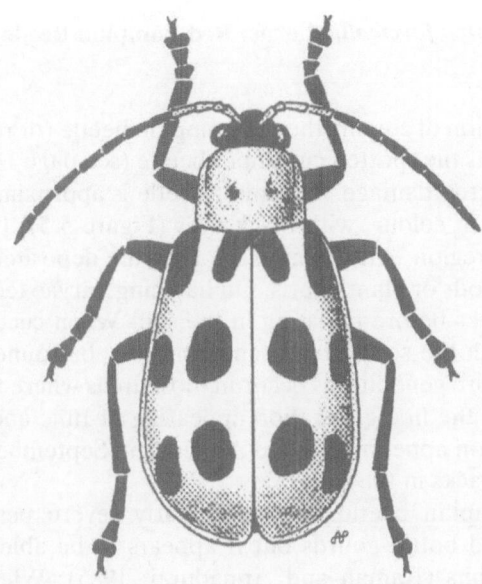

Figure 5.4 Adult spotted cucumber beetle, *Diabrotica undecimpunctata howardi*.
© Purdue University

While the foliar-feeding damage caused by adult spotted cucumber beetles is serious, the beetles' vectoring of bacterial wilt disease of cucurbits is even more serious (see striped cucumber beetle, page 145). Growers of muskmelons generally must tolerate 8–10% diseased plants. Unlike striped cucumber beetles, which feed largely on stems, spotted cucumber beetles feed mostly on leaves. The economic threshold of spotted cucumber beetles on cucurbits to suppress as much as possible the spread of bacterial wilt disease is simply the appearance of beetles in susceptible crops. In south-eastern USA adult spotted cucumber beetles lay eggs in the soil around groundnut plants (see page 198 *et seq.*). The larvae damage shoot tips (pegs) and may prevent pod formation. They also feed on pod tissues and bore into nut kernels. Pod injury may reach 50% in untreated, susceptible groundnut varieties (Grayson, 1947). Pod as well as root injuries also predispose plant tissues to attack by secondary pathogens which produce aflatoxins in the nut kernels.

The same insecticides that control striped cucumber beetles (see page 146) control spotted cucumber beetles. Good results under research conditions have been obtained using baits based on ground dried bitter gourd containing 0.12% cucurbitacins (a feeding arrestant), 1% eugenol (an attractant) and 0.1% methomyl insecticide.

Host-plant resistance has significantly reduced groundnut damage caused by larval feeding of spotted cucumber beetles: cultivar NC6 suffered approximately 85% less pod damage than did the standard cultivar, Florigiant (Campbell and Wynne, 1985).

5.4.3 *Raphidopalpa foveicollis* Lucas: Red Pumpkin Beetle (Coleoptera: Chrysomelidae)

With the exception of colour, the red pumpkin beetle (or red melon beetle) closely resembles the spotted cucumber beetle (see page 146) in morphology and type of crop damage. The adult beetle is approximately 7 mm long and red-orange in colour, with black eyes (Figure 5.5). It is found in the Mediterranean region, Africa and Asia. Eggs are deposited during April in the soil under clods or plant debris. On hatching, larvae feed on plant roots for about 3 weeks before pupating in the soil. When cucurbit fruits come into contact with the soil, larvae can commonly be found feeding on the rind. Two or more generations occur in most areas where the red pumpkin beetle is found, the first generation appearing in June and early July, the second generation appearing in late August and September. Adult beetles overwinter in cracks in the soil.

The red pumpkin beetle is a particularly severe pest of pumpkins, muskmelons and bottle gourds but it appears to be able to feed on any available cucurbits (Raman and Annadurai, 1985). When cucurbits are scarce, it is found feeding on other plant families. Red pumpkin beetles damage most severely cotyledons and young leaves, but they will feed readily on flowers and mature (not senescent) leaves. The beetle appears to be the most serious pest of pumpkins in India. In plant-resistance research the apparent attractiveness of pumpkins to beetles was found to be positively correlated with cucurbitacin content in leaves.

The control of red pumpkin beetles on cucurbits is generally accomplished by spraying insecticides (carbaryl, chlorpyrifos, trichlorfon) when beetles are active on the foliage. Recent research has demonstrated, however, that a single application of carbofuran granules at planting time is as effective as 4–5 applications of carbaryl sprayed weekly on to the foliage. A triterpenoid compound from bitter gourd has been found by Chandravadana and Pal (1983) completely to inhibit feeding by red pumpkin beetle on cucurbit leaves.

5.5 MOTHS

5.5.1 *Diaphania nitidalis* (Stoll): Pickleworm (Lepidoptera: Pyralidae)
Diaphania hyalinata (L.): Melonworm (Lepidoptera: Pyralidae)

The pickleworm and the melonworm are among the most damaging insect pests to cucurbits in the western hemisphere. Both moths are found from southern Canada through the USA and Mexico to Brazil. They are particularly destructive in the Caribbean islands. There has been no evidence of a diapause occurring in any of the life stages of either moth

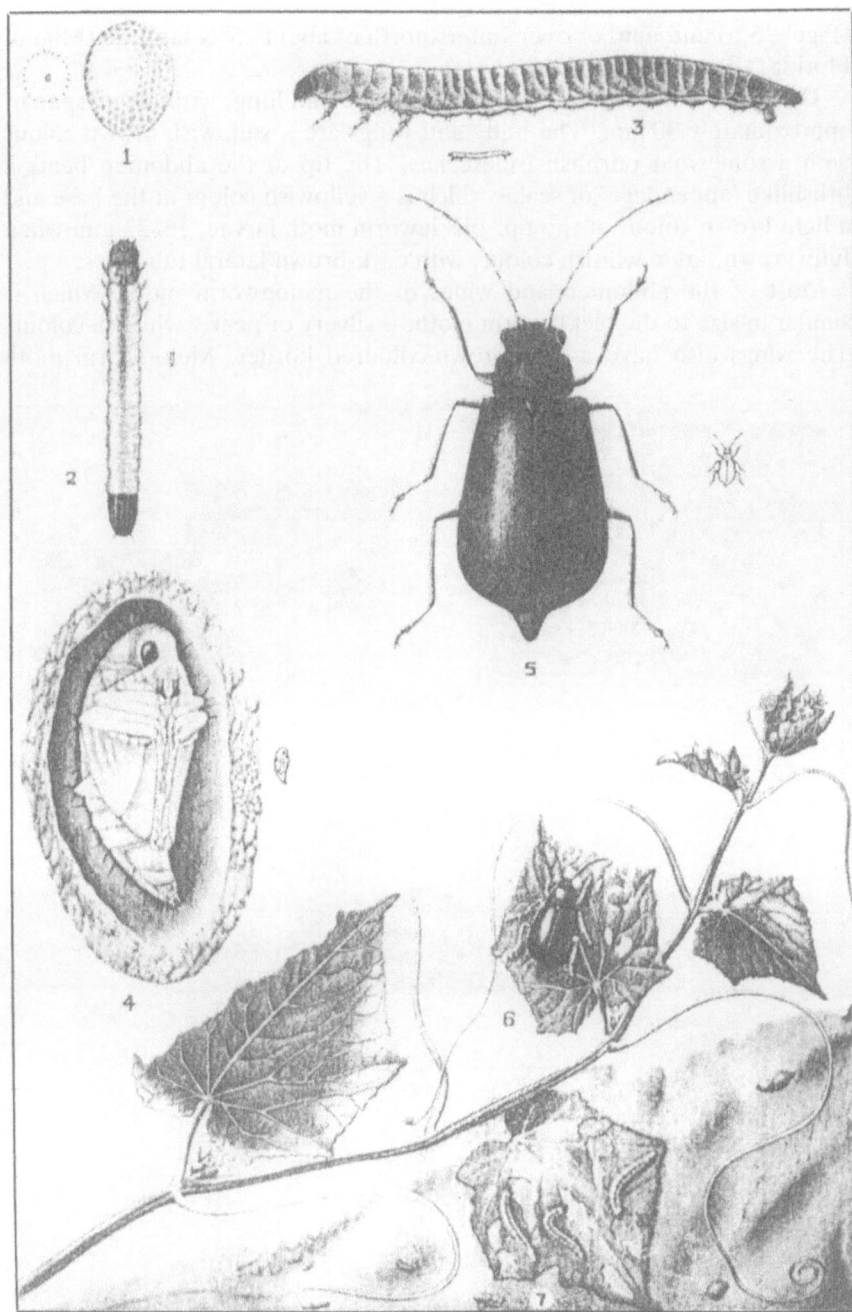

Figure 5.5 Red pumpkin beetle, *Rhaphidopalpa foveicollis*. Reproduced with permission from Ghosh (1940)

(Figure 5.6) and neither overwinters north of about 25°N latitude (Miami, Florida) (Pena *et al.*, 1987a).

The pickleworm moth is approximately 15 mm long, with a wingspan of approximately 30 mm. The body and wings are a yellowish brown colour with a somewhat purplish iridescence. The tip of the abdomen bears a brushlike 'appendage' of scales which is a yellowish colour at the base and a light brown colour at the tip. Pickleworm moth larvae, 16–22 mm when fully grown, are a whitish colour, with dark-brown lateral tubercles.

Most of the abdomen and wings of the melonworm moth, which is similar in size to the pickleworm moth, is silvery or pearly white in colour. The wings also have a light-brown-coloured border. Melonworm moth

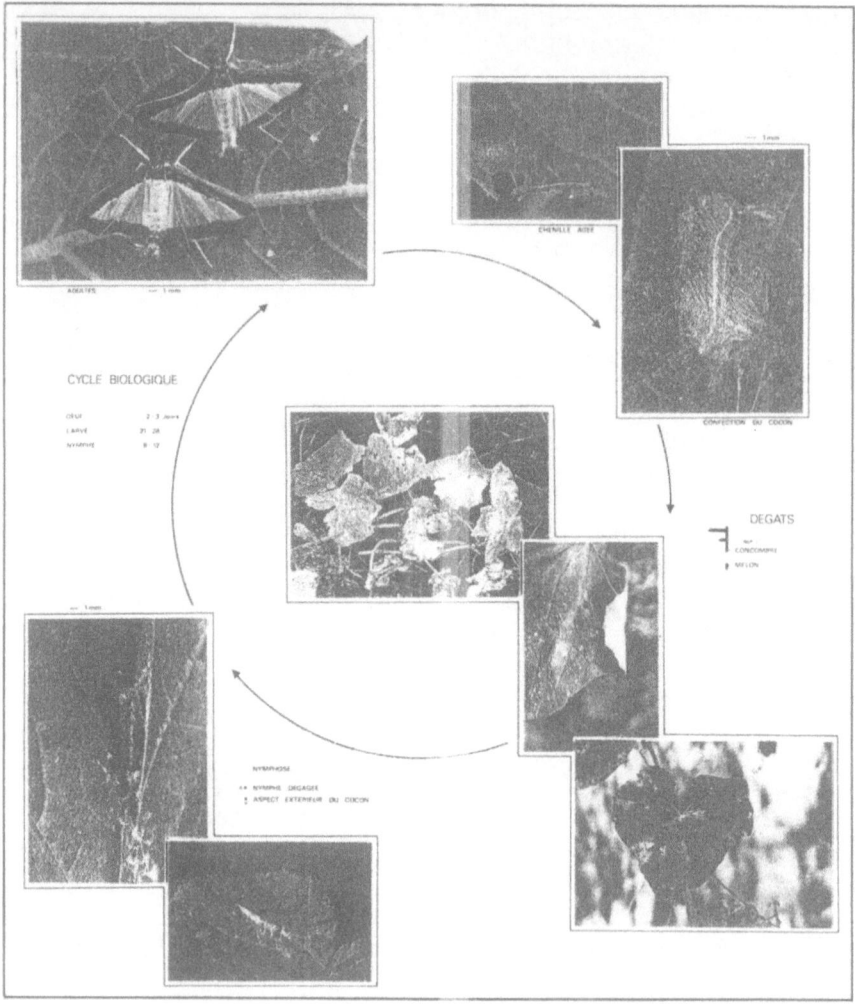

Figure 5.6 Life cycle of *Diaphania hyalinata*. Reproduced with permission from IRAT/CIRAD (1986)

larvae, which, when fully grown, are similar in size to fully grown pickleworm moth larvae, are light green in colour, with dark-brown lateral tubercles.

Emerging from pupae during April, May or June, nocturnally flying female and male moths normally mate within 3 days from emergence. Both pickleworm and melonworm female moths produce a sex pheromone. Although most of the chemical components of the moths' sex pheromones are identical, each pheromone is species-specific. Eggs (0.4–0.7 mm) are laid in clusters of 3–10 on leaves or flower buds. Temperature plays an important role in fecundity—for example, pickleworm female moths lay 300–400 eggs at 27 °C but only 83 eggs at 32 °C. Newly hatched larvae of both species of moth feed on leaves. However, pickleworm moth larvae soon after hatching move to buds or flowers and, on cucumbers, most third- and fourth-stage larvae are found in buds or flowers. Late fourth-stadium or early fifth (last)-stadium larvae begin burrowing into fruit. In cucurbits with large flowers (e.g. squashes) larvae may complete development within flowers. The larval stages require 43 days at 16 °C and 12 days at 35 °C to complete development. The pupal stage lasts 32 days at 16 °C to 8.5 days at 35 °C. At an average daily temperature of 27 °C, larvae require about 13 days and pupae about 9–10 days to complete development. Pupation occurs in folded leaves spun together to form cells. Adults live 35–40 days. In infested areas different generations of moths soon overlap and, as a result, damaging larvae may be continuously present throughout the growing season.

In southern Florida, USA, native (wild) host plants of pickleworm and melonworm moths appear to be *Melothria pendula* L. (creeping cucumber) and *Momordica charantia* L. (wild balsam apple). Both species of moth are supported by these plants when cultivated host plants are not present. Nearly all cultivars of cucumbers, squashes and muskmelons are susceptible to severe damage. Pickleworm is most common on cucumber and zucchini crops, while melonworm is most commonly found on Cuban pumpkin crops. In tropical Mexico melonworm is more common than pickleworm. Fruit losses caused by both moths are high: from 30% of total US production to 66% of Caribbean production. Under experimental conditions larvae of pickleworm moths have been found able to complete their development on snap bean flowers and fruit as well as tomato fruit. Both pickleworm and melonworm moth larvae have been reared on artificial diets (pinto bean), although, after 25 generations on an artificial diet, pickleworm larvae survived better (faster development, greater survival, greater fecundity) when they were transferred to cucurbits (squashes) (Day and Robinson, 1981).

When pickleworm and melonworm moth larvae feed on cucurbit leaves and flowers, fruit set is reduced. As they feed on and in fruit, the larvae reduce quality and introduce secondary fungi and bacteria. Fruit damage from just a single larva can be severe. Because of a low tolerance for fruit damage and infestation by growers and shippers, thresholds for control are very low.

Parasitoids of pickleworm and melonworm moth eggs, larvae and pupae seem to exert little significant control. Egg parasitism ranges from 0 to 69% (*Trichogramma* Westwood spp. (Hymenoptera: Trichogrammatidae)), with parasitism averaging 2–5% in pickleworm eggs and 12–13% in melonworm eggs in Florida, USA (Pena *et al.*, 1987b). In the USA the fireant (*Solenopsis invicta* Buren (Hymenoptera: Formicidae)) does appear to be an important predator of pupae of both species of moth.

In most areas cucurbit growers rely on frequent (5-day intervals) applications of insecticides (organophosphates and carbamates— particularly carbaryl) to control young pickleworm and melonworm moth larvae as they feed on leaves and flowers. Sprays of *Bacillus thuringiensis* Berliner var. *kurstaki* Kurstak have provided only moderate control. For pickleworm moth larvae on summer squashes, Brewer and Story (1987) developed a sequential sampling plan based on staminate green flower buds greater than 5 cm in length. Few cultural controls have proved of value, although on the northern and southern fringes of the moths' range in the USA, early plantings of cucurbits have much lower infestations than do late plantings. Differences in susceptibility do exist between cultivars—for example, in field experiments CVs Butternut and Calabaza squashes were much less suitable for moth oviposition than CV Tablequeen pumpkin. As a rule, adult pickleworm moths lay fewer eggs on glabrous than on pubescent leaves of cucumbers and muskmelons. Furthermore, fruit rinds of CV Calabaza squash appear to inhibit the development of later instars of pickleworm moth larvae (antibiosis). Of 1160 lines of cucumbers tested for antibiosis to pickleworm moth larvae, little or no useful antibiosis was discovered (Wehner *et al.*, 1985). Both gamma radiation and heat have been used in the laboratory to sterilise adult pickleworm moths.

5.5.2 *Melittia cucurbitae* Harris: Squash Vine Borer (Lepidoptera: Sesiidae)

The squash vine borer is restricted in distribution to the western hemisphere from Brazil to southern Canada. The moth is approximately 16 mm long with a wingspan of approximately 35 mm. The forewings of the moth are opaque and coloured olive-brown, appearing somewhat iridescent green, and the hindwings are transparent, each with black-coloured veins and a fringe of brown-coloured hairs. The abdomen of the squash vine borer moth is reddish-orange in colour, with a row of dark-coloured spots along the dorsal, central line. The legs of the moth are orange in colour, with black and white banded tarsi, and the face is white in colour, with yellow palpi. In central and northern USA moths appear from mid-June to mid-July. Generally, the presence of the squash vine borer is not noticed by growers until after damage has been done to crops. The moth is a noisy, daytime flier which deposits eggs (150–200/female) singly on leaf petioles, stems and fruit of susceptible plants. The eggs are small (1 mm), circular, flattened and brown in colour. Hatching from eggs in about a week, larvae begin immediately to bore into plants, leaving behind sawdust-like frass. As larvae feed within stems, frass may fall from larval boreholes in the stems to the soil surface. Often several larvae will be found within a single stem. Larvae feed for 2–4 weeks. When fully fed, they burrow into the soil, where they each create silken-lined cocoons in which to pupate immediately or overwinter, becoming pupae in early spring. A second generation of moths is found in the more southerly areas of its distribution, with mating and oviposition continuing until the onset of frost.

Feeding by squash vine borer larvae within stems of affected plants causes severe wilting and, ultimately, plant death if larvae are not controlled. The squash vine borer is a severe pest of summer and winter squashes: Hubbard, zucchini, yellow and acorn are all badly affected. Squash vine borer attacks on cucumbers, cantaloupes and watermelons are rare.

Growers of small areas of squashes must watch their plants closely from mid-June to mid-August for the first signs of squash vine borer activity. The most common sign of activity is the sawdust-like frass which has been extruded through larval boreholes in the stems to lie on the soil surface. If excrement is seen, plant stems should be split open and any larvae within removed. The length of damaged stems should then be covered with soil. As a result, supplementary roots will often develop and affected plants will survive. More commonly, insecticides are used to protect plants from squash vine borer attack. However, the timing of insecticidal applications is often incorrect, with plants being treated after larval attack.

There is also an additional problem of keeping rapidly growing plant tissue continuously protected with insecticides. Research needs to be done on squash vine borer pheromones, which could help ultimately, perhaps, to time accurately the presence of the moths.

5.6 FLIES

5.6.1 *Dacus cucurbitae* (Coquillet): Melon Fly (Diptera: Tephritidae)

The melon fly is found in Africa, south-east Asia, Japan, Indonesia and the Pacific islands, including Hawaii. It does not occur, as far as is known, in the UK, central Europe and continental USA.

The adult melon fly is 8–10 mm long, including ovipositor, with a wingspan of 12–15 mm. It has a yellow–brown-coloured body, with darker-coloured, transverse, abdominal bands. The melon fly wings are 'horny', each with a dark-brown-coloured costal strip extending to the tip and dark-coloured bands on the radiomedial veins (Figure 5.7). The egg is white in colour, about 1 mm × 0.25 mm in size, and bullet-shaped with a rounded base. The larva, which is about 10–12 mm long and a whitish-yellow colour when fully grown, is a typical acephalic maggot with three thoracic and eight abdominal segments. The puparium is a dull reddish-brown to orange colour, 'barrel'-shaped and 4–5.5 mm long.

While melon flies may mate every 4–5 days, the original mating seems to produce fertile eggs in females for the duration of their lives. Female flies begin laying eggs about 7–10 days after emerging from pupae in the soil as long as they have, in the meantime, fed on protein hydrolysates. Eggs are laid at the rate of 7–10 per female per day. Female melon flies each lay a total of 800–900 eggs with approximately 50% fertility (Vargas *et al.*, 1984). Fertile eggs do not appear to be produced below 13 °C. Female flies seem to require free water for drinking and their longevity is definitely extended if

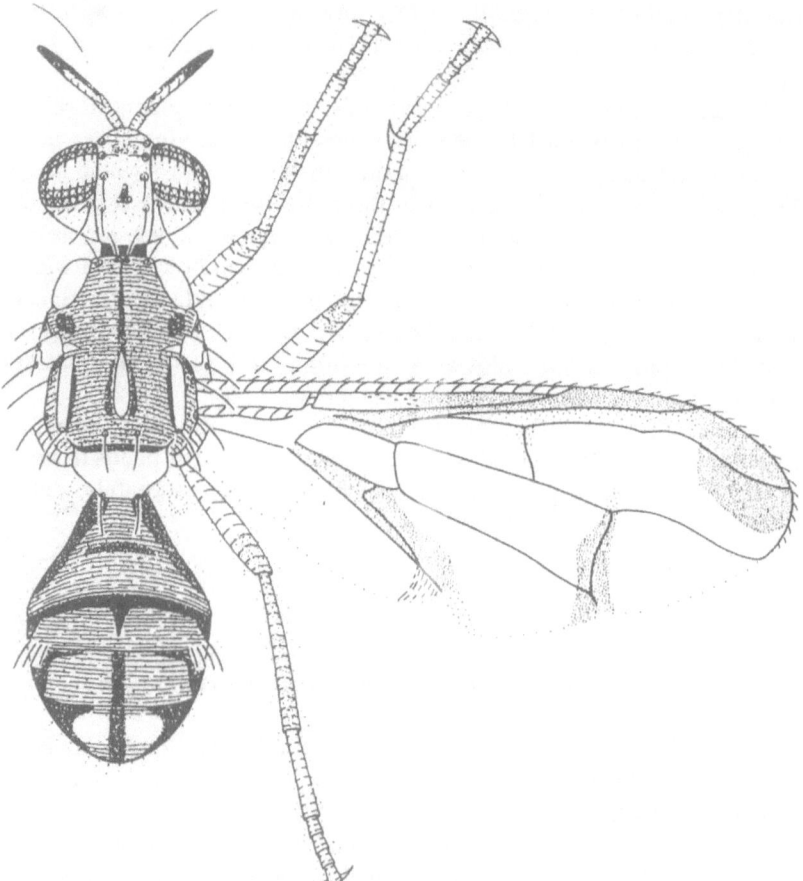

Figure 5.7 The melon fly, *Dacus cucurbitae* (×13). Reproduced with permission from Drew *et al.* (1982)

they are able to feed on carbohydrates. The temperature threshold for egg development is reported to be 10.4°C, with no hatching, pupation or emergence occurring below 10°C (Okumura *et al.*, 1981). Vargas *et al.* (1984) found the egg developmental temperature threshold to be considerably higher at 14.7°C. Eggs are laid by melon flies under the skin of ripe or ripening fruit. The larvae which hatch from the eggs feed on the fruit for three stadia. The fourth larval stage is spent inside a puparium with moulting occurring, eventually to become a true pupa. The rate of larval development is dependent on temperature and host plant. At 25°C larvae complete development in 4–7 days. Pupation, which occurs in the soil, requires 9 days at 25°C. Egg to adult survival rates of between 35% and 85% have been reported under laboratory conditions (Carey *et al.*, 1985).

Adult melon flies readily oviposit in cucumber, muskmelon, watermelon, squash and other cucurbit crops, as well as wild Cucurbitaceae. Between 70 and 100 non-cucurbitaceous fruits and vegetables are also attacked. Larvae tunnelling through fruit cause direct damage. They also cause indirect damage to fruit by contaminating it with frass and accelerating its decomposition by secondary bacteria and fungi. Certain areas of the world (e.g. south-east Asia) report up to 50% of cucurbit production destroyed by melon flies.

The control of melon flies is considered together with the control of oriental fruit flies on pages 156–7.

5.6.2 *Dacus dorsalis* Hendel: Oriental Fruit Fly (Diptera: Tephritidae)

Closely related to the melon fly and found in approximately the same areas of the world, the oriental fruit fly has an even greater host-plant range (Doharey, 1983). Female oriental fruit flies are known to oviposit in more than 150 species of fruits and vegetables, with cucurbits apparently less important hosts to fruit flies than to melon flies. The adult fruit fly is slightly smaller than the adult melon fly and the wings are somewhat less distinctly marked (Figure 5.8; see page 154). The appearance of the eggs, larvae and pupae as well as the timing of the life histories of both flies are similar. Following an 8–12 day preoviposition period, oriental fruit flies normally each lay between 1300 and 1500 eggs over a 70 day oviposition period, but up to 3000 eggs have been reported to be produced under unusual conditions (e.g. when adult females feed on high-protein diets). The egg stage lasts 1–5 days, depending on temperature. At optimum temperatures (about 30°C), three larval instars complete development in 6–7 days, although larvae can need up to 35 days to complete development, depending on time of year. In Hawaii no larval development is reported to take place below 8°C. At 27–30°C the entire life cycle of the oriental fruit fly from egg to adult is reported to require 21 days. Like the melon fly, the fruit fly develops differentially on different host fruit (e.g. papayas,

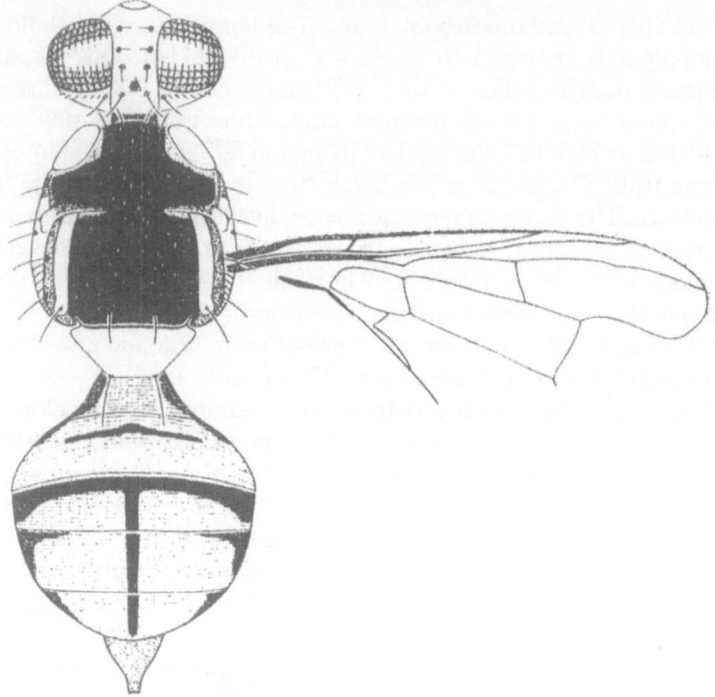

Figure 5.8 Oriental fruit fly, *Dacus dorsalis* (×12). Reproduced with permission from Drew *et al.* (1982)

bananas and mangos are apparently more suitable than watermelons). On Oahu, Hawaii, 95% of the fruit fly population is reported to develop on guavas.

With both oriental fruit flies and melon flies, considerable effort has gone into population surveillance and exclusion from uninfested geographical regions. Upon detection of flies, eradication efforts have been successfully initiated several times in different geographical areas (e.g. USA, Japan). Detection (and in some cases control) has relied on baits comprised of methyl eugenol (97–98%) and naled (2–3%) (Chambers, 1977). Male flies are attracted to baits and killed. In Japan nine traps per hectare with about 33 m between traps was found to be ideal for eradication purposes. Male flies produce a pheromone 'smoke' which is attractive to virgin females. The pheromone constituents differ between oriental fruit flies and melon flies (Fitt, 1981). Even though female flies are not monogamous, sterile male programmes have been effective at controlling populations. Procedures for sterilising have to be applied carefully, as male competitiveness has been shown to be adversely affected. For quarantine purposes, ethylene dibromide and methyl bromide have been used for fumigating fruit prior to shipping. Hot water dipping of fruit (e.g. bananas for 15 min at 50°C) as well as refrigeration have been attempted but fruit can be damaged by high and low temperatures. Refrigeration of infested fruit at 2.8°C for 10 days will kill all oriental fruit fly larvae. To kill all melon fly larvae, infested fruit needs to be refrigerated for 10 days at 0°C. Many fruits and vegetables cannot tolerate the low temperatures needed to kill fruit fly and melon fly larvae. Insecticides have been largely ineffective. Baits comprised of sugar and insecticide applied as large droplets to foliage may help to manage fly populations. Unconventional treatment such as methoprene incorporated into fruit waxes has shown promise but does not kill until pupation, by which time, of course, larval damage has been done. Avermectin B1 is toxic to adult flies and, in addition, sub-lethal dosages reduce significantly both fecundity and fertility. Importation of exotic parasitoids to Hawaii has led to 50% parasitism of the oriental fruit fly population. Crop sanitation by destroying dropped fruit is important to prevent pupation of developing larvae. It is doubtful, however, whether sanitation is a practical method of management in areas where wild hosts are abundant. Cultivation or light tillage of fields after harvest is said to disturb pupae in the soil and, as a result, expose them to predation and parasitism. Host-plant resistance is being researched in many parts of the world and is seen as a critical component of integrated pest management. One or more resistant varieties of cucumber, muskmelon, bitter gourd, pumpkin and bottle gourd have been identified.

5.7 MITES

5.7.1 *Tetranychus urticae* Koch: Two-spotted Spider Mite (Prostigmata: Tetranychidae)

These are discussed in Chapter 8, page 311.

REFERENCES AND BIBLIOGRAPHY

Abdel-Hafez, M. A., Farrag, A. M. I. and Abdul-Ela, M. S. (1982). Relative resistance of some cucurbit varieties to the common spider mite in Egypt. *Agric. Res. Rev.*, **60**, 1–10

Al-Ali, A. S., Al-Neamy, J. K. and Alwan, M. S. (1982). On the biology and host preference of *Aulacophora foveicollis* Lucas (Coleoptera: Galerucidae). *Z. Angew. Entomol.*, **94**, 82–86

Alikhan, M. A. and Yousuf, M. (1985). Effect of host on the oviposition and development and survival of the larvae of *Aulacophora foveicollis* Lucas (Coleoptera: Chrysomelidae). *Can. J. Zool.*, **63**, 1634–1637

Andersen, J. F. (1987). Composition of the floral odor of *Cucurbita maxima* Duchesne (Cucurbitaceae). *J. Agric. Food Chem.*, **35**, 60–62

Basky, Z. (1984). Effect of reflective mulches on virus incidence in seed cucumbers. *Prot. Ecol.*, **6**, 57–61

Bateman, M. A. (1972). The ecology of fruit flies. *Ann. Rev. Entomol.*, **17**, 493–518

Bateman, M. A. (1978). Chemical methods for suppression or eradication of fruit fly populations. In *Economic Fruit Flies of the South Pacific Region* (ed. R. A. I. Drew, G. H. S. Hooper and M. A. Bateman). Watson Ferguson, Brisbane

Blackman, R. L. and Eastop, V. F. (1984). *Aphids on the World's Crops*. Wiley, Chichester

Boudreaux, H. B. (1956). Revision of the two-spotted spider mite complex, *T. telarius* (L.). *Ann. Entomol. Soc. Am.*, **49**, 43–48

Brewer, M. J. and Story, R. N. (1987). Larval spatial patterns and sequential sampling for pickleworm, *Diaphania nitidalis* Stoll (Lepidoptera: Pyralidae) on summer squash. *Environ. Entomol.*, **16**, 539–544

Brewer, M. J., Story, R. N. and Wright, V. L. (1987). Development of summer squash seedlings damaged by striped and spotted cucumber beetles (Coleoptera: Chrysomelidae). *J. Econ. Entomol.*, **80**, 1004–1009

Burnside, J. A. and Barry, B. D. (1976). Destruction of cantaloupe seedlings by cucumber beetles. *Proc. Indiana Acad. Sci.*, **85**, 247

Campbell, W. V. and Wynne, J. C. (1985). Influence of the insect-resistant peanut cultivar NC6 on performance of soil insecticides. *J. Econ. Entomol.*, **78**, 113–116

Carey, J. R., Harris, E. J. and McInnis, D. O. (1985). Demography of a native strain of the melon fly, *Dacus cucurbitae*, from Hawaii. *Entomol. Exp. Appl.*, **38**, 195–199

Chambers, D. L. (1977). Attractants for fruit fly survey and control. In *Chemical Control of Insect Behaviour* (ed. R. R. Shorey and J. J. McKelvey). Wiley-Interscience, New York

Chambers, R. J. (1986). Preliminary experiments on the potential of hoverflies (Diptera: Syrphidae) for the control of aphids under glass. *Entomophaga*, **31**, 197–204

Chandravadana, M. V. and Pal, A. B. (1983). Triterpenoid feeding deterrent of *Raphidopalpa foveicollis* L from *Momordica charantia*. *Curr. Sci.*, **52**, 87–88 (*Current Science*, Bangalore, in Royal Society of Edinburgh Library)

Christenson, L. D. and Foote, R. H. (1960). Biology of fruit flies. *Ann. Rev. Entomol.*, **5**, 171–192

Commonwealth Institute of Entomology (1968). *Distribution Maps of Pests*. Series A, No. 243. London

Cottier, W. (1953). *Aphids of New Zealand*. Bulletin 106, New Zealand Department of Scientific and Industrial Research

Cunningham, R. I. and Suda, D. Y. (1986). Male annihilation through mass-trapping of male flies with methyleugenol to reduce infestation of Oriental fruit fly (Diptera: Tephritidae) larvae in papaya. *J. Econ. Entomol.*, **79**, 1580–1582

Day, A. and Robinson, J. F. (1981). Comparative survival, development, and fecundity of the pickleworm reared on an artificial diet and three cucurbit hosts. *J. Georgia Entomol. Soc.*, **16**, 518–521

Doharey, K. L. (1983). Bionomics of fruit flies (*Dacus* spp.) on some fruits. *Indian J. Entomol.*, **45**, 406–413

Doharey, K. L. and Butani, D. K. (1986). Eco-toxicological studies on *Dacus* spp. *Pesticides*, **20** (10), 14–15

Drew, R. A. I., Hooper, G. H. S. and Bateman, M. A. (Eds.) (1982). *Economic Fruit Flies of the South Pacific Region*. Watson Ferguson, Brisbane

Dupont, L. (1979). On gene flow between *Tetranychus urticae* Koch, 1836 and *Teranychus cinnabarinus* (Boisduval) Boudreaux, 1956 (Acari: Tetranychidae): synonomy between the two species. *Entomol. Exp. Appl.*, **25**, 297–303

Dupree, M., Bissell, T. L. and Beckham, C. M. (1955). The pickleworm and its control. *Bull. Georgia Agric. Expl. Stn*, N.S. 5

Dutt, N. and Dalapati, A. (1977). Dormancy in red pumpkin beetle, *Raphidopalpa foveicollis* (Lucas). *Indian J. Entomol.*, **39**, 70–80

Edelson, J. V. (1986). Comparison of sampling methods for insect pests of cantaloupe. *J. Econ. Entomol.*, **79**, 266–270

Elsey, K. D. (1980a). Pickleworm: effect of temperature on development, fecundity and survival. *Environ. Entomol.*, **9**, 101–103

Elsey, K. D. (1980b). Pickleworm: mortality on cucumbers in the field. *Environ. Entomol.*, **9**, 806–809

Eta, C. R. (1985). Review: Eradication of the melonfly from Shortland Islands, Western Province, Solomon Island. In *Solomon Islands, Agriculture Quarantine Service, 1985 Annual Review*, pp. 14–23

FAO (1989). *Annual Production Yearbook for 1988*, **42**, 183, 185, 192, 201

Fargo, W. S., Renser, P. E., Bonjour, E. L. and Wagner, T. L. (1988). Population dynamics in the squash bug (Heteroptera: Coreidae)–squash plant (Cucurbitales: Cucurbitaceae) system in Oklahoma. *J. Econ. Entomol.*, **81**, 1077–1079

Fitt, G. P. (1981). Inter- and intraspecific responses to sex pheromones in laboratory bioassays by females of three species of Tephritid fruit flies from northern Australia. *Entomol. Exp. Appl.*, **30**, 40–44

Ghosh, C. C. (1940). *Insect Pests of Burma*. Government Printing and Stationery, Rangoon

Grayson, J. M. (1947). Spotted cucumber beetle as a pest of peanuts. *J. Econ. Entomol.*, **40**, 251–256

Gupta, J. N. and Verma, A. N. (1979). Relative efficacy of insecticides, as contact poisons, to the adults of melon fruitfly, *Dacus cucurbitae* Coquillet. *Indian J. Entomol.*, **41**, 117–120

Hall, F. R. and Ferree, C. D. (1975). Influence of twospotted spider mite populations on photosynthesis of apple trees. *J. Econ. Entomol.*, **68**, 517–520

Hill, D. S. (1983). *Agricultural Insect Pests of the Tropics and Their Control*. Cambridge University Press, Cambridge

Hill, D. S. (1987). *Agricultural Insect Pests of Temperate Regions and Their Control*. Cambridge University Press, Cambridge

Howe, W. L. and Rhodes, A. M. (1976). Phytophagous insect associations with Cucurbita in Illinois. *Environ. Entomol.*, **5**, 747–751

Hussey, N. W., Read, W. H. and Hesling, J. J. (1969). *The Pests of Protected Cultivation*. Edward Arnold, London

Iftner, D. C. and Hall, F. R. (1984). The effects of fenvalerate and permethrin residue on *Tetranychus urticae* Koch fecundity and rate of development. *J. Agric. Entomol.*, **1**, 191–200

IRAT/CIRAD (1986). *Annual Report* of the Institute de Recherche Agronomiques Tropicales et des Cultures Vivrières

Jadhav, L. D., Kadam, M. V., Ajri, D. S. and Pokharkar, R. N. (1979). Host preference of leaf-footed plant bug, *Leptoglossus membranaceus* Fabricius. *Indian J. Entomol.*, **41**, 299–300

Jadhav, L. D., Kadam, M. V., Ajri, D. S. and Pokharkar, R. N. (1980). Observations on the biology of leaf-footed plant bug, *Leptoglossus membranaceus* Fabricius, a pest of pomegranate. *Bull. Entomol.*, **21**, 79–82

Janick, J. (1986). *Horticultural Science*. Freeman, New York

Jeppson, L. R., Keifer, H. H. and Baker, E. W. (1975). *Mites Injurious to Economic Plants*. University of California Press, London

Kalshoven, L. G. E. (1981). *Pests of Crops in Indonesia* (revised and translated by P. A. van der Laan and G. H. L. Rothschild). P. T. Ichtiar Baru–Van Hoeve, Jakarta

Khan, M. Z. and Hajely, K. P. (1987). Studies on *Aulacophora africana* Lucas (Coleoptera: Chrysomelidae)—food preference and extent of damage. *Indian J. Entomol.*, **49**, 457–459

Klun, J. A., Leonardt, B. A., Schwarz, M., Day, A. and Raina, A. K. (1986). Female sex pheromone of the pickleworm, *Diaphania nitidalis* (Lepidoptera: Pyralidae). *J. Chem. Ecol.*, **12**, 239–249

Kuba, H. and Koyama, J. (1985). Mating behaviour of wild melon flies, *Dacus cucurbitae* Coq. (Diptera: Tephritidae) in field cage: courtship behaviour. *Appl. Entomol. Zool.*, **20**, 365–372

Lampman, R. L. and Metcalf, R. L. (1987). Multicomponent kairomonal lures for southern and western corn rootworms (Coleoptera: Chrysomelidae:*Diabrotica* spp.). *J. Econ. Entomol.*, **80**, 1137–1142

Latin, R. X. and Reed, G. L. (1984). Effect of root feeding by striped cucumber beetle on the incidence and severity of *Fusarium* wilt on muskmelon. *Phytopathology*, **75**, 209–212

McSorley, R. and Waddill, V. H. (1982). Partitioning yield loss on yellow squash into nematode and insect components. *J. Nematol.*, **14**, 110–118

Marrone, P. G., Ferri, F. D., Mosley, T. R. and Meinke, L. J. (1985). Improvements in laboratory rearing of the southern corn rootworm, *Diabrotica undecempunctata howardi* (Coleoptera: Chrysomelidae), on an artificial diet and corn. *J. Econ. Entomol.*, **78**, 290–293

Messenger, P. S. and Flitters, N. E. (1958). Effect of constant temperature environments on the egg stage of three species of Hawaiian fruit flies. *Ann. Entomol. Soc. Am.*, **51**, 109–119

Metcalf, R. L., Ferguson, J. E., Lampman, R. and Andersen, J. F. (1987). Dry cucurbitacin-containing baits for controlling diabroticite beetles (Coleoptera: Chrysomelidae). *J. Econ.*

Entomol., **80**, 870–875
Miyahara, Y. and Kawai, A. (1979). Movement of sterilized melon fly from Kumi Island to the Amani Islands. *Appl. Entomol. Zool.*, **14**, 496–497
Mollet, J. A. and Sevacherian, V. (1984). Effect of temperature and humidity on dorsal strial densities in *Tetranychus* (Acari: Tetranychidae). *Int. J. Acarol.*, **10**, 159–161
Murray, D. A. H. (1976). Insect pests on passion fruit. *Queensland Agric. J.*, **102**, 145–151
Nechols, J. R. (1987). Voltinism, seasonal reproduction, and diapause in the squash bug (Heteroptera: Coreidae) in Kansas. *Environ. Entomol.*, **16**, 269–273
Nechols, J. R. (1988). Photoperiodic responses of the squash bug (Heteroptera: Coreidae): diapause induction and maintenance. *Environ. Entomol.*, **17**, 427–431
Okumura, M., Ide, T. and Takagi, S. (1981). Studies on the effect of temperature on the development of the melon fly, *Dacus cucurbitae*. *Res. Bull. Plant Prot. Serv. Japan*, **17**, 51–56
Pena, J. E., Waddill, V. H. and Elsey, K. D. (1987a). Population dynamics of the pickleworm and the melonworm (Lepidoptera: Pyralidae) in Florida. *Environ. Entomol.*, **16**, 1057–1061
Pena, J. E., Waddill, V. H. and Elsey, K. D. (1987b). Survey of native parasites of the pickleworm, *Diaphania nitidalis* Stoll and melonworm, *Diaphania hyalinata* (L.) (Lepidoptera: Pyralidae) in southern and central Florida. *Environ. Entomol.*, **16**, 1062–1086
Prokopy, R. J. and Koyama, J. (1982). Oviposition site partitioning in *Dacus cucurbitae*. *Entomol. Exp. Appl.*, **31**, 428
Raina, A. K., Klun, J. A., Schwarz, M., Day, A., Leonardt, B. A. and Douglas, L. W. (1986). Female sex pheromone of the melonworm, *Diaphania hyalinata* (Lepidoptera: Pyralidae), and analysis of male responses to pheromone in a flight tunnel. *J. Chem. Ecol.*, **12**, 229–237
Raman, K. and Annadurai, R. S. (1985). Host selection and food utilization of the red pumpkin beetle, *Raphidopalpa foveicollis* (Lucas) (Chrysomelidae: Coleoptera). *Proc. Indian Acad. Sci. Anim. Sci.*, **94**, 547–556
Renjhen, P. I. (1948). On the morphology of the immature stages of *Dacus (Strumeta) cucurbitae* Coq (the melon fruitfly), with notes on its biology. *Indian J. Entomol.*, **11**, 83–100
Romanow, L. R., Moyer, J. W. and Kennedy, G. G. (1986). Alternation of efficiencies of acquisition and inoculation of watermelon mosaic virus 2 by plant resistance to the virus and to an aphid vector. *Phytopathology*, **76**, 1276–1281
Rose, R. I. and McCabe, J. M. (1973). Laboratory rearing techniques for the southern corn rootworm. *J. Econ. Entomol.*, **66**, 398–400
Roy, D. K. and Behura, B. K. (1983). Notes on host-plants, feeding behaviour, infestation and ant attendances of cotton aphids *Aphis gossypii* Glov. *J. Bombay Natl Hist. Soc.*, **80**, 654–656
Sances, F. V. I., Wyman, J. A. and Ting, I. P. (1979). Physiological responses to spider mite infestation on strawberries. *Environ. Entomol.*, **8**, 711–714
Shaheen, A. H. (1973a). Biological studies on *Aulacophora foveicollis* (Lucas) in Egypt. *Agric. Res. Rev.*, **51**, 91–95
Shaheen, A. H. (1973b). Morphology of adult and immature stages of *Aulacophora foveicollis* (Lucas) in Egypt. *Agric. Res. Rev.*, **51**, 79–90
Singh, D., Narang, D. D. and Chahal, B. S. (1984). Control of red pumpkin beetle, *Raphidopalpa foveicollis* (Lucas) by foliar and soil application of insecticides. *J. Res. Punjab Agric. Univ.*, **21**, 525–532
Ullah, M. (1987). Economic insect pests and phytophagous mites associated with melon crops in Afghanistan. *Trop. Pest Mgmt*, **33**, 29–31
Vargas, R. I., Miyashita, D. and Nishida, T. (1984). Life history and demographic parameters of three laboratory-reared Tephritids (Diptera: Tephritidae). *Ann. Entomol. Soc. Am.*, **77**, 651–656
Vaz Nunes, M. (1986). Some aspects of induction and termination of diapause in a Greek strain of the mite *Tetranychus cinnabarinus* (Boisduval), 1956 (Acari: Tetranychidae). *Expl Appl. Acarol.*, **2**, 315–321
Visalakshi, A., Beevi, S. N., Premkumar, T. and Nair, M. R. G. K. (1980). Biology of *Leptoglossus australis* (Fabr.) (Coreidae: Hemiptera) a pest of snake gourd. *Entomon*, **5**, 77–79
Warrell, E. (1988). *C.A.B. International News*, June, 6

Wehner, T. C., Elsey, K. D. and Kennedy, G. G. (1985). Screening for cucumber antibiosis to pickleworm. *HortScience*, **20**, 1117–1119

Wen, H. C. (1985). Field studies on melon fly (*Dacus cucurbitae*) and attractant experiment in southern Taiwan. *J. Agric. Res. China*, **34**, 228–235

Whitaker, T. W. and Davis, G. N. (1962). *Cucurbits: Botany, Cultivation, and Utilization*. Interscience, New York

CHAPTER 6

Pests of Leguminous Crops

A. J. Biddle, S. H. Hutchins and J. A. Wightman

Peas and Beans

6.1 INTRODUCTION

Pests are capable of causing damage to vegetable legumes in various ways. Direct injury may reduce crop yield and the transmission of virus diseases can generally debilitate crops. An indirect effect of pest attack or infestation may result in a reduction of produce quality. This effect is especially important where the edible parts of the crop are destined for the processing plant, where the physical removal of blemished produce is not possible. In such instances the product quality can be so affected that total crop rejection may be the outcome.

Peas (*Pisum sativum* L.) are grown in Europe, Scandinavia, North America, south Africa and Australasia, mainly as large-scale field crops, and are usually managed with the help of a small labour force and the use of harvesting machinery. The crop may be harvested green as vining peas which are taken directly from the field after harvest to the processing factory. The peas are then either canned or quick frozen. Alternatively, peas of different varietal characteristics may be left in the field until mature and the crop then harvested with a combine harvester, as a grain-legume. These peas (combining peas) are eventually transported to a dry-crops processing plant, where stones, dirt and other unwanted material are removed mechanically before marketing in the dry state either to the consumer for rehydration and cooking at home or to a canning plant where, again, the peas are rehydrated and cooked before resale. A smaller area of peas is grown where the pods are picked fresh for market sales (picking peas).

There are three species of bean grown as vegetables: broad bean; green bean; and runner bean. The broad bean (*Vicia faba* L.) is grown mainly in Europe, with a small area in north-western USA. Broad beans are harvested green to be subsequently frozen or canned or, alternatively, picked fresh. The green, bush, snap or French bean (*Phaseolus vulgaris* L.) is grown in the warmer temperate and tropical areas of the world. It has a wide variety of uses. It can be picked at the green pod stage and sold either fresh, canned or frozen. Some varieties may be left to mature in the field, with the seeds, which are harvested dry, being used in a variety of ways, including canning in tomato sauce (baked beans) or soaking and boiling in stews or in mixed bean salads. *P. vulgaris* is primarily tropical in origin, but varieties have been developed for their cold-tolerance and are grown in more temperate latitudes. The runner bean (*P. coccineus* L.) is almost entirely grown in Europe, where it is a garden or market garden crop for fresh picking.

A variety of pests is able to attack peas and beans at different times, causing problems at various stages of growth. Timing of pesticides is often critical: some pests attack flowering crops when beneficial insects such as bees are present; and other pests attack so close to harvest that unacceptably high residues would be left in the produce.

6.2 APHIDS

6.2.1 *Acyrthosiphon pisum* (Harris): Pea Aphid (Hemiptera: Aphididae)

The pea aphid is found in many temperate countries, including North and South America, Europe, South Australia, Tasmania and New Zealand. It lives on a wide range of leguminous plants, including pea, clover, vetch, sainfoin, bean and broom.

Most of the pea crops in the UK become infested at some time during the growing season. Winged migrants move to crops from mid-May onwards and colonies develop rapidly around growing points. The adult pea aphid is generally bright green in colour, with reddish eyes and long antennae. Body length varies between 2.5 mm and 3.0 mm. Colonies contain several viviparous females with young, feeding on leaf and stem tissues, and are often protected by the leaves surrounding the terminal shoot (Figure 6.1). Colonies may also arise on developing pods. Feeding continues until the pods begin to mature, when winged aphids are produced which migrate to overwintering hosts such as clover and lucerne on which eggs are laid. Adults are able to survive mild winters to infest peas early during the following summer.

Although other species of aphid are found on peas, *A. pisum* is by far the most common and damaging. Infestation may be signalled by flocks of small birds feeding on colonies. Infested plants are retarded, the tops become chlorotic and the production of honeydew by the aphids encourages colonisation by secondary moulds such as *Cladosporium* spp. (Ascomycetes: Dothideaceae) or *Botrytis* spp. (Ascomycetes: Helotiales). Leaves may be puckered and pods severely distorted, coupled with underdeveloped peas. In addition, aphids are efficient vectors of pea enation mosaic (PEMV) and pea seedborne mosaic (PSbMV) viruses, which are very damaging to yield and quality. Symptoms of PEMV include

Figure 6.1   Colony of pea aphid, *Acyrthosiphon pisum*. © PGRO

leaf mosaic and tissue enations on the undersides of leaves and stipules. Pods are often poorly developed, wrinkled and twisted. Symptoms of PSbMV include severe stunting of plants, profusion of axillary shoots and the production of small pods which contain peas with blistered seed coats.

Syrphid flies are important predators. On occasions in the UK, however, contamination of vined peas by syrphid puparia has led to the rejection of crops destined for freezing (Anon., 1970a). In combining peas loss of yield remains the main problem. Aphids are often the cause of the most important of the financial losses suffered by the grower from pest attacks.

Treatment of crops by foliar-applied insecticides is the most widely practised approach to the management of pea aphids. There have been no reports of physiological resistance by pea aphids to any of the commonly used aphicides such as dimethoate, demeton-*S*-methyl or pirimicarb (cf. aphids on potatoes, Chapter 8). There have been various estimates made of the economically important sizes of populations infesting peas, although in vining peas it is common practice to treat crops as soon as aphid colonies are noticed. Recent work by the Agricultural Development and Advisory Service with combining peas in England has shown that significant yield increases can be obtained from aphid control up to the time at which plants have produced four pod-bearing nodes, after which stage no economic benefits are obtained.

6.2.2 *Aphis fabae* Scopoli: Black Bean Aphid (Hemiptera: Aphididae)
Aphis craccivora Koch: Black Legume Aphid (Hemiptera: Aphididae)

A. fabae is mainly a pest of broad beans, but occasionally it can be found in high numbers on *Phaseolus* L. beans. On spinach and sugar beet populations can develop rapidly on the undersides of leaves, which, as a result, become chlorotic and crinkled. *A. fabae* is present in many countries, including Scandinavia, Europe, Asia, Middle East, parts of Africa, and North and South America, but it is not common in the tropics.

A. fabae overwinters as eggs on wild, woody hosts in the UK, especially the spindle tree (*Euonymus europaeus* L.). Wingless females hatch from the eggs in early spring and, after maturing, produce young which, in turn, produce a winged generation of migrating females. Black bean aphids first arrive on beans in Europe during May or June. The aphids are black to dark olive-green in colour, with white markings. The bodies measure 2–2.5 mm in length and are oval in shape. Colonies develop rapidly, particularly in warm, humid weather. The upper part of the stem of the broad bean plant is most commonly infested (Figure 6.2), and on *Phaseolus* beans the undersides of the leaves or the flower clusters may be colonised. Development of colonies reaches a peak in mid- to late July, and then the maturity of the crop and the shortening day length result in the production of winged migrants which return to the overwintering hosts to produce males and egg-laying females (Jones and Jones, 1984). In England the over-wintering egg populations on spindle trees are surveyed annually, so that, if necessary, high-risk regions can be alerted well in advance of the first spring migrants (Cammel *et al.*, 1978).

Severe attacks of black bean aphid cause heavy losses to broad beans, and bean crops growing in areas where regular attacks occur are best protected by routine treatment in advance. Attacks to broad beans which occur before flowering cause most damage.

Damage is typical of aphids: the upper part of broad bean plants becomes stunted and distorted; and honeydew which covers leaf surfaces beneath aphid colonies becomes colonised by black saprophytic moulds. Aphids are important virus vectors to beans. Broad bean leaf roll (BBLRV) and bean yellow mosaic viruses (BYMV) viruses are most damaging, but other viruses may be important in different production areas. Symptoms of BBLRV include tight, upward rolling of new leaves around growing points, leading to subsequent checking of growth and yield loss. Symptoms of BYMV include foliar mosaic, narrowing and twisting of newly developing leaves and stunting and distortion of pods. Although black bean aphid has not been directly associated with virus yellows of sugar beet, it is often found in the company of *Myzus persicae* Sulzer, the peach-potato aphid, which is an important beet virus vector (see Chapters 2 and 8; Chapter 8 discusses importance of *M. persicae* as a virus vector).

Figure 6.2   Colony of black bean aphid, *Aphis fabae*. © PGRO

A wide range of insecticides is effective at controlling *A. fabae*, but, to avoid harming pollinators such as bees, sprays should not be applied during flowering. Systemic granular insecticides such as disulfoton or phorate can be used in advance of flowering to provide persistent control. In *Phaseolus* aphids need to be controlled only if colonies develop on the flower clusters or when varieties are known to lack resistance to the major viruses.

A. craccivora is black in colour and is polyphagous on legumes in many southern temperate countries. The life cycles of *A. fabae* and *A. craccivora* are similar, but the woody, overwintering hosts of *A. craccivora* are different: broom (*Sarothamnus scoparia* Wimm.) and laburnum (*Laburnum anagyroides* Medic.). *A. craccivora* has been shown to transmit 14 viruses to legume crops, including BYMV and bean local chlorosis virus (BLCV) of *Phaseolus* spp. Symptoms of BLCV include severe curling, malformation and mottling of leaves and necrotic streaking of stems and pods.

6.3 BEETLES

6.3.1 *Sitona lineatus* (L.): Pea and Bean Weevil (Coleoptera: Curculionidae)

Probably the most common of pea pests, the pea and bean weevil is found in most of the temperate areas where peas or broad beans are grown, including Europe, Israel and Japan (Hamon and Bardner, 1983). It is also present in the western states of the USA, where it is believed to have been introduced from Canada. Pea and bean weevils have a wide host range, feeding on clovers (*Trifolium* L. spp.), peas, broad beans and *Phaseolus* beans.

The adult pea and bean weevil is about 4–5 mm in length and varies considerably in colour from a light, sandy shade to a much darker brown. There is a faint striping along the length of the back. The adult weevil has a short, blunt rostrum and conspicuous, angled antennae. The larvae are creamy white in colour, wrinkled and fleshy, and when fully grown, reach the same size as the adults (Figure 6.3).

Adult pea and bean weevils overwinter in cereal stubble litter, sides of streams, hedge bottoms and field verges. With the arrival of warm weather in spring, they migrate to pea and bean crops. At temperatures above 18 °C weevils migrate on the wing. Eggs are laid on leaves or soil. Eggs are washed down through the soil by rainfall and the larvae, on hatching 3 weeks after egg laying, make their way to the developing nitrogen-fixing root nodules, into which they tunnel (Figure 6.3) and feed for several weeks. Larval development is completed in 6–7 weeks and, after pupation, a new generation of adults emerges from the soil and feeds on remaining leaf material before migrating to overwintering sites (Figure 6.4). In the western USA there is some indication that the overwintering population may be reduced by natural infection of an entomophagous fungus (*Beauveria bassiana* (Alsamo) Vuill. (Deuteromycetes: Moniliales)) (O'Keeffe *et al.*, 1984).

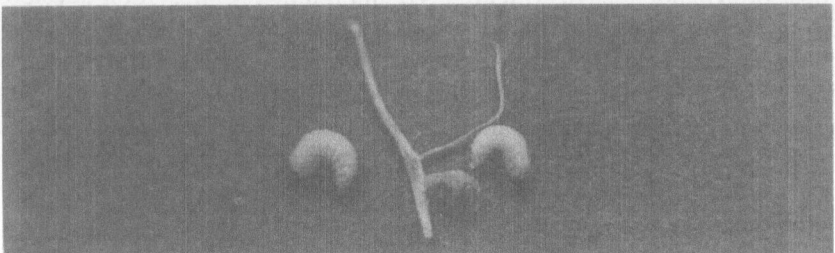

Figure 6.3 Larvae of *Sitona lineatus* from pea root, showing hollowed nodule. © PGRO

Although peas at any stage of their growth may be attacked by pea and bean weevils, damage is most obvious in the early stages of plant development. The adult weevils eat U-shaped notches around the leaf

Figure 6.4   Damage to a pea seedling by the pea and bean weevil, *Sitona lineatus*.
© PGRO

edges of expanded leaves (Figure 6.4), and if a heavy infestation develops
during, for instance, a period of slow growth in cold weather, a considerable proportion of the leaf tissue may be destroyed. Severely affected crops
may exhibit symptoms of nitrogen deficiency. Leaf injury alone is not
thought to be responsible for loss of yield. Experiments using chemicals to
prevent larval attack have provided the greatest yield responses, indicating
that the destruction of nodules has more influence on seed weight than has
defoliation.

The effect of weevil injury on the yield of peas and broad beans can be
similar, but the growth of later-sown bean crops may be so vigorous and
rapid that damage is not noticed and plants may escape injury altogether.

Although foliar sprays of pyrethroid insecticides such as cypermethrin or
deltamethrin or of the organophosphorus insecticide triazophos, applied as
soon as serious levels of leaf damage are seen, will check further damage,

reductions of larvae are seldom achieved, as early treatment is necessary to prevent egg laying. Granular soil-applied insecticides such as phorate or aldicarb have been shown to be effective in reducing adult and larval feeding. Combining pea yield increases have been obtained in the UK when insecticidal granules have been applied as predrilling soil-incorporated broadcast treatments or as 'in-furrow' treatments with the seed (Biddle, 1985). Recently several experimental evaluations have been made of seed treatments which include insecticides (Baughan *et al.*, 1985; Salter and Smith, 1986). While they do significantly reduce leaf attack and larval infestation of roots, seed treatments are still too toxic to birds to be used commercially.

6.3.2 *Apion vorax* Herbst: Bean Flower Weevil (Coleoptera: Curculionidae)

A. vorax and other related species of weevils are present on all continents of the world. Their status as pests of vegetable crops is not very important. However, crop damage can occasionally be severe. The *Vicia* bean is the most commonly affected crop.

The adult weevil is a small (2 mm long), black-coloured beetle, pear-shaped in outline, with a long and downward curving rostrum. The larva is greyish white in colour, with a reddish-brown head, and is similar in length to the adult.

The presence of *Apion* weevils first becomes apparent with the approach of flowering. The adult weevils are frequently found on or near the florets. Eggs are laid singly in the florets over a period of about a month. They hatch in a few days and the larvae burrow through the bases of the florets into the ovaries. The larval and pupal stages last 24 days, when adult weevils emerge to produce a further generation. Adults overwinter in hedges, ditches and field verges.

The principal damage is done by the weevil larvae destroying the developing ovules. The adult weevils, however, are one of the main and most efficient vectors of broad bean stain virus which reduces yield and spoils produce by producing necrotic 'edges' to shelled beans (Cockbain, 1971).

Apion weevil attacks are more likely to occur where beans are growing near to clover crops. Crop siting, therefore, may be a useful means of reducing the risk of attack. Phorate granules, applied just before flowering, prevent weevil infestation and will also control aphids if present.

6.3.3 *Bruchus pisorum* (L.): Pea Seed Beetle (Coleoptera: Bruchidae)

Often referred to as a weevil, the pea seed beetle belongs to the family Bruchidae and is therefore not a true weevil (Curculionidae). Pea seed beetles can be found in all but the coldest countries, including southern Europe, Asia, Australia, USA and Canada, but are not able to sustain populations in more northerly areas. In the USA the beetles are prevalent in the seed production areas of Idaho. Pea seed beetles are pests of the growing crop, and although they are often seen in stores of dried produce as they emerge from the seeds, they are not stored-product pests.

Adult pea seed beetles are small (4–5 mm long), black in colour and oblong in shape, with characteristic white speckling along the elytra. As with other Bruchidae, the elytra are shortened, leaving the tip of the abdomen exposed. The larva is about 5 mm long, creamy coloured, with a brown head.

Adult pea seed beetles overwinter either in stores or in dry sheltered areas around fields or buildings. In the spring they fly to pea fields which have commenced flowering. Adults feed on pollen for 4–10 days. Single eggs are then laid on the surfaces of the young developing pods. After 6–9 days the larvae hatch and burrow through the pod walls into the developing seeds. Only one larva develops in a seed. Each larva develops in a cell close to the testa, which becomes transparent. Infested seeds appear, therefore, to contain circular 'windows'. After 5 weeks the larvae prepare exit holes and pupate. Pupation lasts for 1–3 weeks. After pupating, the adult beetle uses its mouthparts to cut its way out of the seed, leaving a round exit hole. Adult emergence often occurs well after harvest and the beetles can then be found crawling over the seeds in stores.

Apart from some degree of weight loss caused by pea seed beetle injury, damaged peas have to be removed mechanically if the produce is intended for human consumption or seed. Seed peas containing live beetles may also be downgraded by the appropriate seed production control authority.

Control of pea seed beetles in the field is aimed at preventing egg laying and therefore insecticidal sprays must be applied during the flowering and pod setting period. Malathion, parathion and methoxychlor are used in the USA. Because peas are not pollinated by insects, there is little risk of direct injury to bees when insecticides are applied during the flowering period. In stores fumigation of peas with aluminium phosphide or methyl bromide is permitted by some authorities (O'Keeffe and Homan, 1984). However, care is needed in treating seed, as some chemicals under some circumstances (e.g. high seed moisture content, high temperature) may be phytotoxic, impairing germination.

6.3.4 *Bruchus rufimanus* Boheman: Bean Beetle (Coleoptera: Bruchidae)
Acanthoscelides obtectus Say: Dried Bean Beetle (Coleoptera: Bruchidae)

B. rufimanus can build up locally to very high population numbers in areas in all but the coldest countries where dry-harvest *V. faba* beans are grown, including Europe, Asia, Australia, the USA and Canada. In broad beans the main problem occurs when larvae are found in green-harvested seed during canning or freezing. Like the pea seed beetle, the bean beetle adult is oval in shape, about 5 mm long and dark grey-brown in colour, with a slight white flecking on the elytra, which are foreshortened, exposing the tip of the abdomen. Adults fly from overwintering sites in late May and are attracted to bean flowers. Eggs are laid on developing pods and larvae burrow through the pod walls into the developing seeds. Control is carried out as a routine in those areas where processing crops are grown in proximity to dry-harvest beans and where there is a known population of Bruchid beetles. Insecticides (e.g. triazophos) are applied immediately after flowering. Adults overwinter in field edges. Overwinter survival rate is adversely affected by wet weather.

Phaseolus beans are attacked by *A. obtectus*. This insect is widely dispersed throughout the warmer, growing regions of the Mediterranean area, Africa, Japan, Australia and North, Central and South America. It is principally a problem on beans grown for dry harvest. Unlike previously described Bruchid beetles, *A. obtectus* is a pest not only in the field, but also in the store. Damaged seeds are characterised by the circular exit holes left by the emerging adults, or by the testa 'windows' under which larvae are feeding and pupating. The adult dried bean beetle is slightly smaller than the bean beetle. Females lay their eggs in cracks or wounds on pods in the field. Young larvae then penetrate the pods and enter the seeds. After pupating, adult beetles exit from the seeds and immediately begin egg laying on other seeds which may by then be in store. There are several generations a year. Control measures in stores include aluminium phosphide or methyl bromide fumigation or dusting with insecticides. Seed beans may be treated with menazon or cypermethrin sprays (O'Keeffe and Homan, 1984).

6.4 MOTHS

6.4.1 *Cydia nigricana* (Fabricius): Pea Moth (Lepidoptera: Tortricidae)

Pea moths are particularly troublesome in areas of intensive combining pea production, where localised large populations can develop. It is found throughout the temperate areas of Europe including former Eastern Bloc countries. Localised pea moth populations occur in some western states of the USA, southern Canada and Japan.

The pea moth is 4–5 mm in length and silvery brown in colour, with indistinct black and white markings on the leading edge of each forewing. Moths emerge from previous season's pea fields on sunny afternoons from late May to late July and fly to flowering pea crops. Small, flattened eggs are laid singly or in pairs on the leaves, stems and stipules of plants and, after incubation, which is temperature-dependent, larvae hatch and move to young pods, into which they burrow (Lewis and Sturgeon, 1978). Feeding continues for 2–3 weeks, after which time the fully developed larvae cut their way out of the pods, bury themselves in the soil and move to a depth of 20–30 cm, where they spin cocoons for the winter. Pupation commences in early May.

Damage by larvae of the pea moth occurs within pea pods, where feeding takes place on developing seeds (Figure 6.5). Damaged peas exhibit circular or irregular-shaped holes of varying degrees of size, and the pods contain frass which tends to be webbed together in stringy masses. Damage causes most economic losses to crops grown for human consumption or seed. As well as being unsightly, damage may impair seed germination. Damaged peas have to be removed mechanically by merchants before sale, which is both slow and costly, with growers' payments often being penalised. In vining peas, damaged seeds cannot be removed and, therefore, produce showing pea moth damage is likely to be rejected by processing companies. Yield losses in combining peas are rarely significant. Up to 14% loss of weight in a seed is incurred by larval feeding (Bardner, 1978). In the UK it is rare for the pea moth to affect more than 30% of the total production of peas in a crop. For this reason control of pea moths in peas grown for stockfeed in the UK is seldom carried out, as little economic benefit can be obtained. However, unprotected crops do provide a reservoir of moths which can migrate to peas being grown for quality markets.

Figure 6.5   Larvae of the pea moth, *Cydia nigricana*, in a pod. © PGRO

Chemical control is aimed at protecting the young pods by applying an insecticide immediately prior to egg hatch. Triazophos and the pyrethroid insecticides are very effective for this purpose. Timing of sprays is critical. Following identification of a pheromone and subsequent synthesis of an artificial sex-attractant (Wall and Greenway, 1981) which was incorporated into sticky traps, control of pea moth in the UK has been much improved. Two traps (Figure 6.6), each containing lures of the synthesised analogue of the moth's natural sex-attractant pheromone, are placed on adjacent headlands of the pea field as the crop reaches the developing flower bud stage. Traps are examined by growers every two days. When ten or more moths are caught in either of the two traps on two consecutive occasions, calculations are commenced to predict the date of egg hatch, based on accumulated day-degrees. Growers can either record temperatures for themselves or utilise a computed, predicted date supplied by the government advisory services, using data from agrometeorological stations (Biddle *et al.*, 1983). Because egg laying continues for several weeks, a second spray is usually applied 10–14 days after the first spray. The pheromone-based monitoring system serves two purposes: timing of sprays is accurately predicted and, if the necessary number of moths is not trapped, no spraying is required. As any damage is unacceptable in vining peas, a much lower threshold of moths can be tolerated than in combining peas. For this situation, a monitoring system with a more attractive pheromone has been developed (Wall *et al.*, 1986). Growers are able to correlate trap captures with estimates of potential damage, thereby helping them to reach a decision on whether or not to use insecticides. One spray is applied to the vining crop as it reaches the pod set growth stage.

Figure 6.6 A pheromone trap for monitoring the pea moth, *Cydia nigricana*. © PGRO

6.5 FLIES

6.5.1 *Contarinia pisi* (Winnertz): Pea Midge (Diptera: Cecidomyiidae)

Also known as the pea gall gnat, pea midge has been reported as a pest in many northern European countries, including Sweden, Denmark, the UK, Switzerland and the USSR (Geissler, 1966). In England the pest was first reported in 1946 and, with an increasing area of peas grown for canning and freezing during the 1960s, damage became more important. A similar experience has occurred in Sweden. Although pea midge has been recorded in many areas within different countries, populations have reached pest proportions locally only where peas have been grown intensively. Pea midge appears to be specific to peas, although very little work has been carried out to study the host range (Franssen, 1954).

Adult pea midges have a body length of 2 mm and are grey-brown in colour, with a dark head. Wings have simple venation. The antennae of females are shorter than the antennae of males. Adult midges emerge from previous season's pea fields and mate soon after emergence close to the soil surface under this season's cover crop. Shortly afterwards, gravid females fly to this season's pea crops, the peak flight time occurring at the end of the afternoon and in the early evening. Crops are invaded from field edges, where female populations are often high at first. Eggs are laid within the leaves surrounding terminal buds and larvae hatch after about 4 days. Larvae move to the bases of developing flower buds, where they feed, causing deformed buds and sterile flowers to be produced. Galls form and often plant internodes are shortened and 'nettleheads' develop (Figure 6.7). Pods are not produced and during periods of wet weather, damaged tissues become colonised by saprophytic fungi, notably *Botrytis cinerea* Pers. ex. Fr. (Ascomycetes: Helotiales). The principal result of damage is yield loss. After feeding and growing for about 11 days, the fully developed pea midge larvae 'jump' out of the plant shoots by alternately flexing and reflexing their bodies. Upon reaching the soil surface they work their way gradually down the soil profile to spin cocoons within which they remain for the winter before pupating in the spring. Conditions for emergence the following season appear to b.   'tical. Both soil temperature and moisture are important: rising temperature coupled with light rainfall appear to trigger emergence. In some seasons, if these conditions do not occur, emergence of adults is light and prolonged, but, when emergence is at a peak, maximum damage is likely to occur to crops. Although many varieties of combining peas flower over a long period and are able, therefore, to compensate for some loss of flowers, the development of more determinate varieties, especially of vining peas with a shorter flowering period, has resulted in larger numbers of buds being susceptible to damage, with subsequently higher yield losses.

Populations of pea midges build up rapidly in numbers within localised

Figure 6.7 Damage caused by larvae of the pea midge, *Contarinia pisi*. © PGRO

areas, then decline after about 8–10 years. There are at least two species of parasitic wasps reported by Franssen (1954) to be associated with midge: *Pirene graminea* Haliday (Hymenoptera: Lamprotatidae) and *Sactogaster pisi* Foerster (Hymenoptera: Platygastridae). How significant their effect is on pea midge population decline is unknown.

Control by both chemical and cultural means has been practised in several countries. It has been suggested that as pea midges are weak fliers, positioning of pea fields in excess of 100 m from overwintering sites will reduce the numbers of invading female migrants (Meier, 1965; Thygeson, 1971). In England this approach to control is often not possible and the increasing area of peas in the country may result in the pea midge becoming more widespread. Midge damage in Sweden has been reduced as a result of the processing companies' policy of moving pea production away from infested areas for a period of several years. A similar approach has been suggested in Switzerland (Delucchi *et al.*, 1983). Chemicals are generally used to control pea midge in England with varying degrees of success. It is essential to control adults before egg laying commences and, therefore, both the timing and the choice of insecticide are critical. In the UK sprays of either fenitrothion or triazophos + dimethoate mixture are applied to peas at the early green bud growth stage after finding adult midges in terminal shoots. A second spray is applied 5–7 days later. Forecasting midge emergence is difficult and, currently, a pheromone-based monitoring system is being developed in the UK (Anon., 1986).

6.5.2 *Delia platura* (Meigen): Bean Seed Fly (Diptera: Anthomyiidae)

Also known as the seed corn maggot in the USA (Vea *et al.*, 1975), the bean seed fly is almost universally found wherever *Phaseolus* beans are grown. It has virtually a world-wide distribution, extending to the Arctic Circle in Asia, and has been recorded in Greenland and Iceland. It is closely related to the onion fly (*Delia antiqua* L.; see page 243) and will attack onions. A wide range of other crops may be attacked, particularly spring-sown maize and sweet corn, marrows, cucumbers, lettuces, peas and some Cruciferae. Usually, the first sign of attack is a patchy emergence of seedlings. In the case of green beans, attacked plants emerge with damaged plumules and missing cotyledon leaves, damage described as 'baldhead' or 'snakehead'. Infested seeds are eaten through by larvae and often the parts of stems below ground are hollowed out. Emerged, damaged crop plants are often weak and fail to develop into productive plants. Runner bean seeds often fail to emerge. Seeds of other crops are also vulnerable if sown in the late spring and early summer—e.g. sweet corn is often attacked and, in England in recent years, courgettes (which are becoming more widely grown) have been severely damaged. Late-sown peas are also vulnerable.

Adult bean seed flies are inconspicuous, slender flies, about 4 mm long, and grey-black in colour. The females lay eggs on recently disturbed, open soil in late spring and early summer, especially where there are residues of vegetable matter or where large amounts of farmyard manure have been applied. Larvae hatch from the eggs after 2–4 days and burrow into germinating bean seeds, damaging developing plumules, radicles and epicotyls. The small, yellowish-white-coloured larvae feed for 12–16 days before pupating in the soil. Pupation lasts 2–3 weeks. The polyphagous nature of the bean seed fly allows it to survive on any crop debris, particularly root crops or *Brassica* L. vegetable stems. In the UK there are 3–4 overlapping generations each year and elsewhere between 2 and 5. The complete life cycle takes 4–5 weeks in warmer countries. The final generation overwinters as pupae in the soil or may continue feeding on unrotted vegetable residues.

Prevention of damage may be achieved by planting later in the season to avoid first-generation attack. However, seed treatment with insecticides is commonly practised commercially. Green bean seed may be treated with chlorpyrifos (USA), bromophos (UK) or dichlofenthion (Europe not including UK) or other persistent insecticides (King and Biddle, 1973). Runner beans may also be protected in this way, although raising seedlings under glass and then transplanting them are practised in small-scale production.

6.6 THRIPS

6.6.1 *Thrips angusticeps* (Uzel): Field Thrips (Thysanoptera: Thripidae)

Field thrips attack newly emerged pea plants in many areas of Europe, including the UK, France, Holland and Germany, although attacks vary in severity from year to year. On the continent of Europe, where sowings are made earlier than in the UK, *Brassica* L. vegetable seedlings are occasionally damaged by thrips which feed on both growing points and older leaves. This damage appears as surface speckling. Field thrips have been observed regularly attacking peas which follow *Brassica* crops (Gough, 1955), but attacks can also follow a wide range of non-*Brassica* crops, including barley, wheat, rye, beet and onions. Weeds, too, might act as hosts. Damage is especially noticeable on seedlings of early sowings, some seedlings showing symptoms even before they have completely emerged from the soil. Attacks are concentrated primarily on the youngest, enfolded leaflets and may not become fully apparent until the leaflets open. Affected leaflets become characteristically tough and leathery, sometimes deformed, and take on a yellowish, mottled colour which is often the first sign of thrips damage noted by growers. If damaged leaflets are held towards the light, the sites where thrips have been feeding show as translucent spots. In a small number of pea crops patches of stunted plants can appear following thrips attack. Individual plants within the patches exhibit a characteristic proliferation of basal shoots and necroses of growing points and leaf margins. This 'pea dwarfing syndrome', as it is called, can persist throughout the life of the affected crop. The disorder has been noted in England and France (Anon., 1985) and no pathogen has been found to be associated with it. Although pea crops may appear to be retarded by field thrips, most crops will outgrow thrips attack, but in very cold weather when growth is slow, damage may appear to be quite severe, especially where the plants have developed 'dwarfing syndrome'.

Field thrips are dark-brown- or black-coloured insects, about 1–1.3 mm in length, with short legs. As wingless individuals, they live in the soil over winter before laying eggs on host plants in the spring. The almost colourless nymphs which emerge from the eggs feed on seedling leaves for up to 4 weeks before developing into the dark-coloured adults. They then produce the generation that overwinters in the soil (Franssen and Huisman, 1958).

It is often difficult to assess whether or not control measures are warranted, as experience often shows that the crop will recover unless severely attacked (Biddle, 1985). However, thrips tend to be more abundant on chalky or limestone soil types, and growers in these areas often treat crops as a routine before damage becomes severe. If spraying is to be carried out, dimethoate, fenitrothion or triazophos should be applied once as soon as damage is observed on a newly emerging crop.

6.6.2 *Kakothrips pisivorus* **(Westwood): Pea Thrips (Thysanoptera: Thripidae)**

Peas are only seriously attacked by pea thrips where the crops are growing in intensively cultivated ground in close proximity to the previous season's pea crops. The pest is often associated with market garden crops or in home vegetable gardens. During and after flowering, heavy infestations are most likely to occur in humid conditions, especially following a dry spring. The pea thrips has been reported as a pest in the British Isles, much of Europe and Scandinavia. The damage is limited to the external surface of the pod wall, which becomes silver-coloured and distorted. Often the tiny, brownish-yellow-coloured thrips can be found on flowers, pods or foliage.

The adults have elongated bodies, 1.5–1.8 mm long, and have very small wings. The nymphs are bright orange in colour and are found with the adults. Eggs are laid on the stamen sheath and other flower parts, hatching in 7–10 days. The nymphs feed on the flowers and pods and mature in about 17 days. They drop to the soil to complete their life cycle, emerging as adults in the early summer.

As damage is confined to the outer surfaces of pods, crops where appearance is important are at risk. Pea thrips are of no importance to vining or combining peas, but fresh market picking peas or mange-tout or sugar-snap peas may be blemished to an extent where they may be rejected by retailers or processors. Where insecticides are to be used, they should be applied as soon as the plants have produced their first pod.

6.7 NEMATODES

6.7.1 *Heterodera goettingiana* **(Liebscher): Pea Cyst Nematode (Tylenchida: Heteroderidae)**

Pea cyst nematode has been reported in several countries, including the USA, but it is most common in Britain, France and Germany, where it is considered to be a native species. Host-plant range includes broad beans, field beans, vetches, lupins and chickpeas, but the pea crop is the only one to suffer economic losses. It has often been observed that cyst nematode populations are maintained on field beans for many years without perceivable injury, but then when this infested land is cropped with peas, severe damage occurs.

Embedded within the roots of infested plants are tiny (<0.5 mm), lemon-shaped cysts, which begin creamy in colour, but darken to a deep brown with advancing maturity. Nitrogen-fixing nodule production is conspicuously reduced or absent. Initial damage in fields is restricted to small, distinct areas in which plants are stunted, are pale in colour and have leaves tending to point upwards. Flowering commences prematurely and infested plants may die before pods develop. As nematode populations increase with further pea or broad bean cropping, losses can become increasingly more extensive in infested crops.

Infestation of roots begins when larvae hatch from cysts in the soil as a result of chemical stimuli in host-plant exudates. The larvae invade the roots just behind the tips and then move through the tissues until they reach the phloem, where they begin to feed. Larvae thicken and differentiate into males and females. Females remain embedded in the roots (Figure 6.8), while males leave to swim free in the soil pores. The males fertilise the females, which have by now increased in size so much that they have ruptured the root cortex (Franklin, 1951). After fertilisation, the males die and the females continue to swell with eggs. When females die, their skin hardens and darkens to form cysts. There is one generation each year. Cysts may become detached from dead roots and are then dispersed through the soil during cultivations for subsequent crops. Infested soil may be transported inadvertently on implements. The cysts are very persistent in the soil and can survive for long periods in the absence of host crops. Survival of viable populations in the soil has been known to occur for over 15 years.

There is no known pea varietal resistance to pea cyst nematode. In the UK nematicides such as oxamyl or aldicarb are very effective when broadcast and then incorporated into the soil by means of harrows or rotary cultivators immediately prior to sowing (Whitehead *et al.*, 1974).

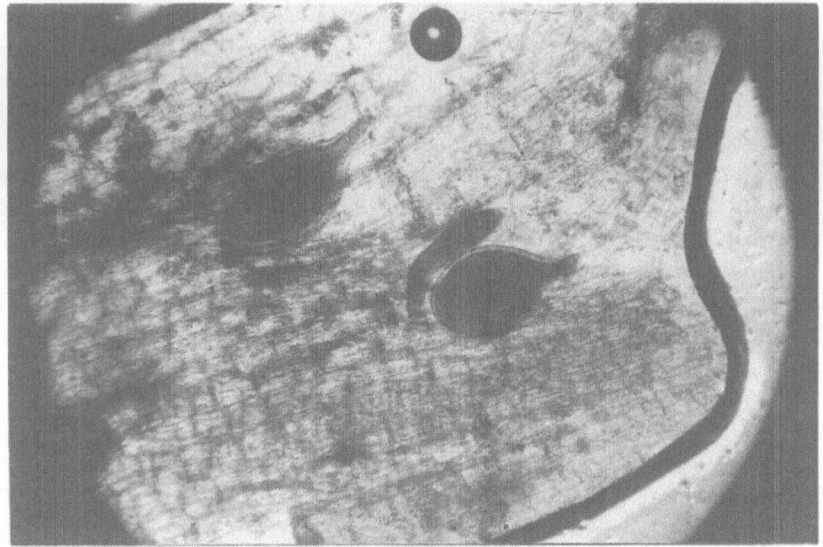

Figure 6.8 A female pea cyst nematode, *Heterodera goettingiana*, within root tissue. © PGRO

6.7.2 *Pratylenchus* Filipjev spp.: Root-lesion Nematodes (Tylenchida: Pratylenchidae)

Root-lesion nematodes are commonly found in all temperate countries, including Europe, the USA and Canada. A wide range of host plants may be attacked and several species of root-lesion nematode may be present in soils. In general, lighter, well-drained soils are more heavily infested than heavier less-well-drained soils. *Phaseolus* beans are attacked by the nematodes, with incidences being more common in warmer production areas of, for instance, the USA. Other important vegetable crops are hosts of root-lesion nematodes, notably tomatoes and *Brassica* crops where they have been growing on lighter, sandier soils (Anon., 1970b).

Both young and adult root-lesion nematodes enter host-plant roots by forcing their way between the cortical cells. They feed on cell contents while migrating through the root tissues. The feeding of the nematodes causes reddish-brown-coloured lesions to appear on roots and partial collapse to occur of sections of rootlets. After fertilisation, adult females each tend to produce single eggs every 24 h and the life cycle is completed in about 50 days. Nematodes overwinter in roots or soil and can survive both drought and severe winters, retaining their normal size.

Several studies examining the role of root-lesion nematodes as a pest of peas suggest that they are not a major pest in their own right. There is, however, a strong possibility that primary nematode damage allows the secondary entry to the root of soil-borne fungal pathogens, particularly *Fusarium oxysporum* Link var. *pisi* (Linford) Snyder and Hansen (Ascomycetes: Hypocreales), which causes pea wilt disease, *F. solani* (Mart.) var. *pisi* (Jones) Snyder and Hansen (Ascomycetes: Hypocreales), which causes a footrot disease, and *Aphanomyces euteiches* Drechsl. (Oomycetes: Saprolegniaceae), which is associated with a root rot disease in the USA, the UK and Scandinavia (Hagedorn, 1984). Because of this complex relationship, direct damage by the nematode only, often cannot be observed. Soil analyses to ascertain populations of root-lesion nematodes are, therefore, of limited value. Because of their wide host range, crop rotation is of little value in reducing root-lesion nematode populations, although marigolds (*Tagetes* L. sp.) will reduce populations in infested soil (Oostenbrink *et al.*, 1957). However, the maintenance of a rotation of not more than one pea crop in a 5 year period is essential to prevent build-up of the secondary fungal pathogens.

6.7.3 *Trichodorus* Cobb spp.: Stubby-root Nematodes (Enoplida: Trichodoridae)

In the UK and Holland light, sandy soils may contain high populations of stubby-root nematodes. These nematodes are vectors of pea early browning virus (PEBV), which can cause severe losses in peas (Bos and Van Der Want, 1962). Symptoms of this virus disease include leaf mosaic, purple discoloration of stems or leaf axils and stunting of plants in distinct patches or generally distributed areas. The growing points may also die: hence the descriptive name of the virus. Many plant hosts, including legumes other than peas, sugar beet and a wide range of weeds, are grazed by stubby-root nematodes: these hosts can be symptomless reservoirs of multiplication of PEBV. Visible injury to root systems is not observed in the field. As the nematodes feed externally, they are not discernible as endoparasites within root tissue. Because of the migratory characteristics of the nematodes up and down the soil profile, the nematodes may escape detection by standard soil sampling and analysis. However, geographical areas and the distinctive soil types favoured by stubby-root nematodes are the key factors which enable potential problems to be identified. Many UK pea growers with a history of nematode-borne virus damage in their fields incorporate aldicarb granules into the soil immediately prior to sowing.

Soybeans

6.8 INTRODUCTION

Soybeans, *Glycine max* (L.) Merr., and soybean products are important sources for edible oil and feed supplements. Indeed, soybean meal and oil represent the most produced, exchanged and utilised meal and oil commodities in the world (Smith and Huyser, 1987). Although native to north-eastern China (presumably between parallels 40° and 50°N), soybeans are cultivated to some degree on all continents. However, current production is concentrated in the USA, Brazil, China and Argentina. These four countries account for 90–95% of world soybean production.

The geographical expansion of soybean production has included subtropical and tropical countries. The arthropod communities associated with particular production centres vary according to geographical location. Because the soybean crop is a relatively new habitat for arthropods outside China (80% of current soybean hectarage has been intensively cultivated for less than 50 years), there are few pests obligatory to this crop. Rather, pests are either facultative colonisers adapting to this crop or generalist feeders (Kogan and Turnipseed, 1987). Moreover, pests of soybeans cultured in temperate regions tend to be sporadic in nature. They are often maintained below economic levels by natural enemies and only infrequently reach outbreak proportions. This natural balance is, however, fragile. The biotic potential of many pest species can be realised when environmental conditions are favourable for pest survival.

Excellent opportunities exist within temperate soybean ecosystems to develop pest management strategies consisting of both preventative and curative tactics. Each tactic within the overall strategy is either a preventative ⌐   ⌐urative control. A preventative control, such as altered planting da⌐   ⌐utumn cultivation or host-plant resistance, is employed because of some advance knowledge about the probability of pest attack. For curative control, however, frequent sampling of soybean pests is necessary to estimate the amount of pest-induced injury. These estimations are evaluated against previously computed economic injury levels to determine whether a given control tactic can be economically justified (Pedigo *et al.*, 1986). Pests which affect soybean physiology in a similar manner can be managed collectively if a broad-spectrum control tactic is used. Indeed, the management of injury guilds allows for the incorporation of different consumption potentials by several insect species, which facilitates the development of multiple-species economic injury levels (Hutchins *et al.*, 1988).

6.9 BEETLES

6.9.1 *Epilachna varivestis* Mulsant: Mexican Bean Beetle (Coleoptera: Coccinellidae)

The Mexican bean beetle is believed to be native to the plateau region of southern Mexico and Guatemala. At present, its range extends from Panama to southern Ontario, Canada (Turnipseed and Kogan, 1976). Within the USA, it was first positively identified in Colorado during 1883, and soon thereafter gained pest status within the semi-arid regions south of Colorado. This distribution of Mexican bean beetles in the USA was extended in 1920, when the species was identified in northern Alabama. The eastward expansion in the USA is believed to have been the result of an accidental introduction of individual beetles in alfalfa around 1918 (Fronk, 1978). From then to the early 1970s, Mexican bean beetles successfully colonised soybeans in the Coastal Plains of the USA and extended into much of the large, midwest, production centre. However, populations high enough to cause economic damage were limited to the eastern seaboard regions. Over the past 10–15 years, population concentrations have shifted west. Currently, damage to soybeans from Mexican bean beetles in the USA is most extensive in areas around southern Indiana (Turnipseed and Kogan, 1976). In addition to soybeans, host crops of the beetles include *Phaseolus* L. spp. beans, cowpeas, peanuts and other vegetable species (Hill, 1983).

Adult Mexican bean beetles are distinctly rounded with eight black-coloured spots arranged in three longitudinal rows on each elytron (Figure 6.9). They are 6–8 mm long, 4–6 mm wide and light yellow to bronze in colour. Males can be distinguished from females by having a

Figure 6.9 Adult Mexican bean beetle, *Epilachna varivestis*, on soybean. © M. Jeffords

small notch on the ventral side of the last abdominal segment. Eggs are 1.2 mm high, with a diameter of about 0.6 mm, and are pale yellow to orangeish-yellow in colour. They are typically found in clusters of 40–60 on the undersides of soybean leaves. Larvae develop through four instars. The size range from first instar to fourth instar larvae is 1.5–10 mm. All larval instars are pale yellow or orange in colour (Figure 6.10) and have branched

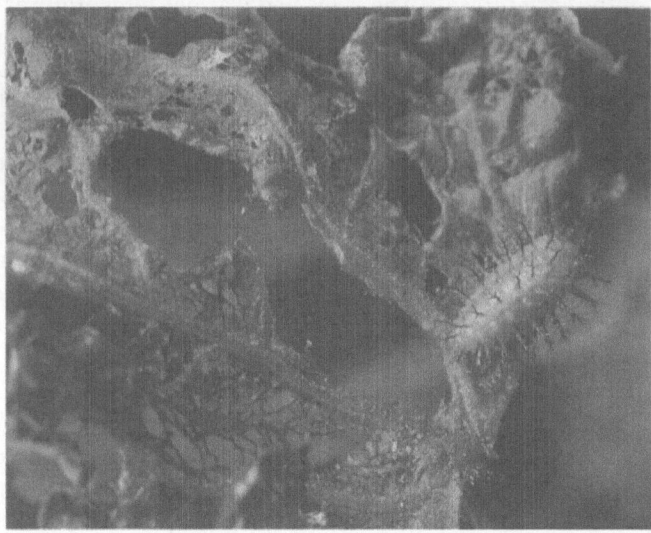

Figure 6.10 Larva of the Mexican bean beetle, *Epilachna varivestis*, on soybean. © M. Jeffords

spines arranged in six longitudinal rows on their backs. The larva has a soft body that tapers posteriorly and has an anal segment having a sucker-like apparatus for attachment to feeding surfaces. Pupae are light yellow in colour, ovoid and partially covered posteriorly with the last larval skin (Figure 6.11). Unlike the larvae, the pupae have a smooth texture.

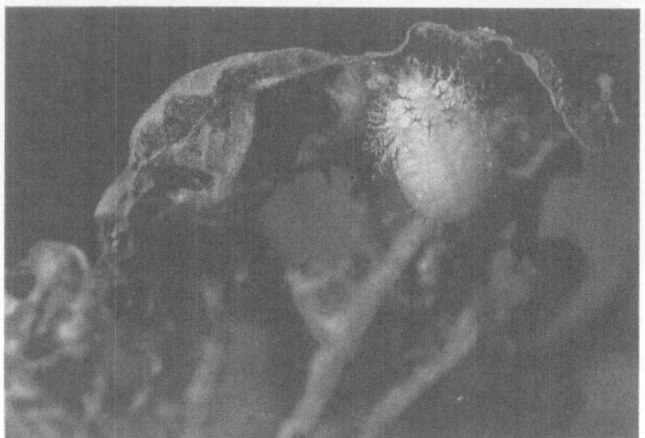

Figure 6.11 Pupa of the Mexican bean beetle, *Epilachna varivestis*, on soybean. © M. Jeffords

Adult Mexican bean beetles (both mated and unmated individuals) overwinter in aggregates in woodlands under pine needles and other debris or within well-drained areas of field margins (Douglass, 1933). Depending on geographical latitude, emergence may begin as early as March and continue into April or May. The rains and warmer temperatures of spring appear to stimulate adult emergence from overwintering. Following emergence, the beetles disperse in search of food sources. Records gathered during the northward expansion in the distribution of the beetles in the USA suggest that they are capable of moving up to 200 km per year (Howard, 1922). Following an initial feeding period of 8–15 days, the adults mate and the females begin to oviposit on the undersides of host-plant leaves. Overwintering females survive and feed for about 48 days and are capable of producing nearly 750 eggs per individual. Eggs hatch after a 5–14 day incubation period. Larvae feed on soybean foliage and pass through four instars. The duration of each instar varies, but ranges from about 4 days to 7 days within a total larval period of 16–27 days. Pupation occurs on the undersides of the leaves following development of the fourth-instar larva. Adult beetles emerge approximately 10 days after pupation. In the south-eastern USA there are three or four generations per year with a great deal of overlap between each. In western and northern USA there is one generation, with a partial second. Adults of the final generation disperse to protected areas in preparation for over-wintering.

The Mexican bean beetle is uncommon among Coccinellid beetles because it is a herbivore. Indeed, except for the squash beetle, *Epilachna borealis* Fabricius, and a few other species of this genus, all other Coccinellids are believed to be predators of small arthropods. As a plant feeder, Mexican bean beetles injure soybeans by removing leaf tissue. Adults feed on the underside of the foliage, using their mandibles carefully to remove leaf epidermis, cuticle and soft mesophyllic tissue. They do not consume leaf tissue *per se*, but plant juices once the tissues have been removed. Feeding is restricted to between the leaf veins, which results in a characteristic skeletonising of foliage. Once tissue is injured, it can no longer maintain a sufficient water balance. This secondary injury results in localised necrosis, with interveinal tissue eventually falling out and leaving a 'net' of leaf veins. From a plant physiological perspective, it is not clear how this skeletonising form of defoliation differs from complete leaf removal typical of other soybean leaf-mass consumers (e.g. green cloverworm: see page 191).

When overwintering adult Mexican bean beetles become active early in the spring while the soybean crop is still in the seedling stage, most serious injuries occur. Because the plants are young and unable, therefore, to tolerate much defoliation, even a small degree of skeletonising may result in considerable stand and yield losses. Skeletonising later in the season has much less impact on final yield of soybeans because the crop is better able then to tolerate or compensate for photosynthetic losses. Once plants begin to direct their carbohydrate energy towards reproductive processes such as flowering and pod development, the continued removal of leaf tissue may result in economic loss. For this reason, when control strategies are deemed necessary, they are frequently directed to populations skeletonising soybeans during later summer months.

Climatic conditions such as intense rainfall or extended drought with high temperatures have been shown to reduce significantly larval and adult populations. In addition, several control tactics have been evaluated for their potential to reduce Mexican bean beetle populations in soybeans. Natural control organisms, including at least 14 species of predators and 10 species of parasitoids, are prevalent in soybeans during the vegetative stages (Howard and Landis, 1936). These beneficial organisms are frequently capable of maintaining pest populations below economic injury levels. To augment natural control, efforts have been made to import natural enemies for mass release in biological control programmes. One larval parasitoid, *Pediobius foveolatus* (Crawford) (Hymenoptera: Eulophidae), has been imported from India and released on an annual basis in Maryland to reduce Mexican bean beetle populations (Fronk, 1978). Extreme caution must be exercised when evaluating biological control organisms for the Mexican bean beetle, to ensure that predatory species of Coccinellidae are not subject to attack.

Cultural control efforts would include destruction of overwintering locations and late planting of the soybean crop. The destruction of overwintering locations increases exposure to inclement weather conditions and can greatly reduce adult numbers the following spring. Another cultural control tactic involves late planting of the crop. Since overwintering adults actively forage upon emergence in the spring, soybeans planted early will attract a disproportionate number of beetles feeding during their preoviposition period. In addition, there is evidence that some soybean varieties adversely affect the behavioural and physiological functions of the Mexican bean beetle (Van Duyn *et al.*, 1971). These varietal differences are being investigated as a basis for developing host-plant resistance in soybeans.

For curative control in outbreak circumstances, several insecticides are currently available. Foliar sprays (e.g. permethrin, tralomethrin) late in the season should be applied as much as possible to the undersides of leaves, where the larvae and adults are actively feeding. In situations where seedling damage is anticipated, a systemic granular insecticide (e.g. phorate) will provide control for a limited period of time. Natural and artificial control tactics should be integrated within an overall strategy which maintains the value of Mexican bean beetle-induced injury in terms of yield losses below the total cost of management.

6.9.2 *Ceratoma trifurcata* (Forster): Bean Leaf Beetle (Coleoptera: Chrysomelidae)

The bean leaf beetle is widely distributed in the western hemisphere. There are several species within *Ceratoma* Gowdey which are capable of feeding on leguminous plants, but *C. trifurcata* is the most widely distributed species, inhabiting most primary soybean production centres in the USA. Specifically, bean leaf beetle populations extend from the south-eastern states of the USA to Kansas, Minnesota and Canada, and as far west as New Mexico.

Adult bean leaf beetles are 4.0–7.0 mm in length, with variable colour and design patterns on their elytra. The most common form has six black spots on buff-coloured elytra (Figure 6.12). There are, however, individuals with buff-coloured elytra without spots (Figure 6.13). Yet another variation, a red form, is characterised by red elytra with six black spots (Figure 6.14). Additional variations in colour and pattern can frequently be found. All polymorphic types have a medial, rearward-pointing triangular spot at the base of the elytra and black wing margins. The body of all forms is black in colour, oval in shape (as seen from above), with a slight narrowing, and moderately convex in form (as seen from the side).

Bean leaf beetle eggs are about 0.8 mm in length and spindle-shaped with a coarse, reticulate surface. They are yellowish-orange in colour and are typically found in clusters of 15–30 just below the soil surface. Larvae develop through three instars in the soil. Each instar is white in colour,

Figure 6.12 Adult bean leaf beetle, *Ceratoma trifurcata*, on soybean. © M. Jeffords

Figure 6.13 The spotless form of the adult bean leaf beetle, *Ceratoma trifurcata*. © M. Jeffords

with black head capsule and anal shield (Figure 6.15). The larvae are slender, with a sub-cylindrical appearance, distinct segmentation and slight posterior narrowing. The thoracic segments have three small pairs of walking legs, and one abdominal segment has a fleshy proleg. The last-instar larva can achieve a total length of 10 mm. Pupae are about 4.5 mm long and white in colour. The developing adult elytra and wings

Figure 6.14 The red form of the adult bean leaf beetle, *Ceratoma trifurcata*. © M. Jeffords

overlap the sixth abdominal segment and the caudal segment has a pair of long spines. The pupae will often be found in earthen cells below the soil surface.

Bean leaf beetles overwinter as adults in crop residues within fields or leaf litter of wood lots, or near field margins. They typically emerge from overwintering when temperatures reach 10–12°C (late March) and move directly to suitable wild leguminous hosts. Following soybean emergence

Figure 6.15 Larva of the bean leaf beetle, *Ceratoma trifurcata*. © M. Jeffords

(May), the beetles frequently relocate to cultivated fields (Kogan *et al.*, 1974). They continue to feed as adults to accumulate energy for oogenesis. Mating eventually occurs and the females begin depositing eggs in clusters in the soil at the bases of soybean seedlings. Females are capable of depositing up to 40 egg clusters containing a total of 175–250 eggs. There is evidence to suggest that the females of the overwintering generation have a higher fecundity than do females of subsequent generations. Following a mean incubation period of about 10 days, the eggs hatch and the first-instar larvae begin to feed on small roots, root hairs and root nodules, which appear to be the preferred feeding sites. Larval feeding continues through the three developmental instars and typically requires 45–50 days under cool spring temperatures. Development may be hastened in more southern latitudes because of higher soil temperatures. Mature larvae construct earthen cells in which the prepupae and then the pupae develop. Total developmental times (egg to adult) under controlled conditions were determined to be about 25 days, 30 days and 47 days at 30°C, 27°C and 21°C, respectively (Isely, 1930). Under field conditions in south-east USA, bean leaf beetle development required about 29 days, with a range of 24–45 days. At these geographical latitudes there are typically three generations. Adult activity peaks generally around the end of May (emerging overwintering adults), mid-July, mid-August and again in mid-September (adults entering overwintering). In the temperate climates of north-central USA there are two generations. Progeny of overwintering adults are active in July or early August and second-generation adults are active in September. These adults feed in preparation for overwintering and leave fields following senescence of the soybean crop.

Because adults and larvae feed on different plant organs, the physiological impact of bean leaf beetle injury on soybean development varies. Adults are leaf-mass consumers and reduce the total, photosynthetic capability of affected crops. Adults emerging early from overwintering, and entering fields soon after soybean emergence, can severely reduce plant stands and final yields. Losses result from cotyledon destruction, which reduces the primary source of energy at that vulnerable stage of plant development. When beetle densities are exceptionally high, they may even clip the stems of seedlings, causing additional stand reductions.

Once soybean crops have safely emerged and are growing vegetatively, bean leaf beetle defoliation can contribute to reduced crop vigour and photosynthetic potential by removing leaf tissue. Damage by bean leaf beetles alone during these relatively tolerant vegetative plant growth stages will rarely cause economic loss. There are, however, several leaf-mass consumers, including the bean leaf beetle (see green cloverworm, page 191), which can feed collectively, with subsequent yield losses. Adult beetles of the final generation, active when soybean plants are beginning to mature, will readily feed on developing flowers and pods. Since soybeans abort many of their blossoms naturally, the impact of beetle feeding on these organs seems minimal. Pod feeding can, however, have a serious impact on the utility of the harvested crop. When pod feeding does occur, the beetles do not penetrate the pod walls to consume the developing seeds. Rather, they remove pod tissue directly above the seeds, to cause circular brown scarring. The extent of quantitative yield loss from this form of injury has not been adequately assessed to date. However, seeds within affected pods are frequently of reduced quality and smaller in diameter than healthy seeds.

In the process of consuming soybean leaf tissue, bean leaf beetles may transmit one or more pathogens—in particular, bean pod mottle virus and southern bean mosaic virus (Walters, 1964). In the case of bean pod mottle virus, *Desmodium paniculatum* L. (stick trefoil species) is an overwintering host for the virus and alternative host plant for the bean leaf beetle (Walters and Lee, 1969).

Bean leaf beetle larvae, feeding in the root zone of plants, decrease turgour by removing root tissue and reduce nitrogen-fixing capacity by consuming root nodules. The nodule destruction is generally considered to be the most significant form of injury resulting from larval feeding. Secondary invasion of roots by pathogens may also occur as a result of mandibular scarring by larvae. Usually the injury caused by the developing larvae is less deleterious than the defoliation and pod feeding caused by the adults.

Adult bean leaf beetles have few known natural enemies. The magnitude of natural control of larvae in the soil is largely unknown. Biological control has not been studied to any great extent. Cultural controls, such as destroying overwintering sites, may be implemented to reduce feeding the following spring.

If outbreak conditions develop, or when soybeans are particularly susceptible to defoliation (e.g. when stressed by drought), insecticides are frequently required to prevent economic losses in yield. When deciding whether an insecticide is necessary, management guidelines typically include estimates of existing defoliation and numbers of adults. Treatment of vegetative soybean crops is often recommended when visual defoliation exceeds 35% and 16 adults or more per metre row are present and actively feeding (Kogan and Kuhlman, 1982). Less defoliation is tolerated during reproductive plant stages. Because bean leaf beetle defoliation may affect soybeans in a similar fashion to other leaf-mass-consuming pests, the total impact of multispecies consumption should be considered when making a management decision.

6.10 MOTHS

6.10.1 *Plathypena scabra* (Fabricius): Green Cloverworm (Lepidoptera: Noctuidae)

The green cloverworm is an occasional pest of soybeans in the USA. It is distributed throughout North America from the Great Plains to the east coast and from the south-eastern areas of Canada to the Gulf of Mexico. Northern regions of its range must be recolonised annually because it is not capable of surviving the winter (Wolf *et al.*, 1987).

Green cloverworm moths hold their wings roof-like over their bodies when they are resting and have distinct labial palps pointing anteriorly to provide a snout-like appearance (Figure 6.16). The hindwings are typically a dark brown to black colour, while the forewings have light-brown markings. In addition, each forewing has an irregular transverse line dividing a dark brown basal area from a lighter apical area. Females generally have a darker and more contrasting transverse line at the edge of each forewing. A more distinguishing characteristic between males and females is the considerably larger compound eyes of males. Wingspan of the moths ranges from 21 mm to 34 mm.

Figure 6.16 A green cloverworm moth, *Plathypena scabra*, on soybean. © M. Jeffords

Eggs of the green cloverworm moth are usually found singly on the undersides of soybean leaves. They are about 0.5 mm in diameter, with ridges and a translucent green appearance. As the head capsule of the larva develops, egg colour will change to light brown about 48 h before hatching. Larvae pass through six developmental instars, with a seventh instar

Figure 6.17   Larva of the green cloverworm moth, *Plathypena scabra*, on soybean.
© M. Jeffords

occurring occasionally. Larval length ranges from 1.5 mm (first instar) to 31 mm (sixth instar) (Figure 6.17). Small larvae are pale yellow to light green in colour, with two white, longitudinal stripes. There are four pairs of abdominal prolegs (including a pair on the last segment), which help to distinguish the green cloverworm from other lepidopterous defoliators that may also inhabit soybeans (especially in southern USA, e.g. velvetbean caterpillar, *Anticarsia gemmatalis* Hubner, and soybean looper, *Pseudoplusia includens* Walker). A behavioural characteristic of green cloverworm larvae is the violent movements when disturbed. The active thrashing movements from side to side are, presumably, intended to be an escape mechanism from predators. Pupae are dark brown in colour and approximately 13 mm long.

Adult green cloverworm moths are capable of overwintering in climates south of 41°N latitude. In the coastal region around the Gulf of Mexico, moths are capable of continuous generations. In the more temperate regions of north central USA, however, annual migrations on southerly winds are necessary to permit colonisation. Adults live 7–16 days and are active at dusk. Each female deposits 150–200 eggs singly on the undersides of alfalfa leaves. Eggs hatch in 2–5 days in warmer weather, but may take longer under cooler, spring conditions. Larvae develop and pupate on the non-soybean host, but emerging adults seem to prefer soybeans for oviposition. These adults oviposit on the undersides of soybean leaves, where larvae develop through six instars (and possibly a seventh), each requiring 2–3 days to complete development. The prepupal and pupal stages develop in earthen cells under the soil surface and require about 12 days before the adult moths emerge. A second generation develops on soybeans, with larval populations usually peaking sometime in August.

Green cloverworm larvae damage soybeans by leaf-mass removal. Developing larvae, either alone or in combination with other leaf-mass consumers, are capable of severely defoliating soybeans. Consumption potential increases exponentially with larval size and development (Hammond *et al.*, 1979). Feeding of very small larvae is restricted to skeletonising the undersides of leaves. As larvae increase in size, feeding sites expand to include all leaf tissue, except for major leaf veins, giving a ragged appearance to the foliage. The extent of damage to soybeans is dependent on the degree of additional stress imposed on the plants.

Research comparing average and stressful growing conditions (drought) has demonstrated that stressed soybean plants are more than 50% less tolerant of insect-induced defoliation (Ostlie and Pedigo, 1985).

Several natural controls may help to maintain the numbers of green cloverworm moths at non-economic levels during many years (Pedigo *et al.*, 1983). Specifically, the entomopathogenic fungus *Nomuraea rileyi* (Farlow) Sampson (Deuteromycotina) has been shown rapidly to reduce populations and, as a result, help to avoid economic losses. Indeed, when outbreaks of green cloverworms occur, the failure of *N. rileyi* to suppress populations is frequently cited (Thorvilson and Pedigo, 1984). In addition to pathogens, green cloverworm larval populations are frequently reduced by parasitoids. A tachinid fly, *Winthemia sinuata* Reinhard, and a braconid wasp, *Rogas nolophanae* Ashmead, have been identified as the most frequent parasitoids in Iowa (Lentz and Pedigo, 1975). Several other parasitoid species utilise green cloverworm larvae or pupae as hosts: in Illinois, USA, *Apanteles marginiventris* Cresson (Hymenoptera: Braconidae) and *R. nolophanae* are the most abundant larval parasitoids; in southern USA production centres, *Chaetophlepsis plathypenae* Sabrosky (Diptera: Tachinidae) is an important larval parasitoid; and among pupal parasitoids are *Brachymeria ovata* Say (Hymenoptera: Chalcididae), *Colpotrochia trifasciata* Cresson (Hymenoptera: Ichneumonidae) and *Vulgichneumon brevicinctor* Say (Hymenoptera: Ichneumonidae). Several insect predators also assist in keeping moth numbers low. Specifically, *Orius insidiosus* Say (Hemiptera: Anthocoridae) and various *Nabis* Latreille spp. (especially *N. roseipennis* Reuter (Hemiptera: Nabidae)) readily feed on small green cloverworm larvae. Other generalist predators (e.g. *Geocoris punctipes* Say (Hemiptera: Lygaeidae) are typically present in soybean fields to suppress further egg and larval development.

Chemical controls represent the most efficient therapeutic control tactic during outbreak situations. Green cloverworm larvae are very susceptible to most insecticides (e.g. permethrin, carbaryl, chlorpyrifos) and control can be achieved at relatively low rates of active ingredient. Deciding if and when chemical controls should be applied has been the focus of much research. Sampling techniques for adults and larvae have been developed and crop damage functions determined under a number of environmental

Table 6.1 Economic injury levels (larvae per metre row) for green cloverworm, *Plathypena scabra*, of different size classes at four different growth stages of soybeans. Based on Hutchins *et al.* (1988)

Larval size	Crop growth stage		
	Late vegetative, Bloom	*Pod set*	*Pod fill*
Small[a]	1250–1300	1450	950
Medium	250–260	290	190
Large	25–26	29	19

[a] Small=instars 1 and 2; medium=instars 3 and 4; large=instars 5+.

conditions. Economic injury levels for different larval sizes and crop
growth stages are presented in Table 6.1. These injury levels reflect the
vast differences in consumption potential of small versus large larvae and
the relative susceptibilities of different soybean growth stages. The economic injury levels for green cloverworm larvae should be combined with
similar data for other soybean defoliators and used to manage multiple
species of leaf-mass consumers in a collective fashion.

6.11 MITES

6.11.1 *Tetranychus urticae* Koch: Two-spotted Spider Mite (Prostigmata: Tetranychidae)

T. urticae is considered in Chapter 8 (see page 311).

6.12 NEMATODES

6.12.1 *Heterodera glycines* Ichinohe: Soybean Cyst Nematode (Tylenchida: Heteroderidae)

The first record of soybean cyst nematode was in Japan during 1915 (Hori,
1916). Reports followed from Korea in 1936 and from China (origin of
soybeans) in 1938. The first record of this pest reaching the USA was in
1954, when it was reported from North Carolina (Winstead *et al.*, 1955).
Since this report, positive identifications have been made in at least 23
other US states, with new records reported continually. Because there are
several races, the range of the soybean cyst nematode will probably expand
to include all intensively cropped soybeans. The reason for the success of
this nematode species is that its races are able to adapt: to diverse
environments; to different cultural practices; and to host diversity. The
soybean cyst nematode, therefore, is currently considered to be a very
serious pest of soybeans in most production centres, including the USA.
    Cysts of *H. glycines* are about 0.5–1.0 mm long and lemon-shaped
(Figure 6.18). The cysts are the expanded abdomens of dead, adult
females. When alive, the female produces eggs within her abdomen and
her body colour is white to light yellow. Following oogenesis, the gravid
female dies and her cadaver becomes dark brown in colour and tough, with
a distinctive zigzag pattern (microscopic). The eggs are protected within
the cyst until it ruptures and releases 200–500 eggs into the soil. The
majority of eggs (about 75%) will either die or hatch soon after being
dispersed from the cyst. Depending on soil conditions and genetic variation, however, some eggs will remain in a dormant state for an extended

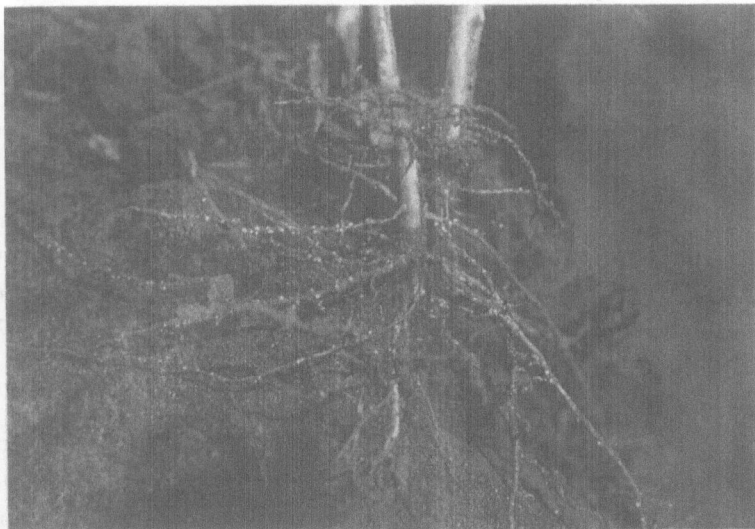

Figure 6.18 Female soybean cyst nematodes, *Heterodera glycines*, attached to soybean roots. © Iowa Agriculture Experiment Station

period of time. The dormancy is believed to be significant as a means of short- and long-term survival. For example, short-term dormancy has been identified with overwintering. Eggs that initiate dormancy in the autumn, when temperatures are cool and soybeans are absent, will typically hatch in the spring, when host plants are available. Hatching is believed to involve adequate soil moisture and a plant-induced hatching stimulant. Long-term egg dormancy can last up to 7 years, allowing the nematode to survive under extended conditions of unfavourable development (Slack *et al.*, 1981). The range of the soybean cyst nematode has expanded as a result of accidental introductions of eggs or cysts on farm machinery, animals or plant material.

First-stage larvae develop inside the eggs. Second-stage larvae, referred to as J-2 larvae, emerge from the eggs and immediately seek out nearby soybean roots. The J-2 larvae penetrate the roots and migrate to the stele region of the cortex to begin feeding. The magnitude of larval penetration appears to be related to temperature. The rate of penetration can vary by as much as a factor of 12, with 28°C providing optimal conditions for penetration. The J-2 larvae feed and grow in length until moulting results in J-3, and subsequently, J-4 larvae. The larvae are generally about 0.5 mm long and each has a thick buccal spear. Adult development follows the J-4 larval stage, but males and females develop differently. Females remain attached to the roots and begin oogenesis. Males become active, dislodge from the roots and fertilise the females. The adult males reach a length of about 1 mm and have very short bluntly rounded tails. A generation (egg–adult) requires about 21–25 days at soil temperatures in the range

24–28°C. Under these temperature conditions, several generations are able to develop during each soybean cropping season.

An important factor affecting degree of soybean cyst nematode infestation is host-plant suitability. Although soybean cyst nematodes display host-specificity, they have been shown to express a high degree of genetic variability for host adaptation. At least five phenotypically distinguishable races have evolved and can be characterised on the basis of their degree of survival on four standard soybean varieties (Epps and Duclos, 1970; Inagaki, 1979). The ability of the soybean cyst nematode to develop rapidly new races when confronted with 'unacceptable' hosts has been documented many times under field conditions (Riggs *et al.*, 1981).

Plant injury by soybean cyst nematodes results in disfiguration or destruction of root tissue, causing poor water and nutrient uptakes. At specific feeding sites injury can be seen as a disruption of cell organisation in and around the vascular elements. Specifically, cell wall perforations occur at feeding sites (presumably arising from salivary secretions). Following disruption of the cell wall, a fluid syncytium forms and unbound cell cytoplasm accumulates at the site of feeding (Jones and Dropkin, 1975). As the syncytium expands, the walls of additional cells break down and their central vacuoles break up into smaller vacuoles. The larvae feed on the solute nutrients within these now dense syncytia and are able to complete their development. Under heavy infestations, the roots become discoloured and necrotic.

Plant leaf area is often significantly reduced, plants are stunted, and their vigour is poor (Figures 6.19, 6.20). The severity of stunting is related to the degree of root destruction, which, in turn, has been correlated with the number of larvae and cysts in the soil (Noel and Stanger, 1982). Leaf chlorosis may appear, but this symptom seems to be associated with feeding by specific races of the soybean cyst nematode. Infested plants are susceptible to secondary invasion by plant pathogens. Indeed, because of the similarity between foliar symptoms and other types of stress, soybean cyst nematode damage may be misdiagnosed as a disease or fertility problem.

Soybean cyst nematode-induced injury may result in significant yield losses, either in seed number or in seed size. The degree of yield loss resulting from nematode-induced stunting and reduction of leaf area is dependent on the severity of other biotic and abiotic stress factors. These factors are almost impossible to separate out from the influence of the nematode alone. Nevertheless, yield losses of up to 50% can occur in heavily infested soils.

Since the cysts of soybean cyst nematode are spread by farm implements moving between fields, strict control of machinery movements will delay introduction into surrounding fields. Thoroughly washing implements may also help. Crop rotation is recommended for severely infested fields: one year of a weed-free, non-host crop can reduce soybean cyst nematode

Figure 6.19   (a) Healthy soybean plants. (b) Stunted soybean plants, showing an open canopy, due to injury by the soybean cyst nematode, *Heterodera glycines*. © Iowa Agriculture Experiment Station

populations by one-half. Because of the extended hatch characteristics of the eggs, it may take 3–4 years of non-host crop rotation before high proportions are reduced to sub-economic levels.

Several varieties of soybeans resistant to cyst nematode attack have been developed and released for use in infested fields. Varietal selections should be based on agronomic characteristics and prior knowledge of which races of the nematode are to be controlled.

Figure 6.20 Soybean field showing injury by the soybean cyst nematode, *Heterodera glycines*. © Iowa Agriculture Experiment Station

Nematicides give one-season control of the soybean cyst nematode. They can be incorporated into the soil in a granular form at planting to provide a zone of protection around the developing root system. Nematicidal protection allows plants to produce more leaf area and generally results in higher final yields. Because the nematicides will eventually biodegrade, late-hatching eggs will survive and successfully develop into adult nematodes which will reproduce and thereby provide adequate numbers of viable eggs for the life cycle to continue next season. Nematicidal control is therefore considered to be an annual practice. Although soil fumigation will destroy the majority of nematodes present in a field, it is frequently too expensive to justify for soybeans.

Groundnuts

6.13 INTRODUCTION

Groundnuts or peanuts, *Arachis hypogaea* L., are grown throughout the warm temperate, subtropical and tropical zones of the world. This section deals with the pests of groundnuts encountered on commercial farms in North America (mainly Virginia, the Carolinas, Georgia, Alabama, Florida, Oklahoma and Texas, but also Ontario, Canada), Australia (Queensland), southern Africa, especially Zimbabwe, and, to a limited extent, North Africa, the Mediterranean and southern Europe.

Groundnut flowers are, for the main part, self-pollinated, so that it is not necessary to specify an insecticide-free period during the prolonged flowering stage of this crop. After fertilisation, the base of the ovary extends to form the gynophore or peg. This stem-like structure grows downwards until its tip is 1 cm or more below the soil surface, where the familiar groundnut pods develop. The entomologist has to deal with pests damaging flowers, foliage, stems, roots, pegs and pods. Insecticides are widely used for killing leaf eaters but, as groundnut plants can withstand considerable amounts of defoliation before suffering yield losses, the application of chemicals may be counterproductive. There are several groundnut genotypes with resistance to a range of pests. This resistance has not been fully exploited. Similarly, the transfer of pest resistance from several wild relatives (*Arachis* L. spp.) to cultivated species has yet to be fully evaluated (Stalker and Campbell, 1983; Amin, 1985a).

Smith and Barfield (1982) have reviewed the management of preharvest pests of groundnuts, with emphasis on the USA. Feakin (1973) has covered the subject, with emphases on the problems of developing and developed countries. Wightman and Amin (1988) and Wightman *et al.* (1990) discussed groundnut pests and their control, with a bias towards the pest problems of farmers in developing countries. Much of the information about groundnut pests in Australia is quoted (personal communication) from an unpublished review and commentary by D. J. Rogers.

6.14 PESTS LIVING ABOVE GROUND

6.14.1 *Helicoverpa (Heliothis) armigera* (Hübner): Cotton Bollworm (Lepidoptera: Noctuidae)

H. armigera is considered in Chapter 7 (see page 220).

6.14.2 *Spodoptera exigua* (Hübner): Beet Armyworm (Lepidoptera: Noctuidae)

S. exigua is considered in Chapter 2 (see page 35).

6.14.3 Various Genera: Thrips (Thysanoptera)

Thrips living on groundnut plants appear to cause considerable damage to the foliage, but there is little evidence to demonstrate any significant crop loss. On the other hand, seven species have been identified as vectors of tomato spotted wilt virus (TSWV), which causes bud necrosis disease

(BND) (Figure 6.21); see Table 6.2. There are other species of thrips living on groundnut plants, e.g. *Megalurothrips sjostedti* Trybom, *F. tritici* Fitch and *F. bispinosa* Morgan, but they are of little economic importance.

Most species live in the flowers and on the foliage, particularly between the unfolded leaflets around terminal buds. Eggs are laid in leaf tissue. There are four nymphal instars: the first two are active, feeding stages; the last two do not feed and are cryptic. They feed by rasping epidermal and mesophyll tissues, causing distortion, shrivelling and brown-coloured lesions to appear on injured leaflets. Badly damaged plants can be stunted. This kind of damage happens most years in the major groundnut-growing areas of the USA. Research by Lynch *et al.* (1984a) showed no yield losses

Figure 6.21 Groundnut plants: the left-hand one is healthy; the right-hand one is infected with bud necrosis disease caused by tomato spotted wilt virus, which is transmitted by thrips. © J. A. Wightman

Table 6.2 Species of thrips identified as vectors of TSWV

Species	Distribution
Thrips tabaci Lind., potato or onion thrips	cosmopolitan
Frankliniella schultzei Tryb., blossom or cotton bud thrips	sub-Saharan Africa, south and south-east Asia, Pacific and Australia
F. occidentalis Pergande, western flower thrips	western north America, but found in Georgia in 1983 (Beshear, 1983); also Europe
F. fusca Hinds, tobacco thrips	widespread
Scirtothrips dorsalis Hood	India
Thrips setosus Moulton	Japan
Thrips palmi Karny	Asia

from thrips damage. *Megalurothrips* Bagnall spp. and, to a greater extent, *F. occidentalis* are flower feeders. Smith and Sams (1977) reviewed research that had been carried out throughout the groundnut-growing areas of the USA between 1945 and 1977. In only two of the total of 14 reports which were reviewed were there significant yield increases attributable to thrips control. Smith and Sams (1977) were not able to detect yield losses in seedling Spanish groundnuts grown in Texas and infested with a high population of some 50 *T. tabaci* per terminal.

Tappan and Gorbet (1979, 1981) reported thrips control with foliar and broadcast granular formulations of systemic insecticides (acephate, disulfoton, carbofuran, aldicarb) without associated increases in yield. Lynch *et al.* (1984a) stated that thrips reduced the profits of groundnut growers in Georgia by $26 million from 1972 to 1981. Of this sum, 76% represented the cost of control agents. A survey of Georgian farmers carried out from 1979 to 1981 showed that about 60% still applied systemic insecticides at planting for thrips control. Owing to declining farm incomes, a more realistic estimate for current systemic insecticide use would be approximately 30% (R. Lynch, personal communication, 1988). Extension officers advised this practice on the grounds that it would prevent the curtailment of early groundnut plant growth by thrips, allowing plants rapidly to cover the soil, thereby reducing subsequently the need for herbicide application. Even though thrips do not apparently cause serious crop losses, they are an economic problem because they induce farmers to apply insecticides for cosmetic reasons. These insecticides can kill not only the thrips, but also their natural enemies (anthocorid and mirid bugs, for instance), as well as the natural enemies of some other potential pests.

The economic importance of thrips has changed rather radically with the occurrence of outbreaks of BND in Australia during the mid-1970s and in the southern groundnut-growing states of the USA during the mid-1980s. Reports of extensive crop losses from TSWV came, in particular, from Texas. The epidemiology and control of TSWV, with emphasis on the role of the thrips vectors, have been reviewed by Reddy and Wightman (1989). TSWV is unusual, but not unique, in that it is transmitted only by thrips. Only nymphs and adults that acquired the virus during their immature stages can be vectors. Farmers wishing to protect their crops from BND might well be tempted to apply one of the many insecticides that will kill thrips. There is evidence to show that this action is counterproductive. Wightman and Amin (1988) showed that high and frequent doses of dimethoate (400 g a.i. ha^{-1} every 3–5 days) were required to protect groundnuts from BND in India. Similarly, insecticides failed to protect Texan groundnuts from a heavy and early infestation of thrips, leading to BND in 1986. Dintenfass *et al.* (1987) described an exponential increase in the density of *F. occidentalis* following insecticidal applications to onions, which indicated that insecticides interfere with natural control processes. Thrips are known to be subject to predation and parasitism but little is known about the dynamics of these processes. Perhaps the best way of overcoming BND in the long term is to exploit the resistance to thrips that exists in several cultivars and genotypes of groundnuts (Campbell and Wynne, 1980; Amin, 1985b; Amin *et al.*, 1985) as well as in several wild *Arachis* species (Stalker and Campbell, 1983). The mechanism of resistance is unknown but may involve the long trichomes on the leaves. It is noteworthy that cultivar Kadiri 3 (Robut 33-1), which is grown extensively in parts of India and elsewhere, has resistance to the thrips vectors of TSWV and is a comparatively rare example of a situation where the effects of a virus disease are minimised by host-plant resistance to the vector.

Reddy *et al.* (1983) provide evidence that sowing crops densely and early in the season, so that they will rapidly form closed canopies before the main thrips migrations occur, are good management options for reducing the losses caused by BND. However, this advice should be accompanied by local knowledge of when the thrips vectors are likely to be flying, supplemented by the possible effects of alternative hosts of TSWV and vectors (a large category that includes many crop plants, weeds and ornamentals) that are growing in the vicinity.

6.14.4 *Empoasca fabae* Harris: Potato Leafhopper (Hemiptera: Cicadellidae)

E. fabae is considered in Chapter 8 (see page 273).

6.14.5 *Tetranychus urticae* Koch: Two-spotted Spider Mite (Prostigmata: Tetranychidae)

T. urticae is considered in Chapter 8 (see page 311).

6.15 PESTS LIVING BELOW GROUND

By way of brief introduction, all groundnut insect pests living below ground could, in theory, be controlled by chlorinated hydrocarbon insecticides, including the cyclodienes (endrin, aldrin, dieldrin). However, the laws of many countries now prevent the use of these insecticides. In the case of groundnuts, the chlorinated hydrocarbon insecticides should not be used, because of their high solubility in the oil that comprises 45–50% of mature seeds. The probability of producing a contaminated product is therefore high if plants are protected from soil insects by these chemicals. This situation is unfortunate, because the chlorinated hydrocarbons are effective and there are no really satisfactory alternatives. The insects that attack the underground parts of groundnut plants are therefore most difficult to control.

6.15.1 *Elasmopalpus lignosellus* (Zeller): Lesser Cornstalk Borer (Lepidoptera: Pyralidae)

The lesser cornstalk borer is restricted to the New World. It is oligophagous, feeding on many leguminous and graminaceous plants, including sugar cane, rice and many kinds of cultivated legumes. Perhaps non-cultivated members of the Leguminosae and the Gramineae act as reservoirs of infestation for the cultivated members.

An adult female lesser cornstalk borer moth is charcoal grey in colour, with brown markings towards the head. The adult male moth is a light buff colour, with a dark-coloured line down the middle of the back and across the posterior border of the wings. Moths of both sexes are 10 mm or a little more long. The lesser cornstalk borer larva is a dark, blue-green, base colour, with brown or purple bands, and reaches a maximum length of 15 mm.

An adult female lesser cornstalk borer moth can lay up to 420 eggs. They are laid singly in the soil at a depth of less than 2 mm under plant canopies. The eggs take 3–5 days to hatch, depending on time of year. The larvae each construct a tunnel lined with silk which connects its underground refugia with its feeding site, which is usually the pegs, pods or flower and stem rudiments concentrated in and around the crowns of plants. The number of larval instars varies between 5 and 9, according to environmental conditions. The larval period lasts for between 33 days and 65 days, with four generations occurring in South Carolina and Mississippi, three generations occurring in Arizona, Texas and Oklahoma, and three generations with a partial fourth occurring in Georgia. The lesser cornstalk borer overwinters as fully grown larvae and/or pupae. There is evidence that it does not enter diapause, but instead slows down its metabolism in response to low seasonal temperatures.

Population outbreaks of the lesser cornstalk borer are most likely to occur during periods of hot, dry weather, because the larvae cannot tolerate much soil moisture. For this reason, the borer is most abundant in areas of deep, freely draining, sandy soil where crops are rain-fed. An analysis of the population dynamics of the lesser cornstalk borer (Smith and Barfield, 1982) showed that there was no density dependence, the implication being that environmental factors were more important population regulators than biotic factors.

Adult lesser cornstalk borers are nocturnal. Females secrete a four-component pheromone that has been identified by Lynch *et al.* (1984b). A synthetic analogue of this pheromone was used by Funderburk *et al.* (1987) to monitor the flight pattern of moths in peanut, soybean, corn (maize), sorghum and wheat fields in northern Florida. Multiple adult generations were active during the vegetative and reproductive stages of peanut and sorghum crops. Two generations were active during the vegetative and early reproductive stages of soybeans. Adults were recovered from seedling wheat crops. Traps in corn fields yielded few moths. The pheromone trap catches indicated that generations did not overlap to any extent, although there were disparities between different crops.

The propensity for lesser cornstalk borer larvae to attack developing tissues in groundnut crowns means that a relatively low density of larvae can cause a lot of damage if the infestation coincides with the period of most active growth and tissue initiation. Crop losses of up to 70% have been attributed to this kind of activity. Berberet *et al.* (1979) found a highly significant relationship between infestation incidence and groundnut yield. There was virtually a 10 kg ha^{-1} reduction in yield for each 1% increase in infestation of non-irrigated, Spanish-type groundnuts. Smith and Holloway (1979) found that groundnut cultivar Starr could withstand damage up to about 1 larva per 10 plants (the damage threshold), but thereafter yield loss was described by a three-parameter, non-linear function. Lynch (1984) described the damage caused by the lesser cornstalk borer to groundnut pods at all stages of development. He found that the smaller (younger) pods suffered considerably more damage than the more mature ones. Older larvae, while more likely to attack older pods, limited damage to scarification. Chemical control measures should, therefore, be aimed at controlling an infestation during pod filling.

The lesser cornstalk borer has many parasites but the degree of attack is not high on groundnuts compared with other host crops. Larvae and eggs are preyed on by carabid beetles, lygaeid bugs and therevid flies. A virulent strain of an entomopox virus attacks the larvae and the pupae.

Insecticidal application can give good control of the lesser cornstalk borer with a recorded yield response of ×2.5 untreated yield (Berberet *et al.*, 1979). A range of insecticides is available: monocrotophos, phorate, carbofuran, etc. Applications of both granular and spray formulations are more effective if they are made directly to the soil surface at the base of plants. Sprays are recommended for rain-fed fields and granules for irrigated fields.

There appear to be no genotypes of groundnuts with outstanding resistance to lesser cornstalk borer, but there is some resistance and some tolerance among several cultivars, e.g. Early Runner. Wild species of cultivated plants are usually a source of resistance, but the wild species of *Arachis* L. did not exhibit enough resistance for continued research (Stalker *et al.*, 1984).

6.15.2 Various Genera: White Grubs (Coleoptera: Scarabaeidae)

White grubs are a particular problem of groundnuts growing in Queensland, Australia, and southern Africa. They are the larvae of melolonthine and ruteline scarabaeid beetles and can achieve a length of 4 cm. In groundnut fields they feed on the roots and developing pods. Where soil water is lacking, root feeding is more important and can lead to the death or stunting of young plants. Damage to the roots and the pods can lead to secondary fungal diseases.

In southern Africa, particularly Zimbabwe, the genera include *Adoretus* Cast., *Schizonycha* Dejean and *Anomala* Samouelle, with several others awaiting description (unpublished). Rogers and Brier (personal communication, 1987) stated that the most abundant species in Queensland is *Heteronyx piceus* Blanchard, with *H. rugosipennis* Macleay and *Sericesthis ino* Blackburn (both identifications are tentative) being less numerous. *Strigoderma arboricola* Fabricius attacks groundnuts growing in deep organic soils in the eastern states of the USA.

The different genera of white grubs have a similar life cycle. The adult beetles emerge from the soil during October and November (just after the start of the rainy season in the semi-arid climates of southern Africa and Australia). They mate and lay eggs in the soil. Adults, which have characteristic lamellate antennae and are usually brown to black in colour (e.g. cockchafers, May beetles, etc.), can cause considerable damage by their feeding activity on the leaves of trees and shrubs near where they have laid their eggs. There are three larval instars, which have tan-to-chocolate-coloured heads and C-shaped white bodies, and can exceed 4 cm in length, depending on species. They usually pupate by the end of the growing season. Some species survive the 'off-season' in the larval stage while others pupate first. Some species remain in the larval stage for two seasons (e.g. *Rhopaea magnicornis* Blackburn in Australia).

There are no data relating groundnut yield losses and white grub densities from southern Africa, but the results of a series of observations in Australia indicate that white grub control can increase yield by between 16% and 35%. Rogers and Brier (personal communication, 1987) calculated that each white grub can consume between 2.5 g and 9.1 g of pods. Yield can be reduced by 52 kg ha^{-1} if there is one white grub per metre row and the rows are 90 cm apart.

In Australia, terbufos (1 kg a.i. ha^{-1}), aldicarb, fensulfothion, isofenphos and phorate granules (all at 3 kg a.i. ha^{-1}) controlled white grubs when placed in-furrow at sowing. Banded applications of insecticidal granules at sowing were also evaluated: carbofuran applied at 1–2 kg a.i. ha^{-1} in 15 cm or 35 cm bands and terbufos applied at 2 kg a.i. ha^{-1} in 15 cm bands were effective. Ethoprophos, fensulfothion and isofenphos granules (all at 3 kg a.i. ha^{-1}) reduced white grub populations when applied as 20 cm bands about 4 weeks after sowing. Although controlling white grubs, the degree of control provided by terbufos and isofenphos was not satisfactory. A series of methamidophos spray applications to groundnut foliage during the adult feeding period resulted in an economic reduction in pod damage.

White grub control on groundnuts in Australia is still at the experimental stage. In southern Africa the estate farmers routinely apply carbofuran granules at sowing as a preventative measure against white grub attack.

6.15.3 *Hilda patruelis* Stal: 'Hilda' or Groundnut Hopper (Hemiptera: Tettigometridae)

'Hilda' or the groundnut hopper is apparently restricted to Africa, south of a line running east–west across the continent, approximately 10° north of the equator. The host-plant range of 'Hilda' is wide and includes a number of cultivated plants such as carrots, potatoes, marigolds, sunn hemp and groundnuts. It feeds on the root system of these plants. Hosts on which it colonises the aerial parts include flowering maple, senna, figs, soybeans, sunflowers and cashew. It is found below and above ground on *Citrus* L. spp. and several kinds of beans (Taylor, 1981). The cashew trees growing along the coast of Tanzania and Kenya may be associated with the high incidence of 'Hilda' in the region. It also lives on many species of weeds.

'Hilda' is distributed over a huge area of Africa, once one of the most important commercial groundnut growing centres of the world and one which should once again come to prominence. Currently, 'Hilda' is of great concern to groundnut farmers in Zimbabwe. It invades a field from the edges inwards, presumably migrating by walking or by low-level flying from alternative hosts. The first sign of invasion is, therefore, dying plants along field borders. If these plants are pulled up, the nymphs and adults will not be found because they will have moved along the rows to feed on the tap roots of neighbouring plants. Only masses of small black ants, *Pheidole megacephala* Fabricius, that attend 'Hilda' for honeydew, will be found on dying plants. 'Hilda' uses the galleries constructed by the ants to move around the root area and perhaps to travel to neighbouring plants.

The 'Hilda' eggs are conspicuous on groundnut tap roots, pods and pegs because of their pale blue coloration. They are about 2 mm long and 0.4 mm wide and are laid in clusters (rafts) of 20–40. The eggs in one batch are always arranged in a row and point in the same direction. Incubation takes about 2 weeks. There are five nymphal instars. The nymphs are dark green to brown in colour. The adults, 4–6 mm long, are a similar colour, but have mottled wings. The complete life cycle requires 30–40 days.

It is not always easy to distinguish groundnut plants that have been killed by 'Hilda' from those plants that have been killed by other insects or pathogens. The brown colour in the vascular system need not necessarily have been caused by a 'Hilda' attack but by fungal attacks. However, the ants and their galleries are clues to the presence of 'Hilda'. Uprooting living plants further along rows from dead and dying plants will probably reveal the hoppers. The brown discoloration of the root vascular tissue which results in rapid death of affected plants has given rise to speculation that 'Hilda' is a vector of a fungal pathogen, because some common root diseases, such as *Fusarium* Link spp. and *Verticillium* Nees spp. (Hyphomycetes), exhibit similar symptoms. However, K. R. Bock (personal communication, 1986) found that when 'Hilda' nymphs were reared on groundnut roots in Petri dishes, the brown discoloration still appeared.

It is possibly caused by a (salivary?) toxin.

Taylor (1981) states that monocrotophos is the best insecticide to apply for 'Hilda' control, although it is not always effective. She suggested that farmers should inspect the vegetation around their groundnut fields for insects, and that they should spray a 3 m wide band around the periphery of each field upon finding 'Hilda' on weeds or already killing plants in the field edges. Spraying the weeds themselves is probably not a good idea, because 'Hilda' eggs are parasitised by an encyrtid wasp, *Psyllechthrus oophagus* Ghesquière (Hymenoptera), and the wild population of hoppers on the weeds will act as a reservoir of hosts for the wasps. Weaving (1980) found that the degree of parasitism was close to 75% for most of the year and that some eggs in all batches were parasitised. However, the degree of parasitism fell off markedly, but not completely, in July–August and in January. There appear to be no specific predators of 'Hilda', the ants, no doubt, giving protection.

6.15.4 *Diabrotica undecimpunctata howardi* Barber: Spotted Cucumber Beetle (Coleoptera: Chrysomelidae)

Diabrotica spp. have been considered in Chapter 5 (see p. 146).

6.15.5 *Graphognathus leucoloma* (Boheman): White Fringed Weevil (Coleoptera: Curculionidae)

G. leucoloma is a pest of groundnuts growing in Australia and the south-eastern states of the USA. The adult weevils are about 10 mm long and grey in colour, with conspicuous white banding along the sides of the elytra. They emerge from the soil throughout the growing season and each lays several thousand eggs. Adult weevils are parthenogenetic. Eggs hatch in 11–30 days. The larvae feed on roots for 2–3 months depending on latitude and each achieves a length of 12–15 mm before pupating in the soil. There appears to be no diapause. There is only one generation per year.

Gross and Harlan (1975) found that it was necessary to apply five sprays of carbaryl insecticide at 2.25 kg a.i. ha^{-1} during the growing season to control adult white fringed weevils, because they took such a long time to emerge. These authors recommended a treatment threshold of two adults per ten plants. A band application of carbaryl at 2 kg a.i. ha^{-1} applied at sowing protected groundnuts from larval attack (Harlan and Gross, 1976).

In southern Africa the species of weevil equivalent to *G. leucoloma* is *Systates* Gerstaecker spp. but little is known about its biology other than the adults appear in sufficiently high densities to suggest that the larvae, which feed on groundnut roots, may cause damage. The adults eat leaf margins but cause little damage to foliage. The life cycles of *Systates* spp. and *G. leucoloma* are similar.

6.15.6 *Meloidogyne* Goeldi spp.: Root-knot Nematodes (Tylenchida: Meloidogynidae)

Meloidogyne spp. have been considered in Chapter 2 (see page 61).

6.15.7 *Pratylenchus brachyurus* (Godfrey) Filipjev and Schuurman-Stekhoven: Smooth-headed Lesion Nematode (Tylenchida: Pratylenchidae)

Feakin (1973) rates *P. brachyurus* as a serious pest of groundnuts in Australia and the USA. All stages of this species are active and can enter or leave root systems, apparently at will. Points of entry appear as brown marks that develop into lesions as a result of toxins produced by the nematodes. In a heavy infestation roots and stems become necrotic. These necrotic areas become the sites of infection for a number of pathogenic and saprophytic fungi.

Other species of nematode that have been reported to attack groundnut crops include *Belonolaimus gracilis* Steiner and *B. longicordatus* Rau (sting nematodes), *Criconemoides ornatus* (Raski) Luc and Raski (ring nematode) and *Trichodorus christei* (stubby-root nematode). All of these species are encountered throughout the tropical and subtropical zones of the world and can cause damage to groundnut plants.

Chemical control of nematodes in groundnuts is usually attempted with in-furrow granular formulations of carbofuran or aldicarb to the soil. Rotation of groundnuts with corn (maize) or small grain (e.g. wheat) crops is a viable method of control in the USA: neither maize nor wheat harbours the same species of nematodes that attack groundnuts (R. E. Lynch, personal communication). Feakin (1973) noted that groundnuts was not a good entry crop to newly cleared land, because of problems created by resident nematode populations.

REFERENCES

Amin, P. W. (1985a). Resistance of wild species of groundnut to insect and mite pests. *Proceedings Int. Wshop Cytogenetics of Arachis*. ICRISAT Center, Patancheru, A.P. 502324, India, p. 57

Amin, P. W. (1985b). Apparent resistance of groundnut cultivar Robut 33-1 to bud necrosis disease. *Plant Dis.*, **69**, 718–719

Amin, P. W., Singh, K. N., Dwivedi, S. L. and Rao, V. R. (1985). Sources of resistance to the jassid (*Empoasca kerri* Pruthi), thrips (*Frankliniella schultzei* (Trybom)) and termites (*Odontotermes* sp.) in ground nut (*Arachis hypogaea* L.). *Peanut Sci.*, **12**, 59–60

Anon. (1970a). *Annual Report of the Pea Growing Research Organisation, UK*, p. 9

Anon. (1970b). *Migratory Root Eelworms*. Short Term Leaflet 113, Ministry of Agriculture Fisheries and Food, London

Anon. (1985). *Annual Report of the Processors and Growers Research Organisation, UK*, p. 7

Anon. (1986). *Annual Report of the Processors and Growers Research Organisation, UK*, p. 33

Bardner, R. (1978). The economic importance of pea moth in the UK. *ADAS Quart. Rev.*, **31**, 159–172

Baughan, P. J., Biddle, A. J., Blackett, J. A. and Toms, A. M. (1985). Using the seed as a chemical carrier. In BCPC Monograph No. 28, Symposium entitled *Application and Biology*, British Crop Protection Council, UK, pp. 97–106

Berberet, R. C., Morrison, R. D. and Wall, R. G. (1979). Yield reduction caused by the lesser cornstalk borer in nonirrigated spanish peanuts. *J. Econ. Entomol.*, **72**, 526–528

Beshear, R. J. (1983). New records of thrips in Georgia (Thysanoptera: Terebrantia: Tubulifera). *J. Georgia Entomol. Soc.*, **18**, 324–344

Biddle, A. J. (1985). Pea pests, yield, quality and control practices in the UK. In *The Pea Crop* (ed. P. D. Hepplethwaite, M. C. Heath and T. D. K. Dawkins). Butterworths, London, pp. 257–266

Biddle, A. J., Blood Smyth, J., Cochrane, J., Emmett, B., Garthwaite, D. G., Graham, J. C., Greenway, A. R., Lewis, T., Maccaulay, E. D. M., Perry, J. M., Smith, M. C., Sturgeon, D. M. and Wall, C. (1983). Pheromone monitoring of pea moth, *Cydia nigricana* F. *Proc. 10th Int. Congr. Plant Prot.*, p. 161

Bos, L. and Van Der Want, J. P. H. (1962). Early browning of pea, a disease caused by a soil and seed-borne virus. *Tijdschr. PlZiekt.*, **68** (6), 368–390

Cammel, M. E., Way, M. J. and Heathcote, G. D. (1978). Distribution of eggs of the black bean aphid on the spindle bush with reference to forecasting infestations on field beans. *Plant Pathol.*, **27** (2), 68–76

Campbell, W. V. and Wynne, J. C. (1980). Resistance of groundnuts to insects and mites. *Proc. Int. Wshop Groundnuts*, ICRISAT Center, Patancheru, A.P. 502324, India, p. 149

Campbell, W. V. and Wynne, J. C. (1985). Influence of the insect resistant peanut cultivar NC 6 on performance of soil insecticides. *J. Econ. Entomol.*, **78**, 113–116

Cockbain, A. J. (1971). Epidemiology and control of weevil transmitted viruses in field beans. *Proc. 6th Br. Insect. Fung. Conf.*, Vol. 1, p. 251

Delucchi, V., Baumgaertner, J. and Bieiri, M. (1983). An integrated control approach for canning pea crop pests in Switzerland. *Proc. 10th Int. Congr. Plant Prot.*, Vol. 3, p. 1188

Dintenfass, L. P., Bartell, D. P. and Scott, M. A. (1987). Predicting resurgences of western flower thrips (Thysanoptera: Thripidae) on onions after insecticide application in the Texas High Plains. *J. Econ. Ent.*, **80**, 502–506

Douglass, J. R. (1933). Habits, life history, and control of the Mexican bean beetle in New Mexico. *US Dep. Agric. Tech. Bull.* 376

Epps, J. M. and Duclos, L. A. (1970). Races of soybean nematode in Missouri and Tennessee. *Plant Dis. Rep.*, **54**, 319–320

Feakin, S. D. (1973). *Pest Control in Groundnuts*, 3rd edn. PANS Manual No. 2, Centre for Overseas Pest Research, London

Franklin, M. T. (1951). *The Cyst Forming Species of Heterodera*. A Technical Manual of the Commonwealth Bureau of Agricultural Parasitology. Commonwealth Agricultural Bureaux, England, pp. 1–147

Franssen, C. J. H. (1954). The biology and control of *Contarinia pisi* Winn. *Meded. Inst. Plziektenk. Onderz., Wageningen*, 71

Franssen, C. J. H. and Huisman, P. (1958). The biology and control of *Thrips angusticeps* Uzel. *Versl. Landbouwk. Onderz. Rijkslandb. Proefstn.*, **64**, 1–104

Fronk, W. D. (1978). Vegetable crop insects. In *Fundamentals of Applied Entomology*, 3rd edn (ed. R. E. Pfadt). Macmillan, New York

Funderburk, J. E., Herzog, D. C. and Lynch, R. E. (1987). Seasonal abundance of lesser cornstalk borer (Lepidoptera: Pyralidae) adults in soybean, peanut, corn, sorghum, and wheat in northern Florida. *J. Entomol. Sci.*, **22**, 159–168

Geissler, K. (1966). Studies on the pea gall gnat. *Arch. Pflanzenschutz.*, **2**, 39–75

Gough, H. C. (1955). *Thrips angusticeps* attacking peas. *Plant Pathol.*, **4**, 53

Gross, H. R. and Harlan, D. P. (1975). Evaluation of preventative adulticide treatments for control of white fringed beetles. *J. Econ. Entomol.*, **68**, 366–388

Hagedorn, D. J. (1984). *Compendium of Pea Diseases*. American Phytopathological Society, St. Paul, Minnesota, pp. 25–28

Hammond, R. B., Pedigo, L. P. and Poston, F. L. (1979). Green cloverworm leaf consumption on greenhouse and field soybean leaves and development of a leaf consumption model. *J. Econ. Entomol.*, **72**, 714–717

Hamon, N. and Bardner, R. (1983). Populations of *Sitona lineatus* on field beans. *Proc. 10th Int. Congr. Plant Prot.*, Vol. 1, p. 162

Harlan, D. P. and Gross, H. R. (1976) Yield of peanuts in soil treated with insecticides for control of white fringed beetle larvae. *J. Georgia Entomol. Soc.*, **11**, 126–130

Hill, D. S. (1983). *Agricultural Insect Pests of the Tropics and Their Control*. Cambridge University Press, Cambridge

Hori, S. (1916). Phytopathological notes. 5. Sick soil of soybean caused by a nematode. *J. Plant Prot.*, **2**, 927–930 (in Japanese)

Howard, N. F. (1922). The Mexican bean beetle in the southeastern U.S. *J. Econ. Entomol.*, **15**, 265–275

Howard, N. F. and Landis, B. J. (1936). Predators and parasites of the Mexican bean beetle in the United States. *U.S. Dep. Agric. Circ.* No. 418

Hutchins, S. H., Higley, L. G. and Pedigo, L. P. (1988). Injury equivalency as a basis for developing multiple-species economic injury levels. *J. Econ. Entomol.*, **81**, 1–8

Inagaki, H. (1979). Race status of five Japanese populations of *Heterodera glycines*. *Jpn J. Nematol.*, **9**, 1–4

Isely, D. (1930). The biology of the bean leaf beetle. *Arkansas Agric. Exp. Stn Bull. 248*

Jones, F. G. W. and Jones, M. G. (1984). *Pests of Field Crops*. Edward Arnold, London

Jones, M. G. K. and Dropkin, V. H. (1975). Scanning electron microscopy of syncytial transfer cells induced in roots by cyst nematodes. *Physiol. Plant Pathol.*, **7**, 259–263

King, J. M. and Biddle, A. J. (1973). Field tests with insecticides for the control of bean seed fly. *Proc. 7th Br. Insect. Fung. Conf.*, Vol. 3, pp. 567–572

Kogan, M. and Kuhlman, D. E. (1982). Soybean insects: identification and management in Illinois. *University of Illinois Agric. Exp. Stn Bull.* No. 773

Kogan, M., Ruesink, W. G. and McDowell, K. (1974). Spatial and temporal distribution patterns of the bean leaf beetle, *Ceratoma trifurcata* (Forster), on soybeans in Illinois. *Environ. Entomol.*, **3**, 607–617

Kogan, M. and Turnipseed, S. G. (1987). Ecology and management of soybean arthropods. *Ann. Rev. Entomol.*, **32**, 507–538

Lentz, G. L. and Pedigo, L. P. (1975). Population ecology of parasites of the green cloverworm in Iowa. *J. Econ. Entomol.*, **68**, 301–304

Lewis, T. and Sturgeon, D. M. (1978). Early warning of egg hatching in pea moth *Cydia nigricana*. *Ann. Appl. Biol.*, **88**, 199–210

Lynch, R. E. (1984). Damage and preference of lesser cornstalk borer (Lepidoptera: Pyralidae) larvae for peanut pods in different stages of maturity. *J. Econ. Entomol.*, **77**, 360–363

Lynch, R. E., Garner, J. W. and Morgan, L. W. (1984a). Influence of systemic insecticides on thrips damage and yield of Florunner peanuts in Georgia. *J. Agric. Entomol.*, **1**, 33–42

Lynch, R. E., Klun, J. A., Leonhardt, B. A., Schwarz, M. and Garner, J. A. (1984b). Female sex pheromone of the lesser cornstalk borer, *Elasmopalpus lignosellus* (Lepidoptera: Pyralidae). *Environ. Entomol.*, **13**, 121–126

Meier, W. (1965). The occurrence and control of pea gall midge (*Contarinia pisi*) in threshed peas. *Mitt. Schweitz. Landw.*, **13**, 43–50

Noel, G. R. and Stanger, B. A. (1982). Estimation of *Heterodera glycines* populations and yield reduction in soybean. *J. Nematol.*, **14**, 461

O'Keeffe, L. E. and Homan, H. W. (1984). *Pea Weevil and Its Control*. Current Information Series No. 475, University of Idaho, Agricultural Experiment Station, USA, pp. 1–2

O'Keeffe, L. E., Homan, H. W. and Schotzko, D. (1984). *The Pea Leaf Weevil*. Current Information Series No. 227, University of Idaho, Agricultural Experiment Station, USA, pp. 1–4

Oostenbrink, M., s'Jacob, J. J. and Kuiper, K. (1957). Tagetes als Feindpflanzen von Pratylechus. *Arten. Nematalogica*, Suppl. 2, 424–433

Ostlie, K. R. and Pedigo, L. P. (1985). Soybean response to simulated green cloverworm (Lepidoptera: Noctuidae) defoliation: Progress toward determining comprehensive economic injury levels. *J. Econ. Entomol.*, **78**, 437–444

Pedigo, L. P., Hutchins, S. H. and Higley, L. G. (1986). Economic injury levels in theory and practice. *Ann. Rev. Entomol.*, **31**, 341–368

Pedigo, L. P., Pitre, H. N., Whitcomb, W. H. and Young, S. Y. (1983). Assessment of the role of natural enemies. In *Natural Enemies of Arthropod Pests* (ed. H. N. Pitre). Southern Coop. Ser. Bull. 285

Reddy, D. V. R., Amin, P. W., McDonald, D. and Ghanekar, A. M. (1983). Epidemiology and control of groundnut bud necrosis disease and other diseases of legume crops in India caused by tomato spotted wilt virus. In *Plant Virus Epidemiology* (ed. R. T. Plumb and J.

M. Thresh). Blackwell Scientific, Oxford, pp. 93–102

Reddy, D. V. R. and Wightman, J. A. (1989). Tomato spotted wilt virus: thrips transmission and management. *Adv. Disease Vector Res.*, **5**, 203

Riggs, R. D., Hamblen, M. L. and Rakes, L. (1981). Infra-species variation in reactions to hosts in *Heterodera glycines* populations. *J. Nematol.*, **13**, 171–179

Salter, W. J. and Smith, M. J. (1986). Establishment pest and disease control in peas by metalaxyl seed coating. *Proc. Br. Crop Prot. Conf.—Pests and Diseases*, Vol. 3, p. 1093

Slack, D. A., Riggs, R. D. and Hamblen (1981). Nematode control in soybeans: Rotation and population dynamics of soybean cyst and other nematodes. *Arkansas Agric. Exp. Stn Rep. Ser.*, **263**, 1–36

Smith, J. W. and Barfield, C. S. (1982). Management of preharvest insects. In *Peanut Science and Technology* (ed. H. E. Pattee and C. T. Young). American Peanut Research and Education Society, Yoakum, Texas

Smith, J. W. and Holloway, R. L. (1979). Lesser cornstalk borer larval density and damage to peanuts. *J. Econ. Entomol.*, **72**, 535–537

Smith, J. W. and Sams, R. L. (1977). Economics of thrips control on peanuts in Texas. *Southwestern Entomol.*, **2**, 149–154

Smith, K. J. and Huyser, W. (1987). World distribution and significance of soybean. In *Soybeans: Improvement, Production, and Uses*, 2nd edn (ed. J. R. Wilcox). American Society of Agronomy, Madison, Wisconsin

Stalker, H. T. and Campbell, W. V. (1983). Resistance of wild species of peanut to an insect complex. *Peanut Sci.*, **10**, 30–33

Stalker, H. T., Campbell, W. V. and Wynne, J. C. (1984). Evaluation of cultivated and wild peanut species for resistance to the lesser cornstalk borer (Lepidoptera: Pyralidae). *J. Econ. Entomol.*, **77**, 53–57

Tappan, W. B. and Gorbet, D. W. (1979). Relationship of seasonal thrips population to economics of control on Florunner peanuts in Florida. *J. Econ. Entomol.*, **72**, 772–776

Tappan, W. B. and Gorbet, D. W. (1981). Economics of tobacco thrips control with systemic pesticides on Florunner peanuts in Florida. *J. Econ. Entomol.*, **74**, 283–286

Taylor, D. E. (1981). Entomology notes *Hilda patruelis*, the ground-nut hopper. *Zimbabwe Agric. J.*, **78**, 177–178

Thorvilson, H. G. and Pedigo L. P. (1984). Epidemiology of *Nomuraea rileyi* (Fungi: Deuteromycotina) in *Plathypena scabra* (Lepidoptera: Noctuidae) populations from Iowa soybeans. *Environ. Entomol.*, **13**, 1491–1497

Thygeson, T. (1971). Pea midge *Contarinia pisi* and other harmful insects in pea crops. *Tidsskr. PlAvl*, **75**, 825–842

Turnipseed, S. G. and Kogan, M. (1976). Soybean entomology. *Ann. Rev. Entomol.*, **21**, 247–282

Van Duyn, J. W., Turnipseed, S. G. and Maxwell, J. D. (1971). Resistance in soybeans to the Mexican bean beetle. I. Sources of resistance. *Crop Sci.*, **11**, 572–573

Vea, E. V., Webb, D. R. and Eckenrode, C. J. (1975). *Seedcorn Maggot Injury*. New York's Food and Life Sciences Bulletin No. 55. New York State Agricultural Experiment Station, Cornell, USA

Wall, C., Garthwaite, D. G., Greenway, A. R. and Biddle, A. J. (1986). Prospects for pheromone monitoring of the pea moth, *Cydia nigricana* (F), in vining peas. *Aspects Appl. Biol.*, **12**, *Crop Protection in Vegetables*, 117–125

Wall, C. and Greenway, A. R. (1981). An effective lure for use in pheromone monitoring traps for the pea moth *Cydia nigricana*. *Plant Pathol.*, **30**, 73–76

Walters, H. J. (1964). Transmission of bean pod mottle virus by bean leaf beetles. *Phytopathology.*, **54**, 240

Walters, H. J. and Lee, F. N. (1969). Transmission of bean pod mottle virus from *Desmodium paniculatum* to soybean by the bean leaf beetle. *Plant Dis. Rep.*, **53**, 11

Weaving, A. J. (1980). Observations on *Hilda patruelis* Stal. (Homoptera: Tettigometridae) and its infestation of the groundnut crop in Rhodesia. *J. Entomol. Soc. S. Africa*, **43**, 151–167

Whitehead, A. G., Tite, D. J., Fraser, J. E. and French, E. M. (1974). Control of pea cyst nematode. *Ann. Appl. Biol.*, **78** (3), 331–335

Wightman, J. A. and Amin, P. W. (1988). Groundnut pests and their control in the semi-arid tropics. *Trop. Pest Mgmt*, **34**, 218–226

Wightman, J. A., Dick, K. M., Ranga Row, G. V., Shanower, T. G. and Gold, C. G. (1990).

Pests of groundnut in the semi-arid tropics. In *Insect Pests of Food Legumes* (ed. S. R. Singh). Wiley, Chichester

Winstead, N. N., Skotland, C. B. and Sasser, J. N. (1955). Soybean cyst nematode in North Carolina. *Plant Dis. Rep.*, **39**, 9–11

Wolf, R. A., Pedigo, L. P., Shaw, R. A. and Newsom, L. D. (1987). Migration/transport of the green cloverworm, *Plathypena scabra* (F.) (Lepidoptera: Noctuidae), in Iowa as determined by synoptic scale weather patterns. *Environ. Entomol.*, **16**, 1169–1174

CHAPTER 7

Pests of Monocotyledon Crops

R. W. Straub and B. Emmett

Sweet Corn

7.1 INTRODUCTION

During the long and complex evolution of maize, periodic mutations produced kernels with endosperms containing high concentrations of sugars. Selection and cultivation of the resulting sweet genotypes eventually led to the product now known as sweet corn, sweet maize or vegetable corn. The sweetness of *Zea mays* var. *rugosa* L. is controlled by a number of recessive genes that also aid in the broad classification of sweet corn cultivars: e.g. *su*, the standard sweet corns; *se*, corns with enhanced sweetness; and sh_2, the recently developed 'super sweets'. The two latter types maintain sweetness, and therefore quality, far longer than the standard *su*s. Although a warm-season, annual crop, and susceptible to frost, sweet corn can be grown in areas having relatively short growing seasons because the edible portion is harvested and consumed prior to physiological maturity. It is grown world-wide for fresh-market consumption and for processing (canned, frozen cut corn and frozen whole ears). The roots, stems, foliage and fruits are attacked by a number of vertebrate and invertebrate pests, as well as many diseases. Home gardeners may accept a great deal of damage from insects, but for commercial purposes, insect-related factors such as feeding on the kernels, insect by-products, or the actual presence of insects or insect parts, may reduce marketability. Because it affects customer appeal, superficial or 'cosmetic' injury to the ear is cause for rejection or downgrading of a fresh-market product. A greater degree of damage may be allowed for processing sweet corn, because some types of superficial injury are eliminated at the processing plant. Insects also cause indirect or secondary effects through their attractiveness to birds, the introduction of decay organisms or the production of honeydew that provides a substrate for the development of black sooty mould fungi on ears and foliage. Insects may also transmit bacterial, virus and mycoplasma-like diseases that affect yield through growth disorders or plant mortality.

7.2 *APHIDS*

7.2.1 *Rhopalosiphum maidis* **(Fitch): Corn Leaf Aphid (Hemiptera: Aphididae)**
Rhopalosiphum padi **(L.): Oat Bird-cherry Aphid (Hemiptera: Aphididae)**
Sitobion (Macrosiphum) avenae **(F.): Grain Aphid (Hemiptera: Aphididae)**
Schizaphis graminum **(Rondani): Greenbug (Hemiptera: Aphididae)**

Aphids are pests of almost every cultivated crop grown throughout the world. Of the numerous species of aphids that infest members of the plant family Gramineae, the above four species are perhaps the most important pests of both dent (feed) and sweet corn. They have overlapping host ranges and infest maize grown on either side of the equator, from approximately 40°S to 55°N. They are also important pests of wheat, *Triticum aestivum* L., oats, *Avena sativa* L., barley, *Hordeum vulgare* L., and other cereal grains.

Published information on description and life cycle for any single species of this aphid complex on corn is incomplete. Members of this complex have characteristic life histories, but for each there may be notable anomalies. The life cycle of aphids demonstrates a type known as heterogamy or cyclic reproduction, in which there occurs an alternation of asexual generations with a sexual generation (compare peach-potato aphid, Chapter 8, page 263). During a complete cycle they pass through egg, fundatrix, viviparae, sexuparae and sexuales (Palmer, 1952). In general, winter is passed as an egg. In spring eggs hatch into fundatrices, reproduce parthenogenetically and give rise to spring migrants that move onto various grass hosts, where successive generations are produced as colonies until autumn, when sexuparae appear and produce true females and males that mate. The females lay overwintering eggs and die. In warmer climates and protected environments such as greenhouses, the sexual part of the life cycle may be omitted. The eggs and sexual forms, particularly males of *R. maidis*, may be rarely found in nature (Cartier, 1957; Orlob and Medler, 1961).

Soft-bodied, apterous viviparae range in length from 1.3 mm for *R. maidis* to 2.5 mm for *S. avenae*, with average body weights of 130 mg and 355 mg, respectively (Palmer, 1952; Malyk and Robinson, 1971). As with most insects, aphid developmental rate is dependent upon temperature. Feng and Yang (1987) determined that each nymphal instar of *S. graminum* required an average developmental time of 4.8 days at 30°C and 16.8 days at 10°C. So, too, the host upon which a species or individual is feeding influences longevity and fecundity. Foster *et al.* (1988) reported that the longevity of apterous *R. padi* viviparae averaged 27.3 days on wheat and 33.1 days on oats. A single apterous *R. maidis* and *R. padi* vivipara may produce 29 nymphs on corn and 39 nymphs on oats, respectively (Ganguli and Raychaudhuri, 1980; Foster *et al.*, 1988). Seedling corn is virtually immune to colonisation by *R. maidis* and *R. padi*, whereas on sorghum, *Sorghum bicolor* L., reproduction by these two species is greater on seedlings than on older plants (Kieckhefer and Gellner, 1988). The genetic composition of an aphid species is often correlated with its host plant. *R. maidis* collected from barley in the northern hemisphere usually has a 10-chromosome karyotype, whereas examples from corn, sorghum and Johnson grass, *Sorghum halapense* L., from all parts of the world commonly have $2n=8$ (Brown and Blackman, 1988). The existence of biotypes and significant variations in responses of aphids originating from different hosts and geographical areas have been established. Four biotypes of *R. maidis* (Pathak and Painter, 1958) and five biotypes of *S. graminum* (Beregovoy *et al.*, 1988) have been reported. There is also evidence that aphid nutrition, behaviour and form determination are strongly correlated to the physiological state of the host plant. Water shortage, for example, provokes various negative responses such as restlessness, slower reproductive rate or precocious production of winged forms (Kennedy and Stroyan, 1959). Branson and Simpson (1965) reported that on sorghum twice the number of *R. maidis* were found on plants receiving nitrogen than on nitrogen-deficient plants, but that deficient plants produced more alatae. Similar results were shown for *S. graminum* on barley (Riedell, 1989). The response of aphids to water and nitrogen deficiencies may be attributable to changes in physiological or chemical properties of the plant sap (Kennedy and Stroyan, 1959).

Aphids readily move on and off host plants as progeny from fundatrices begin to reproduce and excessive crowding occurs (Bakker and Robinson, 1975). Long-range movement or migration is initiated as alatae, apparently excited owing to physiological factors, actively move towards light, and rise out of the slow-moving air among the host plants into the larger-scale, faster-moving air movements above the plants (Moericke, 1955; Taylor, 1958). Once airborne, they are passively transported as the body of each aphid hangs nearly vertically and the wings beat nearly horizontally (Johnson, 1956). Orlob and Medler (1961) concluded that *S. graminum*, aided by strong southerly winds, could migrate 770 km during a 24 h period. Rose *et al.* (1975) reported that an atmospheric convective layer transported large numbers of *R. maidis* some 400 km, from the northern USA into Canada, with an estimated flight duration of 9 h. Passive transport may be interrupted as actively as it is initiated, and well before total exhaustion occurs. Alightments are elicited by response to background colour hues of orange–yellow–green, especially yellow, and are indiscriminate on good and bad hosts (Moericke, 1955; Kennedy and Stroyan, 1959). Upon alighting, the responses of probing, settling to feed and larviposition are readily reversible if the host is not suitable, and the majority of migrants may again take flight.

Although the individual fecundity of a virginopara is modest, it is the geometric rate of production of generations that gives rise to unparalleled numbers. Newton and Dixon (1988) investigated the fecundity of *S. avenae* sexual morphs and concluded that parthenogenetic reproduction presented a 320:1 numerical advantage over sexual reproduction. The basis for this extraordinary productivity is the aphid's feeding process whereby stylets are inserted into the phloem sieve tubes, effectively tapping into the plant's nutrient sap stream, thus enabling uniquely high rates of ingestion relative to the most voracious of chewing insects (Kennedy and Stroyan, 1959).

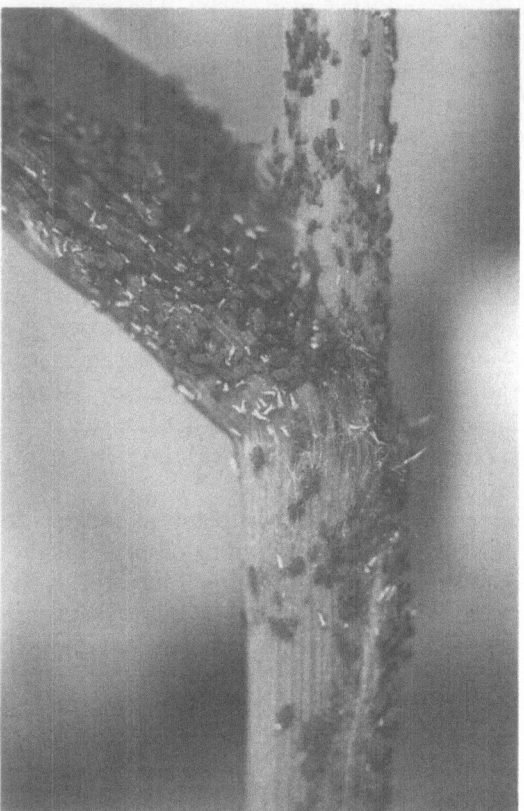

Figure 7.1 Colony of corn leaf aphids, *Rhopalosiphum maidis*, on sweet corn plant. © New York State Agricultural Experiment Station

Because plant sap is a rich source of food, aphids ingest more nutrients than are needed. Excess nutrients are excreted as honeydew, which may comprise free ammonia (Lamb, 1959) and carbohydrates in the form of sucrose or derived monosaccharides (Mittler, 1958). This sticky substance provides a substrate for the development of the black sooty mould fungus, *Alternaria* Nees spp. (Hyphomycetes), which affects plant vigour through a reduction in the efficiency of respiration and photosynthesis. Because of quality standards, sooty moulds on husks and silks may render sweet corn ears unsaleable. Aphids feeding in large numbers (Figure 7.1) cause the breakdown, and release into the sap stream, of materials from tissues not directly attacked. This physiological anomaly may induce premature senescence of tissues, causing localised colour changes—yellowing—and stunting of growth. Although the effects of direct feeding on sweet corn yields have not been well documented, Triplehorn (1959) reported a high incidence of plant barrenness in dent corn due to the confounding effects of *R. maidis* infestations and lack of soil moisture. Similarly, Everly (1960) reported a 53% yield loss in dent corns due to severely infested plants, 44% of which was attributed to plant barrenness. Under high infestation conditions, it is likely that similar effects would occur in sweet corn. The infestation of tassels by high populations of *R. maidis* causes blasting of florets, which seriously interferes with extrusion of anthers and subsequent shedding of pollen (Dicke, 1969). For fresh-market sweet corn, the presence of individual aphids within the silk mass may make ears unsaleable, and the presence of aphid colonies feeding between the layers of husk leaves causes chlorotic or 'bleached' areas that decrease the quality of ears. Similar infestations of processing sweet corn may result in contamination of the product by the presence of whole aphids or parts of aphids in the can.

In many regions of the world, it is as vectors of virus diseases that aphids play their most damaging role (compare peach-potato aphid, Chapter 8, page 263). The two most important virus diseases of sweet corn grown world-wide are maize dwarf mosaic virus (MDMV) and barley yellow dwarf virus (BYDV). All four aphid species transmit both viruses (Orlob and Medler, 1961; Knoke and Louie, 1981) with varying degrees of efficiency. Under natural conditions, the comparative efficiency of a species as a virus vector depends not only upon its relative numbers and probing behaviour, but also on the general condition of the host plant—e.g. whether the plants are chlorotic or previously infected with the virus (Miller and Coon, 1964; Tu and Ford, 1971). MDMV is a flexuous rod particle transmitted in a non-persistent, stylet-borne manner by at least 23 species of aphids, 7 of which transmit it to corn (Knoke and Louie, 1981). Six MDMV strains, designated 'A' through to 'F', have been identified and each may produce slightly different symptoms (Gordon *et al.*, 1981). General symptoms on sweet corn include an initial, light-green-coloured mottle that develops into a mosaic pattern, sometimes with red- or pale-coloured streaks, that may cover the entire surfaces of all leaves. BYDV is an isometric particle transmitted in a persistent, circulative manner by 14 species of aphids (Rochow, 1970). Symptoms include a dark red to purple coloration that develops into an irregular pattern at the tips of leaves, with the oldest leaves showing symptoms first. Some interveinal streaking and malformation of leaves may occur. Under conditions favourable for infection and disease development, either virus may cause severe stunting, poor ear development and barren plants. In south-western

England and western France BYDV infections of sweet corn crops have ranged up to 30% (Pearson and Robb, 1984; Henry and Dedryver, 1989). In eastern USA results from plots sown sequentially from June 6 to July 20 showed that MDMV infections ranging from 7% to 100% caused yield reductions of 10–88%, respectively (Straub and Boothroyd, 1980). In western USA massive flights of *S. graminum* into late-season sweet corn fields contributed to extreme levels of MDMV infection and 75% reductions in yield (Forster *et al.*, 1980).

The seasonal occurrence of alate aphid flights can be monitored by suction traps (Taylor, 1977), by wind traps (Orlob and Medler, 1961), by vertical boards painted yellow in colour and coated with a sticky substance (Blair, 1970) or by pans, yellow in colour, containing water (Moericke, 1950). Evans and Medler (1967), employing water-pan traps, concluded that aphid colonies on sweet corn tended to be largest approximately 10 days preceding large collections of alatae in the traps. Once aphid populations have been established, management can be accomplished by applications of insecticidal emulsions, often in combination with treatments used to manage lepidopteran pests. Granular formulations of systemic insecticides applied in-furrow can significantly reduce the numbers of aphids per plant (Onazi and Wilde, 1974). Although insecticidal treatments may reduce the spread of circulative viruses, they are ineffective against the spread of stylet-borne viruses, because transmission requires such a short period of time that the effects of toxins are nullified (Zitter and Simons, 1980). Furthermore, insecticidal treatments do not prevent infections by immigrating, viruliferous alatae from outside habitats (Schepers, 1972). Certain insecticides may, in fact, contribute to the spread of virus diseases within crops by altering the normal arresting and probing behaviour of vectors (Shanks and Chapman, 1965; Gonzalez and Rawlins, 1969). Some species of aphids, including *S. graminum*, exude droplets of an alarm pheromone when attacked by predators or when disturbed by various physical stimuli. The odour, produced by *trans*-β-farnesene (Bowers *et al.*, 1972), appears to excite or repel other aphids for up to 60 min at a distance of 1–3 cm from the site of secretion (Nault *et al.*, 1973). They suggested that alarm pheromones might be useful for the management of plant virus diseases. Field trials with periodic releases of *trans*-β-farnesene have not been effective, however, in reducing virus transmission by alatae (Zitter and Simons, 1980). Applications of mineral oils to sweet corn plants have been effective in reducing transmission of MDMV in laboratory tests, but are ineffective in controlling the spread of this disease in field situations (Szatmari-Goodman and Nault, 1983).

Although information on natural control of aphids infesting sweet corn is incomplete, the relative abundance of information derived from research on cereals and sorghum may be applicable. According to Jackson *et al.* (1970), three primary hymenopterous wasp parasitoids have been recovered from mummies (mummified dead bodies of parasitised viviparae) of *S. graminum*, *R. maidis* and *R. padi*: the braconid *Lysiphlebus testaceipes* Cresson; the eulopid *Aphelinus nigritus* Howard; and the eulopid *A. varipes* Forester. *L. testaceipes* is native to North America and *A. varipes* has been recorded from most European countries and Egypt. Four hymenopterous hyperparasitoids from the families Encyrtidae, Pteromalidae and Cynipidae have been recovered from these three species of primary parasitoids. Diurnal arthropod predators such as coccinellids, nabids, syrphids and anthocorids may account for significant reductions of aphid numbers (Ohnesorge, 1988). Rice and Wilde (1988) recorded high incidences of predation in sorghum and wheat in the USA by the coccinellids *Coleomegilla maculata lengi* Timberlake, *Hippodamia convergens* Guerin and *H. tredecempunctata tibialis* Say. Malyk and Robinson (1971) reported that *H. convergens* and *H. tredecempunctata tibialis* were more effective as predators of grain aphids in Canada than other species of coccinellids. Insecticides are generally incompatible with efforts to establish or maintain an equilibrium between aphid pests and their natural enemies. For example, applications of pyrethroid insecticides to control European corn borer, *Ostrinia nubilalis* Hubner, contribute greatly to outbreaks of aphids in sweet corn (Moreau, 1987; Cananettes, 1988; Ohnesorge, 1988).

Numerous dent corn genotypes having some resistance to the development of aphid colonies have been found (Neiswander and Triplehorn, 1961; Milinko *et al.*, 1983; Rabichuk, 1985), but similar searches for resistant sweet corns have been uncommon. Huber and Stringfield (1942) found a high correlation between aphid and European corn borer infestations of dent corn strains and suggested that the relationship could be used to index plants for susceptibility to the lepidopteran pest. Long *et al.* (1977) reported up to 21% mortality in *R. maidis* that had fed on an artificial diet containing 2,4-dihydroxy-7-methoxy-1,4-benzoxanzin-3-one (DIMBOA), an active agent in the resistance of corn to *O. nubilalis* (Klun *et al.*, 1967). The correlation of DIMBOA concentrations and resistance to aphids has been further substantiated by Beck *et al.* (1983), Zungia *et al.* (1983) and Awadallah and Hanna (1985). Because of the shortcomings of chemical and biological control methods to control aphid vectors, plant resistance is generally recognised as the most promising strategy for the suppression of virus diseases of sweet corn. Most resistant sources, however, are derived from dent corn and efforts to incorporate resistance factors into sweet corn have been complicated by the difficulty of retaining the high ear-quality factors necessarily associated with a marketable sweet corn product. In the USA there has been evident progress in the development of MDMV resistance. Dale *et al.* (1982) evaluated numerous sweet corn hybrids and found a wide range of resistance, but concluded that there was a general lack of quality in the higher-yielding hybrids. Straub (1984) reported that commercial hybrids were generally quite susceptible to MDMV, but the best experimental MDMV-resistant hybrids yielded up to 71% marketable ears in spite of severe disease incidence.

Generally speaking, moderate infestations of sweet corn by any of the four aphid species do not appreciably affect yield. As secondary pests or as potential contaminants, moderate to severe infestations that occur close to harvest are important and usually require control measures. Methods to combat the spread of virus diseases are at present limited to the use of resistant cultivars and the avoidance of late-season plantings.

7.3 *MOTHS*

7.3.1 *Agrotis* Ochsenheimer spp., *Euxoa* Huebner spp.: Cutworms (Lepidoptera: Noctuidae)

Discussed in Chapter 9.

7.3.2 *Helicoverpa* Heliothis Hardwick spp.: Corn Earworm Complex (Lepidoptera: Noctuidae)

Taxonomically, *Helicoverpa* spp. is called the corn earworm complex, consisting of *H. zea* Boddie in North America, and *H. armigera* Hübner, *H. assaulta* Gn. and *H. punctigera* Wallengren in the old world. The corn earworm (also tomato fruitworm, cotton bollworm, gram podworm, etc.) is a polyphagous pest that attacks a wide range of outdoor crops throughout the tropical and temperate regions of the world. Primarily a pest of numerous vegetable crops, other hosts include: cotton; field maize; tobacco; grain sorghum; and a variety of ornamental plants.

Insects of the family Noctuidae undergo complete metamorphosis. The four-winged adult moth is approximately 30 mm in length and the anterior wings are covered with tannish-brown-coloured scales and marked with dark-grey, irregular lines. The irregular lines often shade into an olive green colour. Distinctive eyes are a light-green colour. The moths are relatively strong fliers, dispersing widely within areas containing host plants. They can also be dispersed by winds accompanying weather fronts and be carried great distances (Rainey, 1973). Preparatory to copulation, females attract males by emitting sex pheromones (Kou and Chow, 1987). Females oviposit single, white-coloured hemispherical eggs, 0.75 mm in diameter, on the foliage of growing maize plants, but more commonly on the silks of developing ears, to which they are attracted by volatile components (Flath *et al.*, 1978; Cantelo and Jacobson, 1979). Eggs hatch in 7–10 days at 18°C and in 2.5 days at 28°C (Hminina, 1981; Ongoren *et al.*, 1977). Average egg deposition on maize silks by *H. zea* is approximately 5.5 per day, but can range as high as 35 per day over the 7–14 day life of an individual female moth (Phillips and Barber, 1936). Egg density is significantly correlated with sweet corn maturity stage and proximity to the nearest previous field of sweet corn (Wiesenborn and Trumble, 1988). Upon hatching, neonate larvae are negatively phototaxic and crawl away from light to seek refuge under leaves, in whorls or within silk masses. The mature larva is approximately 37 mm long, with lengthwise, alternating, light- and dark-coloured stripes (Figure 7.2). Larvae vary greatly in colour, ranging from light green or pink to brown, the colour phases being

Figure 7.2 Corn earworm, *Helicoverpa* sp., in ear of sweet corn. © New York State Agricultural Experiment Station

influenced by dietary carotenoids (Ramos and Morallo-Rejesus, 1978). Generally, larvae pass through five moults in 26 and 14 days at 22 °C and 28 °C, respectively (Singh and Singh, 1977). However, average duration of the larval stage is influenced by diet and ranges from 15 days to 31 days on soybean and corn leaf roll diets, respectively (Gross and Young, 1977; Doss, 1979). Physiologically mature larvae descend to the soil and prepare tunnels and earthen pupation cells 5–10 cm below the surface (Barber and Dicke, 1939). The larvae metamorphose into copper-coloured pupae approximately 20 mm long. Duration of the pupal stage varies with temperature and time of season, lasting 19 days at 22 °C to 10.5 days at 28 °C (Stoeva, 1973). *Helicoverpa* spp. pupae rarely survive prolonged overwintering temperatures below −12 °C (Barber and Dicke, 1939). The temperature over winter determines whether *Helicoverpa* spp. are indigenous or migratory pests. Barber and Dicke (1939) showed that *H. zea* could consistently overwinter at 40° N in the USA. Overwintering at more northerly latitudes would be unusual. The number of generations completed per year is also dependent on latitude, and subsequently temperature. In the Cotton Belt of the USA the number of *H. zea* generations ranges from 6 at approximately 31° N to 3 at approximately 38° N (Dicke, 1939).

Because of the propensity of *Helicoverpa* spp. to oviposit on sweet corn during peak silking periods, the greatest economic damage results from larval feeding at this plant growth stage. Extensive feeding on the silk itself can cause failure of kernel development over the distal 1–5 cm of the cob (Vargas and Nishida, 1978). More commonly, larval feeding on developing or 'milk-stage' kernels causes an unsightly ear that reduces marketability. A significant amount of damage to processing sweet corn can be removed at the processing plant. Corn earworm damage on fresh-market sweet corn is relatively more serious. Sweet corn fields in Florida, USA, that reach silking during intensive flight periods may sustain 12–60% damaged ears, even with 20 insecticidal applications (Mitchell, 1978). In more northerly regions where *H. zea* are immigrants and infestations are less severe, fewer insecticidal treatments are necessary. Where severe damage to fresh-market sweet corn occurs, the affected fields are often ploughed under or utilised as ensilage for livestock.

Successful management of corn earworm has been achieved with various organophosphate, carbamate and pyrethroid insecticides applied during the 2–3 week period when the ears of sweet corn are susceptible to damage. After silks have been pollinated and have dried, they are no longer attractive as ovipositional sites for *Helicoverpa* spp. Since corn earworms enter the ear through the silk channel soon after hatching, it is imperative that insecticidal treatments be applied before neonate larvae enter the silk channel, where they are protected from toxin. The proper timing of applications is aided by a number of techniques: (1) regular scouting of fields, as in most pest management programmes (Straub and VanKirk, 1987); (2) black-light-trap networks (Prostak, 1973); (3) pheromone-trap networks (Wiesenborn and Trumble, 1988); and (4) correlations of likely growth of insect populations with degree-days (Scott, 1987). In general, because of high initial toxicity and field persistence, the pyrethroids (permethrin, cypermethrin, fenvalerate, esfenvalerate, etc.) have become the most widely used corn earworm insecticides (Hamilton and Muirhead, 1981; Linduska *et al.*, 1988). The continual use of broad-spectrum insecticides for the control of *Helicoverpa* spp. has caused secondary resurgence of other sweet corn pests, notably two-spotted spider mites, *Tetranychus urticae* Koch (Pike and Allison, 1987) (see page 311). Because toxicants are widely employed to manage *Helicoverpa* spp. on sweet corn and on many other crops, physiological resistance to the pyrethroids and other insecticides has become problematic throughout the world. Future control programmes will have to address the management of resistance.

Insecticides containing strains of the bacterium *Bacillus thuringiensis* Berliner are efficacious (McLeod, 1989), but these biological insecticides are generally more effective when used in conjunction with biological control techniques, or with synthetic insecticides. Similarly, commercially produced nuclear polyhedrosis viruses are efficacious against *H. zea*, but repeated applications at short intervals are necessary for adequate control (Oatman *et al.*, 1970). In areas of consistently high infestation levels, biological insecticides are less effective than synthetic insecticides when used on comparable application schedules.

Infestations of a particular sweet corn field may be greatly influenced by planting date—i.e. plantings made early in the season sustain less damage from corn earworm than do later plantings. Early seeding is a more effective cultural practice in areas where *Helicoverpa* spp. populations are migratory than in areas where they overwinter. Altieri (1981) showed that *H. zea* infestations of tomatoes were reduced in fields that contained or were bordered by weeds. In maize ecosystems the insect was not greatly affected by weed diversity, and only strip-cropping of maize with soybeans significantly reduced corn earworm damage (Altieri and Whitcomb, 1980).

A number of biological control agents are active against corn earworm on sweet corn, and can be used to augment conventional management systems that rely primarily on insecticides. Oatman (1966) reported 38% parasitism of *H. zea* eggs by the hymenopterous wasp *Trichogramma pretiosum* Riley. At this rate of parasitism, however, 99% of the ears were infested with an average of 3 corn earworm larvae per ear. Under favourable summer conditions in Hawaii, USA, egg parasitism by *T. chilonis* Ishii reached 95%, but averaged only 32% during the remainder of the season (Vargas and Nishida, 1982). In China field releases of *T. pretiosum* and *T. evanescens* Westw. provided egg parasitism rates of 75% and 72%, respectively (Yin and Chang, 1987). The most effective release rates of egg parasitoids for control of corn earworm in sweet corn are not well documented, but considerable work has been done on tomatoes. Natural agents that parasitise larvae may also be effective control factors. In the USA *H. zea* larvae collected from crops of sweet corn yielded 0.8–2.9% parasitisation by: the ichneumonid wasps *Campoletis* Foerster sp., *Hyposoter* Foerster sp. and *Meloborus fuscifemora* Graf; the braconid wasp *Chelonus texanus* Cresson; and the tachinid flies *Eucelatoria armigera* Coquillett and *Lespesia archippivora* Riley (Oatman and Platner, 1970). Mortality of eggs and larvae is also caused by predacious insects. Common predators include: the anthocorid bug *Orius insidiosus* Say; the coccinellid beetle *Coleomegilla maculata* DeGeer; the lacewing fly *Chrysopa oculata* Say; the geocorid bug *Geocoris punctipes* Say; the nabid bug *Nabis ferus* L.; and several species of syrphid fly larvae (Oatman, 1966; Harrison, 1960). Generally, parasitisation of larvae is a minor factor in the overall natural control of the corn earworm on sweet corn when compared with parasitisation and predation of eggs (Oatman and Platner, 1970). Effective management of *Helicoverpa* spp. on sweet corn by biological agents necessitates minimal usage of insecticides, as most insecticides are generally incompatible with efforts to establish or maintain an equilibrium between pest and parasitoid or predator.

The development of resistant cultivars of sweet corn has received considerable research emphasis since Collins and Kempton (1917) reported correlations between husk length and decreased damage by corn earworms. Husk tightness and 'silk balling' are also related to the resistance of some genotypes and hybrids (Walter, 1962; Widstrom *et al.*, 1970). The silks of Zapalote Chico (Pl 217413), a dent corn line from Mexico, have antibiotic properties (Walter, 1962; Straub, 1968). The antibiotic factor in silks was later identified as maysin, a flavone glycoside (Waiss *et al.*, 1979). Efforts to transfer and utilise this antibiotic mechanism of resistance from dent corn to sweet corn have been generally unsuccessful, primarily because sweet corn quality factors are too greatly altered. Within sweet corn cultivars, however, there are wide-ranging degrees of susceptibility to corn earworm damage. McMillan *et al.* (1977) evaluated 61 hybrids and found that Silver Queen and Burpee Snow Cross sustained significantly less ear damage than other hybrids, and that damage was not correlated with husk tightness or husk extension. Story *et al.* (1983) also reported that Silver Queen possessed resistance to corn earworm. At present, the degree of resistance found in these cultivars of sweet corn is not sufficient to eliminate the need for supplementary insecticidal treatments.

Current methods employed for the management of *Helicoverpa* spp. on sweet corn generally rely on multiple insecticidal treatments to maintain damage below threshold levels. Owing to the severity of infestations in many areas where this vegetable crop is grown in monoculture, insecticides will continue to be a necessary and major production tool. Future pest management strategies are likely to integrate the use of computer-based biological monitoring and pest event scheduling systems (Gage *et al.*, 1982); early planting dates that allow ears to ripen before peak infestation periods; cultivars with some degree of tolerance; as well as indigenous or released parasitoids and predators that are allowed to proliferate through the judicious use of selective insecticides.

7.3.3 *Spodoptera frugiperda* J. E. Smith: Fall Armyworm (Lepidoptera: Noctuidae)

The genus *Spodoptera* Guenée contains three species that are major pests of agricultural crops throughout the world: *S. exigua* Hubner, known as the beet armyworm in America, but occurring throughout the world (see page 35); *S. exempta* Walker, an African armyworm that also occurs in North America; and *S. frugiperda*, the fall armyworm (Rose *et al.*, 1975). The fall armyworm is a polyphagous pest that attacks over 60 wild and cultivated plant species. Cultivated host crops, other than sweet corn, attacked by the fall armyworm include: alfalfa, Bermuda grass, cotton, millet and soybean. It is indigenous to tropical–subtropical areas of North, Central and South

America, some of the West Indian Islands, and the Gulf states of the USA (Wood *et al.*, 1979). Although the fall armyworm has no diapause mechanism and, thus, rarely overwinters at temperatures less than 10°C, it often damages cultivated crops at locations far north of overwintering sites. Large numbers of moths may be transported great distances on an atmospheric convective layer (1500–2000 m altitude): the origin of a sudden infestation of fall armyworm moths occurring at Sault Ste. Marie, Ontario, Canada (approximately 45° N) was traced, by 'back-track' analysis of synoptic weather data, to northern Mississippi, USA (approximately 35° N), some 1600 km to the south, a distance which was travelled by the moths in 30 h (Rose *et al.*, 1975). However, short-range migrations account for the general dissemination of *S. frugiperda* throughout areas in which it is a pest of sweet corn. Adults migrate an uncertain distance, alight and produce a generation. The subsequent adult generation migrates further north and the reproduction process is repeated. In this fashion, the moths may spread approximately 580 km per generation (Luginbill, 1928).

Insects of the family Noctuidae undergo complete metamorphosis. The four-winged adult fall armyworm moth is approximately 20 mm in length, and 8 mm wide when the wings are held in the resting position. Anterior wings of the female moth are dark grey in colour, are mottled with lighter and darker splotches in a somewhat mosaic pattern and have a noticeable whitish spot near the extreme tip of each wing. The male moth has anterior wings that are a lighter grey in colour, with a more distinctive mosaic pattern. The posterior wings of both sexes are coloured light grey without distinctive markings. The nocturnal moths feed on the nectar of various plants. After feeding, the female moth sits near the top of vegetation or crop canopy, extends her ovipositor and initiates the process of reproduction by emitting a sex pheromone, (*Z*)-9-dodecen-1-ol acetate (Sekul and Sparks, 1976; Sparks, 1979). Mated females each deposit 7–9 egg masses that average 143–243 eggs per mass (Luginbill, 1928). The moth's wing scales, loosened during the act of oviposition, usually cover each egg mass, making it appear as though covered with mould. On sweet corn plants egg masses are most abundant on the underside of leaves in the region proximal to the stalk (Thompson and All, 1984), but when populations of mated females are high, oviposition may occur indiscriminately over most of the plants (Sparks, 1979). The rate of egg development is dependent upon prevailing temperatures: in south-eastern USA the time from oviposition to eclosion varies from 3 days during July and August to 3.5 days during September and 6 days during October (Luginbill, 1928).

Upon hatching, each colony of neonate fall armyworm larvae feeds for a short time at the oviposition site, and then seeks shelter among the whorl-leaves of young sweet corn plants. On mature plants the larvae enter the ears through the silk channels. In general, larvae pass through 5 moults in 14–18 days at 27°C (Luginbill, 1928). Fully grown larvae are approximately 25 mm in length. The body colour ranges from light tan to nearly black, with three whitish dorsal and subdorsal stripes, which are very prominent on the prothoracic plate, running the entire length of the body. Laterally, there is a dark-coloured stripe and a somewhat wavy, yellow-coloured stripe running the entire larval body length. The larval head capsule is dark in colour, with a prominent epicranial suture appearing as an inverted 'Y'. Physiologically mature larvae descend to the soil, burrow to a depth of 2.5–7.5 cm and metamorphose into copper-coloured pupae approximately 17 mm long. Duration of the pupal stage is greatly influenced by mean soil temperature: in south-eastern USA the number of days required for emergence varies from 6–13 during July to 7–12 during August and 16–27 during September and October (Luginbill, 1928). During warm weather throughout the south-eastern USA, the entire life cycle of *S. frugiperda* requires approximately 28 days. During colder seasons a generation may take as long as 80–90 days (Sparks, 1979).

The primary damage to sweet corn caused by *S. frugiperda* is through direct feeding of larvae on or in the ears (Figure 7.3). Extensive feeding upon the silks results in poor pollination, and damage to kernels or husk

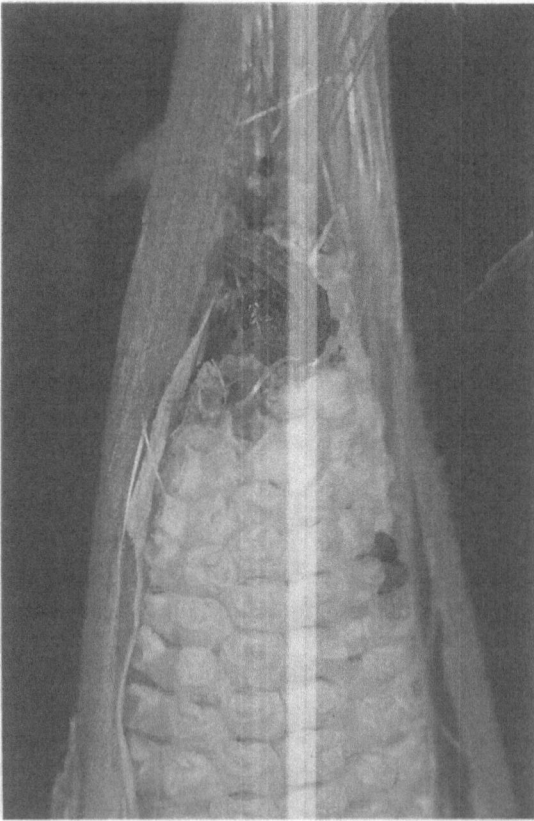

Figure 7.3 Fall armyworm, *Spodoptera frugiperda*, in ear of sweet corn. © New York State Agricultural Experiment Station

leaves causes unsightly ears that reduce marketability. Much of the ear-tip damage to processing sweet corn can be removed at processing plants, but damage on lateral portions is more difficult to detect and eliminate. Any damage to the fresh-market product is poorly tolerated and the acceptance threshold is usually set at 5% damaged ears. Fall armyworm larvae are voracious feeders—during the six stadia of growth, a single larva is able to consume 138 cm^2 of leaf tissue (Luginbill, 1928). Fields containing small plants can be severely damaged or completely consumed by larval infestations (Brett, 1953). Negative effects on grain and silage yields owing to fall armyworm infestations of whorl-stage dent corn are well documented (Straub and Hogan, 1974; Cruz and Turpin, 1983; Hruska and Gladstone, 1988). Similarly, yields of sweet corn may decrease linearly with increasing defoliation (McMillian *et al.*, 1980). As a result, fall armyworm infestations often have to be managed not only during the silking stage, but also during the vegetative stages of plant growth.

Although fall armyworm infestations of sweet corn can be significantly reduced by growing corn in polyculture with beans (*Phaseolus vulgaris* L.) sown 20–40 days earlier than the corn (Altieri *et al.*, 1978), in comparison with other lepidopteran pests, few cultural practices are employed to combat *S. frugiperda*. Successful management has been achieved with various organophosphate, carbamate and pyrethroid insecticides applied to sweet corn ears, starting at pre-silk and continuing until harvest (All *et al.*, 1986; Straub, 1986). This approach often requires that ears be sprayed at frequent intervals to ensure that larvae are continually exposed to fresh insecticidal residues. Foster (1989) showed that maintaining the crop relatively free of fall armyworm infestations during the whorl stage of growth reduces the number of insecticidal treatments needed during the silking period. Since larvae feed deep within the whorl tissues and are thus shielded from direct contact by insecticides and encounters with residues, efforts to maximise the amount of toxicant reaching the site of larval activity are likely to augment control. Because applications of granular formulations effectively place the toxicant near the primary point of feeding, where it persists for a relatively long period of time (Cox *et al.*, 1956), granular treatments are generally more effective than equivalent rates of emulsifiable insecticides (Whalen and Vanderhoef, 1985; Ghidiu *et al.*, 1988).

Effective management of the fall armyworm, particularly with migrant populations, requires locally diligent monitoring. Scouting for larval feeding activity on sweet corn plants, with treatment decisions based on established thresholds (Straub and VanKirk, 1987), remain reliable techniques in most pest management programmes. Eggs are deposited in a non-random fashion and damage is therefore usually localised or clustered within fields with several consecutive plants infested. Under low infestation conditions, the sample size on which to base management decisions is often increased to ensure accuracy of quantitative estimates of infestation. In

contrast to many Noctuidae, *S. frugiperda* adults are relatively unrespon-
sive to black-light traps. Traps baited with fall armyworm pheromone
(Sparks, 1979; Barfield and Stimac, 1980; Starratt and McLeod, 1982) and
sugarline sampling (i.e. a row in the field is treated with sugar-baited
insecticidal solution, and dead moths along the treated row subsequently
counted) (Boyer *et al.*, 1966; Chowdbury *et al.*, 1987) have been employed
to monitor populations of adults in sweet corn fields. At present neither
method presents a defined estimate of the quantitative relationship between moths captured or killed and oviposition. However, Thompson and
All (1984) reported efficient sampling of *S. frugiperda* egg masses by
red-coloured vinyl flags placed within sweet corn crops and egg counts
correlated with subsequent infestation levels.

Luginbill (1928) listed numerous natural enemies of the fall armyworm,
and surmised that local outbreaks are possible only when environmental
factors have reduced natural enemies, permitting the armyworm to develop during early spring. Fifty-three species of hymenopterous parasitoids
from 43 genera and 10 families have been reared from larvae of the fall
armyworm (Ashley, 1979). Ashley (1979) indicated that importations from
Central and South America, primarily of the braconid wasps *Apanteles
marginiventris* Cresson and *Chelonus texanus* Cresson, significantly reduce
overwintering populations of *S. frugiperda* in Florida and Texas, USA.
Maggots of the tachinid fly *Archytas marmoratus* Townsend, a primary
endoparasitoid of noctuid larvae and pupae in North and South America,
have been mechanically extracted from fecund flies, suspended in an
aqueous carrier and successfully reared in a mass production process
(Gross and Johnson, 1985). Applications of 28 extracted tachinid larvae
per sweet corn plant reduced subsequent *S. frugiperda* adult emergence by
67% (Gross *et al.*, 1985). Survival of maggots, however, declines significantly after 72 h in the field (Gross, 1988). Isenhour (1985) and Isenhour
and Wiseman (1987) reported that parasitisation by the ichneumonid wasp
Campoletis sonorensis Cameron caused significant reductions in larval
weights and leaf-area consumption. Agents other than arthropods may also
be effective in biological control. A naturally occurring, specific nuclear
polyhedrosis virus causes varying degrees of larval mortality in the field
(Hamm and Wiseman, 1986). Landazabal *et al.* (1973) reported that a
nematode, *Neoaplectana* Steiner sp. (Rhabditida: Steinernematidae), applied at 4000 larvae per plant, elicited up to 60% mortality of fall
armyworm larvae after 96 h.

Wiseman *et al.* (1966) noted that, from the point of view of studying
insect–plant relationships, the fall armyworm has been studied less than
have other insect pests of sweet corn. Most plant-resistance characteristics
known at present can be traced to Antigua corns and to similar Coastal
Tropical Flints from Antigua and other islands of the West Indies
(Wiseman and Davis, 1979). Scott *et al.* (1977) released the first corn
germplasm with known resistance to *S. frugiperda*: dent corn genotype

MP496. It has Antigua Gp2 in its background and withstands the feeding effects of extremely high larval populations. More recently, both antibiotic and non-preference types of resistance have been reported in the dent corn hybrid MpSWCB4 (Wiseman *et al.*, 1981), and in the silks of Zapalote Chico, a flint type from the dry coastal region of Oaxaca and Chimpas, Mexico (Wiseman and Widstrom, 1986). Silks from Zapalote Chico were previously shown to be antibiotic to larvae of the corn earworm, *Helicoverpa zea* Boddie (Walter, 1962; Straub, 1968) (see page 220). Efforts to transfer resistance to the fall armyworm from dent corn to sweet corn have been unsatisfactory, because numerous sweet corn quality factors are too greatly altered in the process. Within existing sweet corn hybrids, resistance (undefined mechanisms) has been reported in Silver Queen and Bonanza (Mitchell, 1978). Because of extremely high densities of fall armyworm larvae in many production regions, resistance in sweet corn is typically associated with vigorous plant growth and the ability of a cultivar to withstand the stress of severe damage (Brett and Bastida, 1963).

Current methods employed to manage *S. frugiperda* on sweet corn generally rely on multiple insecticidal treatments to maintain damage below threshold levels. Future management programmes are likely to integrate novel classes of long-residual and environmentally safe insecticides: insect growth regulators (Linduska *et al.*, 1985); botanical antifeedants, e.g. azadirachtin from the neem tree, *Azadirachta indica* A. Juss (Redfern *et al.*, 1984); a complex of mass-produced and released parasitoids; sprays of biologically based kairomones that attract parasitoids and predators (Gross *et al.*, 1985); genetically engineered pest-resistant hybrids that derive from closely related species, e.g. *Tripsacum dactyloides* L. (Gama grass) (Wiseman *et al.*, 1967); mass releases of pheromone-communication disruptants, e.g. synthetic acetates (Mitchell *et al.*, 1974); and geographically specific, computer-based pest management models (Barfield and Jones, 1979).

7.3.4 *Ostrinia nubilalis* (Hübner): European Corn Borer (Lepidoptera: Pyralidae)

The European corn borer is a polyphagous pest that attacks over 200 wild and cultivated plant species throughout the world (Hodgson, 1928; Hudon and LeRoux, 1986a). Host crops include: sweet corn; beans; cotton; potatoes; wheat; rhubarb; shoots of fruit trees; etc. The genus, *Ostrinia* Hübner spp., contains four species that have overlapping distributions in Europe, north-west Africa, western Asia and North America: *O. nubilalis*, *O. furnacalis* Guenee, *O. obumbratalis* Led. and *O. zaguliaevi* Matuura and Munroe. The European corn borer, long recognised in Europe as an insect pest of economic importance, was accidentally introduced into North America during the early 1900s and has become a serious pest of sweet corn and many other crops. Dispersion of *O. nubilalis* within a geographical area is relatively slow. It required 21 years to spread approximately 900 miles from the point of introduction into the USA (eastern seaboard) to the western shore of Lake Michigan (Caffrey and Worthley, 1927; Anon., 1989).

Insects of the family Pyralidae undergo complete metamorphosis. The four-winged adult female European corn borer moth is approximately 20 mm in length and the anterior wings are covered with yellowish-brown-coloured scales. Each anterior wing is marked by two diagonal, dark-coloured serrate lines. The slightly smaller male moth has anterior wings that are reddish-brown in colour, with similar, but darker, line markings. The posterior wings of both sexes are lighter in colour than the forewings and are unmarked.

European corn borer moths aggregate in grass habitats near cornfields, where females attract males by emitting a sex pheromone. *O. nubilalis* populations comprise two strains having distinct pheromone communication systems: a *Z* strain using a 97:3 mix of (*Z*)- and (*E*)-11-tetradecenyl acetates, and an *E* strain using a 1:99 blend of (*Z*)- and (*E*)-components. The *Z* strain is widely distributed, representing nearly all populations in Europe and North America, whereas the *E* strain is found primarily in north-eastern USA and Italy (Klun *et al.*, 1973, 1975a; Kochansky *et al.*, 1975; Roelofs *et al.*, 1985). Peak sexual activity is between 2400 and 0100 hours. Female moths normally mate when 48 h old, following which they leave the aggregation site and deposit egg masses of 15–30 eggs on the underside of sweet corn leaves. Each mated female moth is capable of depositing an average of 2 egg masses per night and fecundity ranges from 50 to 190 eggs (Hudon and LeRoux, 1986b; Anon., 1989). Egg development is completed in 5–7 days at average temperatures of 17–20°C (Hudon and LeRoux, 1986a). Upon hatching, neonate larvae are negatively phototaxic and seek refuge in the sweet corn plant's whorls or wander off the plant and perish. Temperature-related climatic variables such as moisture stress and evaporation will cause up to 68% mortality of neonate larvae (Anon., 1989). Fully grown larvae are approximately 25 mm long. The body is flesh-coloured, ranging from light grey to faint pink, with dark-brown spots on each body segment. A distinct reddish-brown-coloured stripe extends the entire length of the larval body (Andaloro *et al.*, 1983). Larvae pass through four moults in 21–25 days at 27°C (Straub, unpublished). Larvae feeding on sweet corn plants pupate inside the stems or ears. The reddish-brown-coloured pupa is approximately 18 mm long. After about 14 days, the imago emerges, to begin another generation.

The number of generations of the European corn borer on sweet corn ranges from one to four per year, depending on the length of season: in North America, for instance, *O. nubilalis* has a single generation in Alberta, Canada, and four generations in south-eastern USA (Anon., 1989; Hudon *et al.*, 1989). However, differences in voltinism among European corn borer populations occur in identical environments, and such variability is known to have a strong genetic basis involving sex-linked recessive inheritance (Showers, 1981). There is also evidence to suggest that voltinism is correlated with photoperiod–temperature interactions (Showers, 1979; Anon., 1985). On the basis of larval diapause characteristics, three biotypes exist in North America: northern univoltine (Minnesota and Quebec), central bivoltine (Iowa, Nebraska and Ohio) and southern multivoltine (Alabama, Georgia and Missouri) (Hudon *et al.*, 1989).

O. nubilalis overwinters as diapausing fifth-instar larvae. Cold-hardy larvae are able to withstand temperatures of −20°C for up to 3 months, despite the formation of ice in their body tissues (Hanec and Beck, 1960). This physiological capability of *O. nubilalis* larvae to withstand low temperatures results in their being distributed over a wide range of latitudes—i.e. approximately 30° N to 55° N (Hudon and LeRoux, 1986a).

The primary damage to sweet corn caused by *O. nubilalis* is direct feeding on or in the ear (Figure 7.4). Because of local, regional or federal requirements, sweet corn for fresh-market has a low insect-damage

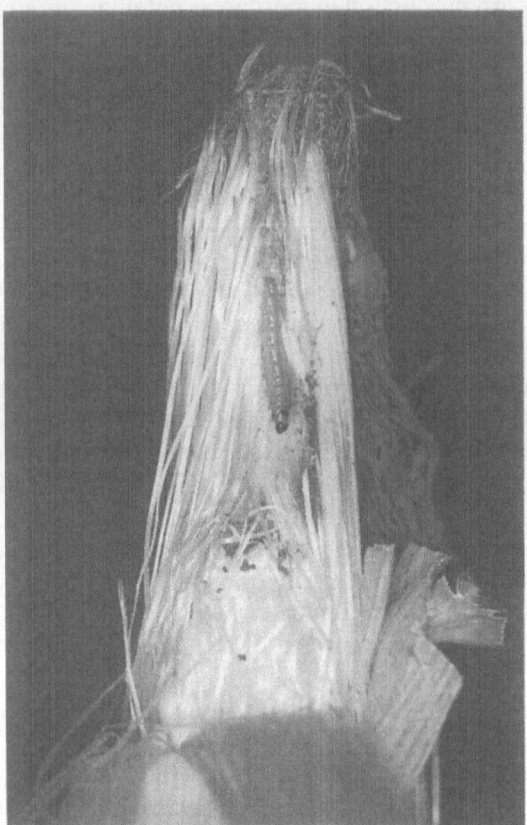

Figure 7.4 European corn borer, *Ostrinia nubilalis*, in ear of sweet corn. © New York State Agricultural Experiment Station

threshold with an acceptance level between 95% and 100% damage-free ears (Straub, 1983). The European corn borer is a particularly serious problem for processors because of the difficulty of detecting infested kernels on the sides of ears and, consequently, the increased likelihood of contamination from insect parts or frass (Anon., 1989). First-generation larvae feed within plant whorls and on leaves, subsequently boring into the stalks and eventually migrating into the shank portion of the ears (Straub, 1977). Straub (1983) found that the principal mode of ear entry was through the silk channel, either by larvae migrating from opening tassels or by direct oviposition on or near the ear. Second- and succeeding-generation larvae feed initially on pollen that accumulates in leaf axils and on sheath and collar tissues, subsequently attacking the ear and the shank before boring into the stem (Hudon *et al.*, 1989). Leaves in the ear region appear to be preferred for oviposition. Windels and Chiang (1975) found that 34% of all egg masses were located on the three leaves in closest proximity to the ear, and 17% were located on the flag leaves of primary and secondary ears.

With sweet corn production in most areas entailing progressive plantings at regular intervals, synchrony between plantings and bivoltine and multivoltine biotypes of the European corn borer may exist at any plant growth stage throughout the season. As a result, management programmes must be exercised on a field-by-field basis, the specific approach being determined by geography and *O. nubilalis* behaviour. Successful management has been achieved with various organophosphate, carbamate and pyrethroid insecticides (Whalen and Breeding, 1988; Bowman and Barry, 1989) applied to sweet corn plants during the whorl stage and at tasselling, or to ears during the approximate 30 day period in which they are susceptible to oviposition and damage. McClanahan and Founk (1972), examining 17 insecticides for ovicidal activity, found that parathion-ethyl was effective against eggs, but otherwise the larva was the most susceptible life stage to all insecticides. Given the propensity of *O. nubilalis* to oviposit on the underside of leaves, and, thus, the low probability that eggs will be exposed to a toxin, larvae are the primary target of insecticidal treatments. Infestation of ears by first-generation larvae depends upon the degree of prior whorl infestation (Straub, 1983). As a consequence, a single treatment of organophosphate or carbamate granules applied to whorls may be as effective as multiple foliar and silk sprays (Straub, 1983, 1986; Whalen and Breeding, 1989). The numbers of adult European corn borers, and, thus, the numbers of egg masses deposited in corn fields, can be greatly reduced by treatment of the adult aggregation sites bordering fields (Showers *et al.*, 1980).

Because changes in *O. nubilalis* behaviour can occur with shifting weather patterns (Chiang *et al.*, 1961) or invasion by new strains (Chiang, 1972), effective management of this pest with insecticides requires the diligent monitoring of local populations. Field-scouting for egg masses and larval activity is the most reliable method in most pest management programmes (Anon., 1985; Straub and VanKirk, 1987). Combined data derived from regular field-scouting, black-light traps and traps baited with European corn borer pheromone have been used in the USA and Europe to determine optimum treatment periods based on damage thresholds (Showers *et al.*, 1974; Katona and Cziklin, 1984; Ferro and Fletcher-Howell, 1985). Temperature summations (heat units, day-degrees, etc.) have been correlated with adult activity (Apple, 1952; Clement *et al.*, 1981; Straub, 1986) and have been employed to detect variable voltinism and identify biotypes (McLeod, 1976; Eckenrode *et al.*, 1983). Anderson *et al.* (1982) utilised post-diapause developmental rate and temperature curves to construct a mathematical model that provided good prediction of all percentiles (10%, 50% and 90%) of spring emergence of adults. They, as well as Eckenrode *et al.* (1983), point out, however, that when weather-driven programmes are applied over large regions, inaccuracies in predictions for localised environments are often unavoidable. In local situations, phenological events of selected woody and herbaceous perennial plants have been used to forecast the occurrence of bivoltine life stages (Straub and Huth, 1976).

Cultural control measures have long been employed as aids to the management of European corn borer. Such methods for the most part modify or manipulate the environment to the disadvantage of the pest. Clean ploughing is the classic tactic recommended for cultural control (Crawford, 1924; Caffrey and Worthley, 1927). Although ploughing may be somewhat effective in reducing overwintering populations within sweet cornfields, the multitude of alternative hosts available to the European corn borer ensures a high degree of survival outside cornfield sites. Intercropping of sweet corn with red clover, *Trifolium pratense* L. (Lambert *et al.*, 1987), potato, *Solanum tuberosum* L. (Anderson *et al.*, 1984), and other susceptible corn hybrids (Derridj *et al.*, 1988) significantly reduces numbers of egg masses. If sweet cornfields are maintained relatively free of grass and dense stands of weeds, and field edges are mowed, *O. nubilalis* adults can be forced into predesignated patches of grass and subsequently killed with non-persistent organic or biological insecticides (Anon., 1989).

Although the intensive production of sweet corn requires regular insecticidal applications to manage *O. nubilalis*, insecticidal programmes could in less intensive production systems be augmented by an array of natural enemies. In North America two species of the hymenopterous wasp, *Trichogramma* Westwood (Trichogrammatidae), occur naturally and parasitise *O. nubilalis* eggs. Generally, rates of parasitism by *T. minutum* Riley are low, a result thought to arise from the wasp's poor synchrony with its host (Hudon and LeRoux, 1986c; Jarvis and Guthrie, 1987). Parasitisation of European corn borer eggs by *T. nubilale* Ertle and Davis ranged from 5% in June to as high as 70% in September (Ertle and

Davis, 1975), but this species of wasp has spread slowly from its original site of discovery in Delaware, USA. Difficulty in the development of mass rearing techniques has stalled commercial efforts to produce *T. nubilale* for mass release. In Europe, however, mass releases of the indigenous egg parasitoid *T. evanescens* Westwood have proved effective against *O. nubilalis*. Neuffer (1982) concluded that releases of between 240000 and 300000 *T. evanescens* per hectare at first appearance of the European corn borer were needed to protect sweet corn ears in heavily infested fields. Subsequently, Hassan *et al.* (1986) reported that 76–89% borer-free ears could be achieved with 150000 parasitised European corn borer eggs per hectare. Similar results have been achieved with mass releases of *T. maidis* Pintureau and Voegele in Austria (Ravensberg and Berger, 1988).

Diurnal arthropods such as coccinellid beetles, chrysopid lacewings, nabid bugs, syrphid flies and anthocorid bugs may account for some degree of reduction in *O. nubilalis* infestations, but there is great variation in predator effectiveness from season to season (Frye, 1972). *Coleomegilla maculata* DeGeer (Coccinellidae) is abundant in some sweet corn eco-systems and accounts for significant egg mortality (Andow and Risch, 1985; Hudon and LeRoux, 1986c), but its effectiveness is decreased by the braconid wasp *Perilitus coccinellae* Shrank, a natural enemy of adult coccinellid beetles (Risherson and DeLoach, 1973). The protozoan *Nosema pyrausta* Paillot (Nosematidae) occurs wherever *O. nubilalis* is found and, at times, produces epizootic infections (Hill and Gary, 1979). It is, however, an obligatory, intracellular parasite which varies in natural incidence each year (Burbutis *et al.*, 1981) and is difficult to disseminate by artificial means (Lewis *et al.*, 1983). Birds are conspicuous predators on insect pests of sweet corn, but are rarely mentioned as biological control agents. In the USA, the downy woodpecker, *Dendrocopos pubescent* L., and the red-winged blackbird, *Agelaius phoeniceus* L., are efficient pre-dators of overwintering *O. nubilalis* larvae (Frye, 1972; Whitman, 1975; Bendell *et al.*, 1981). Red-winged blackbirds can be pests of sweet corn, however, as they are attracted to ears containing *O. nubilalis* larvae and their foraging activity usually causes extensive damage (Straub, 1989).

Research to identify dent corn genotypes resistant to *O. nubilalis* has been extensive. Resistance to leaf feeding by the first generation has been relatively easy to find, while resistance to stalk boring by the second generation has proved more difficult to find (Gracen, 1989). Resistance has been related to the presence of an aglycone (DIMBOA), silica and lignin in leaf tissues (Hudon *et al.*, 1989). Efforts to transfer resistance mechan-isms from dent corn to sweet corn have been generally unsatisfactory, because numerous sweet corn quality factors are too greatly altered in the process.

Current methods employed for the management of the European corn borer on sweet corn typically rely on multiple insecticidal treatments to maintain infestations below established threshold levels. Owing to the severity of infestations in most areas where this vegetable crop is grown in monoculture, insecticides will probably continue as vital components in production systems. Future strategies for management of this very important pest are likely to integrate the use of cultivars having resistance to leaf feeding by first- and second-generation larvae; mass-produced releases of *Trichogramma* spp. parasitoids; selective insecticides having minimal toxicity to natural control agents; inundative releases of synthetic pheromones that disrupt long- and short-range communications between the sexes (Klun *et al.*, 1975a, b); applications of botanical antifeedants such as azadirachtin from the neem tree, *Azadirachta indica* A. Juss (Meisner *et al.*, 1985) or feeding deterrents such as DIMBOA (Robinson *et al.*, 1982); and general usage of computer-based biological monitoring and pest scheduling systems (Gage *et al.*, 1982).

7.4 NEMATODES

7.4.1 *Ditylenchus dipsaci* **(Kuhn) Filipjev: Stem Nematode (Tylenchida: Tylenchidae)**

D. dipsaci is discussed in Section 7.9; see page 248.

Onions and Leeks

7.5 INTRODUCTION

The onion (*Allium cepa* L.) is a vegetable of great antiquity and is now unknown as a wild plant. It is believed to have originated in south-eastern Europe and is traditionally associated with Turkey. It is the most widely grown vegetable in the world and is now cultivated from the coldest areas of the temperate zones to the tropics. Onions are biennial, herbaceous plants with tubular leaves and a swollen, pithy stem base which functions as a drought-resisting organ. When allowed to grow naturally from over-wintered seed, the onion bulb develops during the late spring and summer; the plant then flowers, sets seed and dies in the following year.

The crop is cultivated in many different ways. Seed may be sown in spring or autumn or specially prepared small bulbs (sets) may be planted for development of one or several daughter plants. They may be grown singly or in clumps; harvested immature or fully bulbed; stored for long periods or cooked immediately after harvest. Onions are not highly nutritious but have health-giving properties partly associated with their high sulphur content. They are the most popular flavouring vegetable and may be used either raw or cooked. They have a wide range of culinary uses and are an essential ingredient of many soups, sauces, pickles and preserves. Onions are also cooked as a main-course vegetable. Bulbs and immature (spring) onions are also widely used, uncooked, in salads.

Onions are attacked by many pests that seriously reduce yields by stunting or killing the plants. These include the stem nematode, onion fly, bean seed fly and cutworms. Crop quality is frequently impaired by thrips or aphids.

The closely related leek (*Allium porrum* L.) is less popular than the onion world-wide, being a winter-hardy plant, mainly of Great Britain and Northern Europe. It forms a basis for soups and special leek dishes. Essentially, the same pests attack this crop, but generally to a lesser extent than onions. The most damaging insect pests are onion thrips and cutworms.

7.6 THRIPS

7.6.1 *Thrips tabaci* Lind.: Onion Thrips (Thysanoptera: Thripidae)

Onion thrips are commonly found insects throughout the UK. They infest many crops both outdoor and under protection, including onions, leeks, leaf and root brassicas, spinach and sugar beet. Tomatoes may be attacked as well as marrows and courgettes. The onion thrips is considered to have originated around the eastern Mediterranean, but it is now distributed almost world-wide, extending north as far as Scandinavia, Finland and Canada (Hill, 1985) and south as far as Chile and New Zealand. In warm temperate and tropical areas the crops attacked, in addition to those mentioned above, include garlic, tobacco, cotton, pineapples and legumes. Onion thrips are polyphagous and have been recorded on more than 300 species of plants.

Description

Members of the Thysanoptera are small, four-winged insects, known commonly as thrips, thunder-flies or thunder-bugs. Thrips develop through two nymphal instars, a quiescent prepupal form and one or two more or less immobile pupa-like stages. These developmental stages are of similar general appearance to the adults but without functional wings.

The adult onion thrips is 1–1.3 mm long. The head is rounded, with dark-coloured eyes and short, 7-segmented antennae. The mouthparts consist of paired, tubular, asymmetrical stylets adapted for rasping and sucking. The thorax carries two pairs of long, narrow wings fringed with long hairs. The legs are devoid of obvious hairs, but each has a protrusible bladder at the tip of the tarsus which improves adherence to smooth surfaces. The abdomen is elongated and pointed at the tip. It has a continuous row of setae on the eighth abdominal tergite. At the tip of the female abdomen is a saw-like ovipositor. Females tend to be larger and darker in colour than males. The body colour of adult onion thrips varies from pale yellow to dark brown.

The onion thrips egg is minute (0.3×0.1 mm) and kidney-shaped, and varies in colour from white to yellow. The nymph is 1.4 mm long, similar in appearance to the adult but without wings and almost transparent when newly hatched, with a disproportionately large head. The nymph develops a pale cream or yellow body colour with conspicuous reddish eyes. The prepupa is 0.9 mm long and white in colour and has short wing-sheaths and antennae which are held straight in front of the head. The pupa-like stage is 1.1 mm long and brownish-coloured with longer wing-cases and antennae than in the prepupa. The 'pupal' antennae are held back over the thorax.

Life History

Adult onion thrips emerge from overwintering in early spring, having spent the winter in a quiescent condition in herbage and plant debris. They fly weakly, but may be carried great distances on wind and air currents. On reaching host plants, each female lays up to about 100 eggs in slits made by the ovipositor in the plants' soft tissues. The head end projects slightly to facilitate nymphal emergence. On onions in Italy (Ferrari, 1980) the egg stage lasted 5–10 days and the nymphal stage lasted for 10–20 days. Other workers have suggested shorter nymphal periods, even for cooler climates—e.g. about 5 days in the UK (Edwards and Heath, 1964) or 4–7 days (Hill, 1985). Nymphs puncture plant cells and suck out the contents, moving frequently to new locations to make more feeding incisions. After the second ecdysis, nymphs develop the short wing-sheaths of the prepupal stage. The nymphs usually conceal themselves among debris or in the surface layers of the soil before changing to prepupae but may sometimes remain on host plants. After about 2 days, prepupae change into 'pupae'. The onion thrips has only one pupa-like stage, which remains where the prepupa developed. These resting stages do not feed but are capable of slow movement in response to suitable stimuli—e.g. they tend to avoid light. Prepupae and 'pupae' together require about 6 days for complete development, after which the adults emerge. Thrips are gregarious insects and large numbers are often found together on single leaves. The number of generations varies greatly. In the UK, although each generation may be completed in less than 3 weeks, there are usually two generations only each year (Edwards and Heath, 1964). In Iraq Al-Faisal and Kardou (1986) recorded a higher egg hatch-rate, a shorter life cycle, a longer adult life span and a greater female fecundity on younger than on older cotton plants. Temperatures above 35 °C were more favourable to thrips development than was a temperature of 20 °C. Many workers draw attention to the importance of warm, dry weather for rapid thrips development. Population dispersal and damage decline in response to wet conditions. Ferrari (1980) reports 70% mortality following heavy rainfall in Tuscany, Italy. In warm temperate or tropical climates favourable seasons lead to the development of 10 or more generations a year. In New York State, USA, North and Shelton (1986) found onion thrips overwintering in winter wheat and lucerne. The adult females were capable of ovipositing on these plants the following spring.

Damage

Onion and leek plant tissues attacked by onion thrips develop white or silvery blotches (Figure 7.5). These blotches later turn yellow in colour and

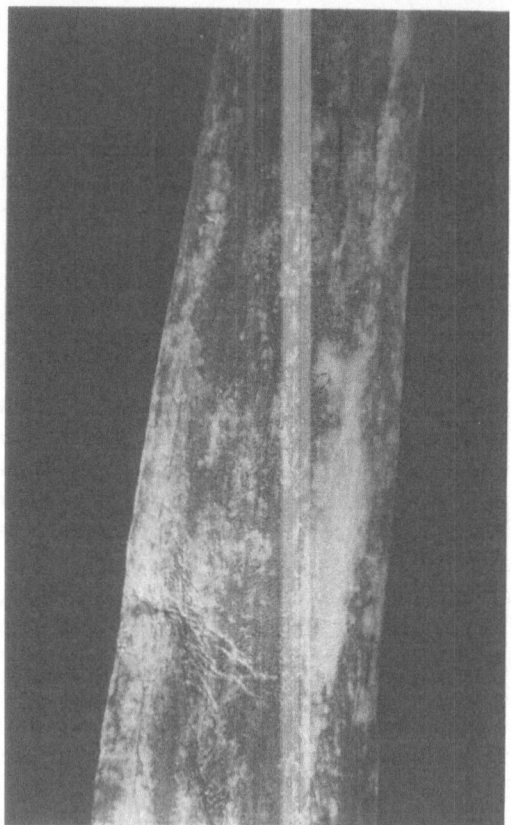

Figure 7.5   Damage to onion stem by the onion thrips, *Thrips tabaci*. © MAFF

affected tissues dry out or become mottled with fungal growth. Leaves may twist, crinkle and curl, or even die if heavily infested. In southern Europe,

India and Africa crops are usually sown or transplanted early, to avoid damaging populations of thrips. In Sudan light infestations of thrips on late transplantings of onions led to yield losses of 40%, while severe attacks reduced yields by almost 60% (Kisha, 1977). In the UK mid-season or autumn sowings of salad or bulb onions and leeks are worst affected when the weather is warm and dry. When young plants are severely damaged, the loss of photosynthetic area can cause death.

The onion thrips is a vector of various diseases in tobacco, tomato, cotton, pineapple and other crops, and Ferrari (1980) suspects that it transmits onion yellow dwarf virus in Italy. In India *T. tabaci* is implicated in the development of 'purple blotch' of onions. The causal organism (*Alternaria porri* (Ellis) Cif (Hyphomycetes)) develops rapidly when foliage is injured by insect feeding (Bhangale and Joi, 1983). In the UK onion thrips can transmit spotted wilt virus of tomatoes (Edwards and Heath, 1964).

When outdoor tomatoes and cucurbits are attacked by onion thrips, the symptoms of attack on these crops are similar to the symptoms of attack on onions—i.e. the tissues around feeding punctures become desiccated, giving the affected leaves or petals a flecked appearance; on rapidly expanding leaves of cucumbers these flecks enlarge into blotches of necrotic tissue, which may rupture to form star-shaped or irregular holes; in the case of cucurbits, thrips usually lie along prominent leaf veins so that damage becomes concentrated where the principal veins radiate from the petioles; damage to cucurbit fruits is rare; tomato fruits, on the other hand, are damaged by surface speckling and small necrotic patches. The large leaves of cucumbers cover the soil surface, so providing shelter for thrips populations from adverse weather conditions.

Cultural Control

Wherever possible, the timing of sowings should be adjusted so that crops mature before damaging onion thrips populations occur. Control by rotation or spatial separation from previous crops is usually not feasible because of the wide distribution and dispersal of thrips. Uvah and Coaker (1984) reduced attacks from onion thrips by intercropping carrots with onions.

Onion breeding programmes and field evaluations have evaluated commercial cultivars for resistance to onion thrips attack (Saxena, 1975; Coudriet *et al.*, 1979). Saxena (1975), working with five cultivars, found that CV Bombay White had fewest larvae/plant, while CV Pusa Red had most. In similar work by Coudriet *et al.* (1979) the cultivar Nebuka had fewest thrips/plant. The length of sheath and shape of leaves were not correlated with numbers of larvae/plant. In America germplasm is available for breeding resistant onion cultivars (Schalk and Radcliffe, 1977). Molenaar (1984) compared varietal phenotypes and found that breeding lines with glossy foliage were more resistant because thrips appeared to prefer non-glossy plants.

Pest Monitoring

Although white and, more particularly, yellow sticky traps attract thrips, this technique has apparently been used for monitoring onion thrips only in America (Coudriet *et al.*, 1979). The number of nymphs per plant is considered now to be a more reliable index of activity. Plant assessments in Sudan led to a recommendation to spray when a mean of 5–10 nymphs were present on each onion plant (Kisha, 1979).

Chemical Control

In India good results were achieved—up to 90% onion thrips mortality—when insecticides were applied to onion crops between 0500 and 0800 hours or between 1700 and 1900 hours (Mote, 1981). In temperate climates there may be no benefit from these spray timings, even though the Indian work was done during the cold season. Insecticides shown to be effective in large numbers of Asian and Indian trials include acephate, carbaryl, carbofuran, cypermethrin, deltamethrin, endosulfan, fenitrothion, malathion and quinalphos. In most cases repeat sprays were made at 10–15 day intervals. Yield increases of up to 36% have been reported.

In Italy acephate and etrimphos sprays were recommended for control of onion thrips by Ferrari (1980). In the UK malathion and nicotine only are currently approved for use. Onion crops for processing require a 7 day harvest interval after application of malathion. Triazophos, when used for the control of cutworms, will also give some incidental control of thrips populations.

For control of onion thrips on cucurbits and tomatoes, all insecticides listed above, if approved in country of intended use, are likely to be effective. In the UK, deltamethrin, diazinon, fatty acids and heptenophos are recommended for use on cucumbers, and certain off-label uses are permitted where western flower thrips (*Frankliniella occidentalis* Pergande) is also present. For tomatoes in the UK carbaryl, diazinon, fatty acids, gamma-HCH and malathion are approved. Gamma-HCH may also be used on cucumbers but only as a bed treatment or as a spray to young plants before flowering.

Integrated Control

In Jaipur, India, Saxena (1975) recommended a combination of resistant onion cultivars (e.g. Bombay White) and releases of predators (*Chrysopa* Leach sp. (lacewings)) to suppress onion thrips populations with judicious applications of insecticides (monocrotophos, formothion, dimethoate or thiometon) during peak periods of infestation.

7.7 MOTHS

7.7.1 *Acrolepiopsis assectella* (Zeller): Leek Moth (Lepidoptera: Yponomeutidae)

The leek moth occurs throughout most of western Europe—particularly France, Belgium, Italy and Spain—and extends eastwards to Siberia. It is also recorded from Hawaii (Carter, 1984). Leek moth larvae feed mainly on leeks but also attack chives, garlic, onions and shallots. Plants in families other than the Amaryllidaceae are never infested.

Description

The adult leek moth has a 15 mm wingpsan. Its forewings are slender and narrow and have a variable brown colour, with paler scales near the apices. There is a conspicuous, white-coloured triangular mark, half-way along the rear margin of each forewing. The rear margins of the forewings are fringed with pale-coloured hairs. The hindwings are pale in colour and, like the forewings, fringed with hairs along the rear margins.

The egg of the leek moth is about 0.4 mm in diameter, oval-shaped, white in colour and iridescent, with raised sculpturing. The larva is 13–14 mm long and has a brown- or yellow-coloured head and a yellowish-green body with inconspicuous grey-brown patches, especially around the spiracles, and yellow plates on the first and last segments, freckled with brown. The pupa is 5.5–6.5 mm long and chestnut brown in colour, and has abdominal spiracles which are raised on tubercles and a rounded cremaster with hooked setae (Carter, 1984).

Life History

Adult leek moths emerge from overwintering during April and are nocturnal. Eggs (up to 100 per female) are laid singly on leaves towards the base of food plants. They hatch in about a week into larvae which feed during May and June. At first the larvae mine leaves, leaving the epidermis of each leaf intact. After about 5 days, the caterpillars bore through the folded leaves of each plant to feed near the centre. Around 3 weeks after hatching from the eggs, the caterpillars pupate within flimsy, silken cocoons attached to the dead leaves of host plants (Edwards and Heath, 1964). A new generation of adult moths emerges from the cocoons after 2 weeks and gives rise to further caterpillars. If larvae experience short day lengths and low temperatures, the subsequent adults undergo reproductive diapause. This diapause results in delay and reduction of female reproductive potential (Abo Ghalia and Thibout, 1982). There are up to 6 generations each year in Hawaii and 4 or 5 in southern Europe. In northern Europe there may be only 2 generations, with second-generation larvae present during August and September. Adult moths emerge during September and October and overwinter among plant debris until the following spring. More usually in northern Europe, leek moths overwinter as pupae.

Damage

Mining of the leaves of leeks by young leek moth larvae results in patches of papery, necrotic tissue, which is of minor importance only. Older larvae make 'shot holes' in folded leaves. Severely attacked leek leaves sometimes rot and, if the rotting is extensive, the affected plants will wilt and die. When onions are attacked, the larvae feed inside the hollow leaves, where they cause little damage. The larvae may also bore into the onion bulb or attack the flowering shoot, resulting in severe rotting and loss of seed, respectively (Jary and Rolfe, 1945). Later generations of the leek moth are usually more damaging than earlier ones and in most years in the district of Lusia, Italy, the fifth-generation larvae cause severe economic damage to leeks (Scaltriti *et al.*, 1981). In Brittany, France, later-generation larvae excavate feeding galleries in mature bulb onions during September (Rahn, 1982a).

Cultural Control

Crop rotation and siting susceptible crops well away from infested soil are necessary to prevent reinfestation by leek moths. Burning crop debris will kill any pupae present, but overwintering adults will not be affected. Whenever possible, crops should be harvested early, to prevent damage from later generations.

Biological Control

Leek moths are attracted to host crops by volatile chemicals, especially sulfinothioates. Investigations at Tours, France, aim to use the compounds to lure moths away from leek crops (Auger and Thibout, 1981). In Poland nine species of ichneumonid and braconid wasp parasitoids were shown between them to limit leek moth populations, especially the third generation of the moth, which was found to have between 74% and 89% parasitism (Legutowska and Plaskota, 1986). The bacterial insecticide *Bacillus thuringiensis* Berliner (Eubacteriales: Bacillaceae) is used to control leek moths in Europe.

Monitoring and Forecasting

A synthetic pheromone ((Z)-11-hexadecenal) attracts male leek moths and has proved effective in France for monitoring moth activity and giving early warning of attack (Rahn, 1982b). Traps based on this pheromone are used in France and Spain as part of a supervised control programme.

Chemical Control

When leek moth infestations are under way, the application of insecticides is usually of very limited value, because affected plants will have begun to rot. A control strategy should be based on pest monitoring followed by judicious use of insecticides. In France at least 13 insecticides are recommended for use (Goix, 1986). These include azinphos-methyl, carbaryl, deltamethrin, malathion, permethrin and phosalone. Where leek moth attacks are sporadic (e.g. UK), insecticidal control is unnecessary.

7.7.2 *Agrotis* Ochsenheimer spp., *Euxoa* Huebner spp.: Cutworms (Lepidoptera: Noctuidae)

These pests are discussed in Chapter 9; see pages 352–364, inclusive.

7.8 FLIES

7.8.1 *Delia antiqua* (Meigen): Onion Fly (Diptera: Anthomyiidae)

The onion fly is distributed throughout Europe, throughout much of Asia, including Japan and Korea, and across northern USA and Canada. In addition to salad and bulb onions, it attacks chives, garlic, leeks and shallots. Plants in families other than the Amaryllidaceae are never infested.

Description

The adult onion fly is pale grey in colour, hairy and 5–7 mm long. It is a robust-looking fly with reddish-brown-coloured, hairless eyes. The thorax has five, longitudinal bands of backward-curving, dark-coloured bristles and the abdomen has a darker grey-coloured, central line. The female fly has widely separated eyes and a swollen, barrel-shaped abdomen. The male's eyes are set more closely together than the female's and, unlike the female's swollen abdomen, the abdomen of the male is parallel-sided with the terminal genitalia, visible as claspers, folded forward against the ventral surface.

The egg of the onion fly is elongate (1.1–1.2 mm long), is white in colour and has a convex surface as well as a slightly concave surface. The chorion of the egg is finely reticulated with longitudinal ridges giving the appearance of broken ribbing. A pair of ridges extends along the concave surface of each egg for about a quarter of its total length. The larva is a whitish-coloured, legless maggot which is 9–10 mm long when mature. The head, which is very small and retracted into the prothorax, bears a pair of mandibles, robust, black-coloured hooks mounted on the stem of a Y-shaped armature. The posterior tip of the larval body is triangular in profile, with four distinct, conical tubercles below a pair of dark spiracles. The two anterior spiracles each have 10–13 finger-like processes. This feature is the only diagnostic character which could be used to separate onion fly larvae from bean seed fly larvae (Miles, 1953) (see page 176). The pupa is an oval puparium, 6–7 mm long, and reddish-brown to dark brown in colour, with the larval tubercles visible at the posterior end.

Life History

Adult onion flies emerge from overwintering pupae during May or June. Males usually emerge first but are not ready to mate for several days after emergence. They are then able to copulate repeatedly but females usually mate only once. Females are crepuscular, avoiding host crops for much of the day and returning to them from adjacent vegetation in the evening (Finch *et al.*, 1986b). Both sexes feed at simple flowers that have accessible nectaries—e.g. Umbelliferae. These food plants are usually located in headlands adjacent to crops. As a consequence, crop damage is sometimes found more in headland areas than in central areas.

First eggs of the onion fly are laid usually towards the end of May. Adult females are attracted to onion plants by volatile chemicals and by aromas produced by bacteria associated with plant decomposition (Dindonis and Miller, 1981). Oviposition, which begins about 8 days after emergence of adult female flies, tends to occur in sandy soils. Eggs are laid in batches of up to 30 in soil adjacent to host plants or in leaf sheaths. Given the choice, onion flies appear to prefer high-density to low-density plantings for oviposition. In captivity, egg batches are laid at one- or two-day intervals for several weeks. The average life span of female onion flies in a Canadian mass-rearing exercise was 48 days, with some flies still ovipositing after 66 days: fecundity ranged from 114 eggs to 491 eggs per female (Vernon and Borden, 1979). Variations in diet and environment may result in different fecundities in the field. The rate of egg production decreases as females age, and later-emerging females suffer from reduced fecundity and are prone to mortality (Goth *et al.*, 1983). Host-plant attractiveness declines with age of plant, and crops, after 12 weeks of growth, are rarely attacked unless they have been physically injured. Larvae hatch from eggs within a few days of oviposition, move through the soil and burrow into the bases of plants, where they feed on soft tissue until fully fed 2–3 weeks later. Younger larvae appear to prefer decomposing tissue, with bacteria playing a major role in their development. Older larvae appear to prefer to feed on fresh onion tissue, especially the interior sections of the bulb, in which they develop quickly (Zurlini and Robinson, 1978). Large numbers of onion fly larvae feed in the same plant and there may be insufficient food for all. Larger larvae can move to other plants, where they resume feeding until mature.

Pupation of onion fly larvae takes place in soil near to host plants at a depth of 4–8 cm. Adults of the second generation emerge in 2–3 weeks, with a peak population in early to mid-July. In cool locations (e.g. UK), a small proportion of flies may remain within their puparia until the following spring. Second-generation larvae feed until mid-August, when they pupate. The onion fly usually overwinters as second-generation pupae, but in southerly areas, sheltered locations or during exceptionally warm summers, a partial third generation can occur, with larvae feeding long into the autumn. Pupae remain in the soil until May of the following year.

Damage

Onion seedlings attacked by onion fly larvae quickly collapse and die. The outer surface of each affected plant usually remains intact but it may be eaten through near the base. When present in large numbers, onion fly maggots can destroy a large proportion of young plants, resulting in patchy crops. Older plants wilt: foliage becomes flaccid and discoloured before drying out or starting to decompose. Larger bulb onions withstand attacks with few symptoms above ground at first. The outer leaves then turn yellowish-white in colour and wilt as the infested bulbs begin to rot. If these plants are pulled, the leaves easily break away from the stems. Eventually, the foliage of each plant will die and fall to the soil surface. When the bulbs are cut open, the extent of the damage will be revealed, the larvae visible within the unmarketable onions. The presence of maggots prevents confusion with damage by stem nematode (see Section 7.9, page 248).

Cultural Control

Crop rotation is essential to prevent onion fly reinfestation, and susceptible crops should be situated as far apart as possible and well away from infested land. Attacked plants should be burned to kill maggots. When removing badly decayed plant material, care is needed to avoid leaving maggots or puparia in surrounding soil. Substandard onions should be burned, to reduce build-up of onion fly populations.

The most important cultural consideration is avoidance of physical damage to onion crops. The use of implements that damage plants can greatly increase the likelihood of attacks from onion fly. Similarly harvesting of crops should be done in such a way as to minimise physical damage so that flies are not attracted to any injured bulbs lying in windrows.

Plant breeding programmes to develop onion cultivars resistant to onion fly have met with only limited success (Ellis *et al.*, 1986), and no commercial cultivars with significantly reduced attractiveness to egg-laying flies are yet available. The most promising approach for the future is likely to be the development of cultivars with mid-season resistance, reducing the period of susceptibility to attack (Finch *et al.*, 1986a).

Biological Control

Studies of beneficial insects in Canada (Tomlin *et al.*, 1985) identified several that exert some influence on populations of onion flies. Adults of the staphylinid beetle *Aleochara bilineata* Gyllenhal, together with at least 41 other staphylinids and 20 carabid beetles, accepted immature stages of the onion fly as food. The braconid wasp *Aphaereta pallipes* Say and larvae of *A. bilineata* were significant mortality agents, together parasitising 38% of the observed onion fly population. In Romania another braconid of the genus *Aphaereta* Foerster constituted 95% of the total parasitoid population (Gherasim and Lacatusu, 1977). In the UK, and probably much of Europe, chalcid wasps of the genus *Trybliographa* Foerster (possibly *T. rapae* Westwood) often infest a large proportion of onion fly larvae and pupae. Although these beneficial insects cannot give adequate control of damaging onion fly populations, they none the less afford considerable protection and their safety should be considered in the formulation of pest control strategies.

Genetic manipulation of onion fly populations and radiation pathology are reported from Canada and Holland. X-ray irradiation of pupae 2–3 days before adult emergence causes sterilisation of both sexes (Theunissen, 1977). As optimum dose rates differ for male and female onion flies, puparia are sexed before treatment. Irradiated sterile males are released annually onto Dutch onion crops, and a high degree of sterility in field populations is claimed for various release rates and techniques. Heemert (1977) introduced mutant flies bearing compound chromosomes which conferred sterility on crosses with the wild type or on hybrids from such crosses. Chromosomal aberrations that increase the incidence of gynandromorphs and other individuals with low reproductive potential are also under investigation (Vosselman, 1978; Reid and McEwan, 1979).

Monitoring and Forecasting

Adult onion flies are attracted to adhesive traps of various designs and colours which can be used to form the basis of a pest monitoring and forecasting system. Vernon and Bartel (1985) concluded that white was the most attractive colour to monitor onion flies. They found that blue and violet colours were more attractive than saturated hues and that yellow was a relatively unattractive colour. Traps baited with lacerated whole onions, macerated onion tissue or sulphur-bearing chemicals attract ovipositing flies. Surrogate onions—i.e. artificial plants of various colours (Harris and Miller, 1983)—have also been tested. The most attractive colours or scents are not always the most environmentally acceptable, because some will also catch butterflies and other 'amenity' or beneficial insects.

The thermal constants associated with the adult emergence of the various generations of the onion fly have been investigated. Andaloro *et al.* (1984) established that 50% of first- and second-generation flies are caught at 427 and 1073 day-degrees above 4°C, respectively. Whitfield *et al.* (1985) worked to base 4.4°C and timed the emergence of each of the 3 generations of the onion fly at 399, 1142 and 1840 day-degrees. These models of onion fly development can be used to trigger the start of monitoring programmes, avoiding unnecessary expenditure of time and resources before flies are active. In practice, simultaneous and continuous monitoring of several insect pests, not just onion fly, is required. In these circumstances, Vernon and Borden (1983) considered yellow to be the most suitable, general-purpose, trap colour.

Chemical Control

Before 1952, onion fly attacks in the UK were controlled by calomel dust applied as a seed treatment. Later, dieldrin gave excellent control (Carden, 1960), but by 1969 resistance of the onion fly to dieldrin was reported from Bedfordshire and Norfolk (Gostick *et al.*, 1971). Resistance to organochlorine insecticides has also been recorded in Canada, the Netherlands and France (Missonnier *et al.*, 1966). The sale and use of dieldrin were not permitted in the UK after 1980 and it is now banned in most of the Western World. In the search for alternative insecticides, Thompson and Wheatley (1973) showed that seed treatments of diazinon, pirimiphos-ethyl and isofenphos were effective, but were less effective than dieldrin. Saynor and Hill (1977) confirmed these findings and also showed that chlorpyrifos, trichloronate, ethion, iodofenphos, chlorfenvinphos and triazophos were effective. Emmett and Savage (1980) evaluated treatments for plants grown in peat blocks and established the efficacy of iodofenphos seed treatments. Emmett (1981) drew attention to the value of incorporating iodofenphos into seed-coating compounds and the dangers of phytotoxicity when combined with some fungicides.

Insecticides recommended in the UK for the control of onion fly in bulb onions are soil-incorporated carbofuran granules and bromophos or iodofenphos as seed treatments. For salad onions, these seed treatments as well as diazinon wettable powder are recommended, but carbofuran must not be used. Bromophos only is approved in the UK for onion fly control on leeks. When using seed coating compounds, iodofenphos should be the insecticide of choice. This insecticide may be mixed with the fungicides metalaxyl and benomyl but must not be used with thiram or iprodione, because of phytotoxicity problems. Aldicarb granules or chlorpyrifos, when used to control other pests, may also give incidental control of onion fly.

Integrated Control

Integrated control programmes based on previous onion fly damage, cropping history, day-degree models and peaks of adult activity can reduce the need for insecticide by 30–100% (Madder, 1981). The absence of resistant onion cultivars is a barrier to the integrated control of the onion fly. The use of mid-season insecticidal treatments can suppress beneficial insects and cause physical injury to onion plants, leading to an increase in onion fly incidence (Talman *et al.*, 1981).

7.8.2 *Delia platura* (Meigen): Bean Seed Fly (Diptera: Anthomyiidae)

This pest is discussed in Chapter 6. See page 176.

7.9 NEMATODES

7.9.1 *Ditylenchus dipsaci* (Kuhn) Filipjev: Stem Nematode, Stem and Bulb Nematode, Onion Nematode (Tylenchida: Tylenchidae)

The stem nematode is distributed in most temperate areas of the world. It is reported, mainly as a pest of garlic and onions, in Europe, USSR, North America, Brazil, Chile, Israel, Africa and Thailand. In spite of the importance of onions and garlic in the Indian subcontinent, there is a paucity of published information on the stem nematode as a pest in India.

Stem nematodes damage many different crops. Onions are the most seriously and widely damaged vegetable crop, but all alliaceous plants— e.g. shallots, chives, garlic and leeks—may become infested. Other vegetables, including carrots, parsnips, beans, peas, rhubarb and sweet corn, are also attacked. The stem nematode occurs as 'races' or strains which differ in the range of host plants they attack. Alliaceous and other vegetable crops are infested by a race that also attacks strawberries, oats and beets. Both this oat–onion race and a so-called 'giant race' may infest field and broad beans. A narcissus race can also infest onions. Several races infest flower bulbs and other races affect various legumes, including clover. Races can be distinguished only by their ability to multiply on specific host plants and by the damage they cause to those plants. However, attacks are not necessarily limited to the most suitable hosts. Stem nematodes of most races may survive in many kinds of plants, including a wide range of weed species.

Description

Nematodes are unsegmented worms and are usually circular in cross-section. Plant-parasitic species possess a mouth (or buccal) spear used to puncture plant cells. Some, like the stem nematode, are internal parasites while others are free-living in the soil. Stem nematodes are colourless and slender and cannot usually be seen without magnification. The adult body is 2.2 mm long, filiform and tapering at each end, with the internal organs visible through the cuticle.

Life History

Stem nematodes live as parasites within plant tissues, where both males and females often occur in enormous numbers. They feed on parenchymatous tissues in stems and bulbs, causing the breakdown of the middle lamellae of cell walls. Mating is necessary for reproduction. Females lay up to 500 eggs and can live for more than 2 months. There is a total of four moults, with the first completed within the egg. The life cycle in onion plants growing at 18°C occupies about 21 days. Breeding is continuous. The stem nematode can survive, mainly as desiccated fourth-stage juveniles, for several years. Fourth-stage juveniles tend to aggregate at the surface of heavily infested plant tissue, to form clumps of 'nematode wool'. They may sometimes become attached to the seeds of host plants. The nematodes become active again only when moistened. In wet soil they can remain alive in the absence of host plants for a year or longer, according to soil type. They persist longer in clay than in sandy soils. Stem nematodes may be able to reproduce in soil fungi, but weeds are important reservoir hosts for some races.

Damage

Onion plants may be attacked by stem nematodes any time after germination. Infested seedlings or young plants become greatly swollen at their bases and have malformed and twisted leaves (Figure 7.6). Infested tissue has a characteristic loose, puffy texture and an epidermis which is generally dull in appearance, a condition known as 'bloat'. Rotting then occurs at soil level, so that badly infested plants can be easily pulled to leave most or all of their roots in the soil. Eventually, affected plants die and rot. Slight infestations may pass unnoticed at harvesting, but onion bulbs may deteriorate during air-ventilated storage, especially if the temperature is too high or the bulbs are too moist. Garlic plants infested with stem nematodes develop similar symptoms to nematode-infested onions (garlic

Figure 7.6 Onions damaged by the stem nematode, *Ditylenchus dipsaci*.
© MAFF

bloat) with a thickening of the pseudostems and a yellowing of the leaves. Infested garlic bulbs are usually pale yellow in colour and necrotic, but some cultivars show no symptoms even when heavily infested. On leeks, symptoms of stem nematode attack are similar to nematode-infested onions, but damage is usually less severe.

When sugar beet seedlings are attacked, the tissues become swollen and characteristically spongy. The stems, midribs and main veins of leaves are affected and, sometimes, distinct galls are formed. The growing points of plants may become malformed or killed. If they are killed, fresh growing points develop, leading to a multiple-crowned ('many-necked') condition. During summer, healthier growth is produced and the earlier damage becomes less apparent. When infestation occurs in autumn, the roots at around soil level may become cracked. Sub-epidermal tissues become soft and brownish in colour and a dry, spongy rot gradually extends inwards. This dry rotting may be followed by a darker-coloured, wetter, bacterial rot until eventually root structure breaks down. In advanced cases the crown and leaves of sugar beet plants infested by stem nematodes can be easily pulled away from the roots.

Broad bean plants infested with stem nematodes develop reddish-brown discoloured stems that darken with age, starting at the bases and extending upwards to the pods. Symptoms are most readily seen in July, when pods are still green in colour, but eventually the symptoms become masked by the general brown colour of crop ripening. Stems infested with the oat–onion race of stem nematodes are seldom distorted, but are often thinner and shorter than the stems of uninfested plants. The 'giant race' severely damages stems, which often become twisted and blistered.

Infestations in the pod-bearing regions are more common than stem infestations, and leaf petioles and pods may become distorted. Secondary rotting at the base may cause collapse of the whole plant. Ripe, infested pods contain many nematode larvae which are attached to each seed, especially in the area of the slit of the hilum. Sometimes these nematodes are massed together to form 'nematode wool'. Beneath the seed coats of heavily infested seeds, discoloured patches or lesions containing many nematodes can often be seen.

French bean plants attacked by stem nematodes can become severely stunted with the leaves bunched around each growing point.

Stem nematode damage to beans usually shows as a stem discoloration: reddish-black in broad beans; and reddish-brown in runner beans and French beans.

Cultural Control

In the Netherlands the stem nematode was found to persist in heavier soils, even when host plants were not frequently grown. Conversely, in light soils the nematode became important only where there was over-cropping with host plants. Opinions regarding optimum intervals between host crops vary from 2 years to 6 years. It has been suggested that, as long as it remains economic to do so, onions be grown every year in the same soil. The use of nematode-free seed was recommended, but it was considered that the nematode would eventually still be introduced. Once it is introduced, the stem nematode spreads in soil, whether or not onions are grown, at about the same rate. Continuous onion growing in a field avoids rapid contamination of other fields.

Preventing Spread

Nematodes do not move actively in soil more than a few centimetres each year and most movement is passive. They may be transferred in soil or in host plants, including the seed. When onions are grown for seed, stem nematodes may be carried upwards with the growing flower stems, so that they infest the flowers. The seed from these plants is often heavily infested. Stem nematodes are probably most commonly introduced to fields by sowing infested onion seed. They may also be spread by rain, flood water and wind. National and international traffic in plants and seeds renders the spread of the stem nematode inevitable. Small numbers of nematodes introduced into a previously clean field seldom cause damage to the crop grown from the infested seed. However, the nematodes can multiply to serious proportions on subsequent crops. It is then impossible to prevent local spread between fields, farms and cultivars, especially by the movement of contaminated soil. Ideally, there are three steps which should be taken to ensure the production of 'clean' seed:

1. Only nematode-free seed should be used for growing seed crops.
2. This seed should be sown in soil where the nematode is known not to exist.
3. The crop should be inspected for the presence of the stem nematode, preferably in the early spring after growth commences.

Chemical Control

Stem nematodes are difficult to control with nematicides, because they can multiply up to 1000-fold on susceptible plants. Soil fumigants sometimes prevent serious injury but give inadequate control of populations. Aldicarb, oxamyl, fenamiphos and thionazin have protected onion seedlings from injury and greatly increased yields. Row treatments are better than broadcast applications, because stem nematodes are not soil-borne, affecting vegetative shoots only, and row treatments concentrate nematicides close to plants. In the UK only aldicarb, carbofuran and oxamyl are approved for use with bulb onions. Treated onion crops must not be harvested until they have reached the mature bulb stage. Control of the stem nematode with carbofuran is partial. Oxamyl may be used for pickling or silverskin onions. There are no recommendations for control of the stem nematode on salad onions.

In the UK aldicarb is recommended for use against nematodes on broad beans. There are no recommendations for chemical control in sweet corn, French bean, runner bean or sugar beet crops. The recommendations are intended for soil migratory nematode control and are unlikely to be fully effective against stem nematodes.

Physical Control

With varying degrees of success various workers have investigated the use of hot water treatment of shallots and onion sets for stem nematode control.

Legislative Control

Within the European Economic Community (EEC) a directive on protective measures against the introduction of harmful organisms or plants or plant products prohibits the importation of the stem nematode into member states. Plant Health Certificates are issued by exporting countries to confirm compliance with the importing country's requirements.

REFERENCES

Abo Ghalia, A. and Thibout, E. (1982). *Annales de la Societe Entomologie de France*, **18**, 173–179

Al-Faisal, A. H. M. and Kardou, I. K. (1986). *Journal of Biological Sciences Research, Iraq*, **17**, 9–19

All, J. N., Javid, A. and Guillebeau, P. (1986). Control of fall armyworm with insecticides in North Georgia sweet corn. *Fla Entomol.*, **69**, 598–602

Altieri, M. A. (1981). Weeds may augment biological control of insects. *Calif. Agric.*, **35**, 22–24

Altieri, M. A., Francis, C. A., Schoonhoven, A. and van Doll, J. D. (1978). Effects of bean association on yields and yield components of maize. *Field Crop Res.*, **1**, 33–49

Altieri, M. A. and Whitcomb, W. H. (1980). Weed manipulation for insect pest management in corn. *Environ. Mgmt*, **4**, 483–490

Andaloro, J. T., Muka, A. A. and Straub, R. W. (1983). Cornell University, *Fact Sheet* (European Corn Borer)

Andaloro, J. T., Rose, K. B. and Eckenrode, C. J. (1984). *Search Agriculture*, **29**, 5–10

Anderson, T. E., Kennedy, G. G. and Stinner, R. E. (1982). Temperature-dependent mode for postdiapause development and spring emergence of the European corn borer, *Ostrinia nubilalis* (Hubner) (Lepidoptera: Pyralidae) in North Carolina. *Environ. Entomol.*, **11**, 1307–1311

Anderson, T. E., Kennedy, G. G. and Stinner, R. E. (1984). Distribution of the European corn borer, *Ostrinia nubilalis* (Hubner) (Lepidoptera: Pyralidae), as related to oviposition preference of the spring-colonizing generation in Eastern North Carolina. *Environ. Entomol.*, **13**, 248–251

Andow, D. A. and Risch, S. J. (1985). Predation in diversified agroecosystems: relation between a coccinellid predator *Coleomegilla maculata* and its food. *J. Appl. Ecol.*, **22**, 357–372

Anonymous (1985). Corn pest management for the Midwest. Ohio Agric. Exp. Stn, *N. Cent. Reg. Pub. No. 98*

Anonymous (1989). European corn borer: development and management. Iowa Agric. Exp. Stn, *N. Cent. Reg. Extn Pub. No. 327*

Apple, J. W. (1952). Corn borer development and control on canning corn in relation to temperature accumulation. *J. Econ. Entomol.*, **45**, 877–879

Ashley, T. R. (1979). Classification and distribution of fall armyworm parasites. *Fla Entomol.*, **62**, 114–123

Auger, J. and Thibout, E. (1981). *Comptes Rendus Hebdomadaires des Seances de l'Academie des Sciences*, **29**, 217–220

Awadallah, W. H. and Hanna, L. I. (1985). Evaluation of certain US corn inbred lines for their resistance to infestation with the corn leaf aphid, *Rhopalosiphum maidis*, in Egypt. *Ann. Agric. Sci.*, **23**, 327–333

Bakker, T. and Robinson, A. G. (1975). Movements of English grain aphids on barley plants. *Manitoba Entomol.*, **9**, 9–12

Barber, G. W. and Dicke, F. F. (1939). Effect of temperature and moisture on overwintering pupae of the corn earworm in the northeastern states. *J. Agric. Res.*, **59**, 711–724

Barfield, C. S. and Jones, J. W. (1979). Research needs for modeling pest management systems involving defoliators (e.g. the fall armyworm) in agronomic crop systems. *Fla Entomol.*, **62**, 98–114

Barfield, C. S. and Stimac, J. L. (1980). State-of-the-art for predicting damaging infestations of fall armyworm. *Fla Entomol.*, **63**, 364–375

Beck, D. L., Dunn, G. M., Routley, D. G. and Bowman, J. S. (1983). Biochemical basis of resistance in corn, *Zea mays* L., to the corn leaf aphid, *Rhopalosiphum maidis. Crop Sci.*, **23**, 995–998

Bendell, B. E., Weatherhead, Pl. J. and Stewart, R. K. (1981). The impact of predation by red-winged blackbirds on European corn borer populations. *Can. J. Zool.*, **59**, 1535–1538

Beregovoy, V. H., Starks, K. J. and Janaradan, K. G. (1988). Fecundity characteristics of the greenbug biotypes C and E cultured on different host plants. *Environ. Entomol.*, **17**, 59–62

Bhangale, G. T. and Joi, M. B. (1983). *Journal of Maharashta Agricultural Universities*, **8**, 299–300

Blair, B. D. (1970). Aphids collected from a Scioto County, Ohio, corn field and areas bordering the field. *J. Econ. Entomol.*, **63**, 1099–1101

Bowers, W. S., Nault, L. R., Webb, R. E. and Dutky, S. R. (1972). Aphid alarm pheromone: isolation, identification, synthesis. *Science*, **177**, 1121–1122

Bowman, J. S. and Barry, D. W. (1989). Control of first-brood ECB, 1988. *Insect. Acar. Tests*, **14**, 115

Boyer, W. P., Lincoln, C. and Philips, J. R. (1966). Surveys for *Heliothis* moths in Arkansas. *US Dep. Agric. Coop. Econ. Insect Rep.*, **16**, 950–951

Branson, T. F. and Simpson, R. G. (1965). The effects of a nitrogen-deficient host and crowding on the corn leaf aphid. *Proc. N.C. Branch, Entomol. Soc. Am.*, **20**, 56

Brett, C. H. (1953). Fall armyworm control on late-planted sweet corn. *J. Econ. Entomol.*, **46**, 714–715

Brett, C. H. and Bastida, R. (1963). Resistance of sweet corn varieties to the fall armyworm, *Laphygma frugiperda*. *J. Econ. Entomol.*, **56**, 162–167

Brown, P. A. and Blackman, R. L. (1988). Karyotype variation in the corn leaf aphid *Rhopalosiphum maidis* (Fitch) species complex in relation to host plant and morphology. *Bull. Entomol. Res.*, **78**, 351–363

Burbutis, P. P., Erwin, N. and Ertle, L. R. (1981). Reintroduction and establishment of *Lydella thompsoni* and notes on other parasites of the European corn borer in Delaware. *Environ. Entomol.*, **10**, 779–781

Caffrey, D. J. and Worthley, L. H. (1927). The European corn borer: its present status and methods of control. *US Dep. Agric. Farmers Bull. No. 1548*

Cananettes, J. P. (1988). Maize aphids, a major problem in 1987. *Phytoma*, No. 398, 20–21

Cantelo, W. W. and Jacobson, M. (1979). Corn silk volatiles attract many pest species of moths. *J. Environ. Sci.—Health (Part A)*, **14**, 695–708

Carden, P. W. (1960). *Bulletin of Entomological Research*, **50**, 795–807

Carter, D. J. (1984). *Pest Lepidoptera of Europe with Special Reference to the British Isles*. Junk, Dordrecht

Cartier, J. J. (1957). On the biology of the corn leaf aphid. *J. Econ. Entomol.*, **50**, 110–112

Chiang, H. C. (1972). Dispersion of the European corn borer (Lepidoptera: Pyralidae) in Minnesota and South Dakota, 1945 to 1970. *Environ. Entomol.*, **1**, 157–161

Chiang, H. C., Jarvis, J. L., Burkhardt, C. C., Fairchild, M. L., Weekman, G. J. and Triplehorn, C. A. (1961). Populations of European corn borer, *Ostrinia nubilalis* (Hubner), in field corn, *Zea mays* L. Missouri Agric. Exp. Stn, *N. Cent. Reg. Pub. No. 129*

Chowdbury, M. A., Chalfant, R. B. and Young, J. R. (1987). Comparison of sugarline sampling and pheromone trapping for monitoring adult populations of corn earworm and fall armyworm (Lepidoptera: Noctuidae) in sweet corn. *Environ. Entomol.*, **16**, 1241–1243

Clement, S. L., Rubink, W. L., Rings, R. W. and Casey, M. A. (1981). Predicting flight activities of the European corn borer. *Ohio Rep.*, **66**, 3–4

Collins, G. N. and Kempton, H. H. (1917). Breeding sweet corn resistant to the corn earworm. *J. Agric. Res.*, **11**, 549–572

Coudriet, D. L., Kishaba, A. N., McCreight, J. D. and Bohn, G. W. (1979). *Journal of Economic Entomology*, **72**, 614–615

Cox, H. C., Lovely, W. G. and Brindley, T. A. (1956). Control of the European corn borer with granulated insecticides in 1955. *J. Econ. Entomol.*, **49**, 834–838

Crawford, H. G. (1924). The status of the control practice for the European corn borer in Ontario. *Ann. Rep. Entomol. Soc. Can.*, **54**, 78–82

Cruz, I. and Turpin, F. T. (1983). Yield impact of larval infestations of the fall armyworm (Lepidoptera: Noctuidae) to midwhorl growth stage of corn. *J. Econ. Entomol.*, **76**, 1052–1054

Dale, J. L., McFerran, J., Wann, E. V. and Bona, R. L. (1982). Evaluation of sweet corn hybrids for yield and ear quality when grown under virus disease conditions. *Arkansas Agric. Exp. Stn Rep. Ser.*, pp. 27–28

Derridj, S., Lefner, H., Argendre, M. and Dukrand, Y. (1988). Use of strips of *Zea mays* L. to trap European corn borer, *Ostrinia nubilalis* (Hubner), oviposition in maize fields. *Crop Prot.*, **7**, 177–182

Dicke, F. F. (1939). Seasonal abundance of the corn earworm. *J. Agric. Res.*, **59**, 237–258

Dicke, F. F. (1969). The corn leaf aphid. *Proc. 24th Ann. Corn and Sorghum Res. Conf.*, **24**, 61–70

Dindonis, L. L. and Miller, J. R. (1981). *Journal of Chemical Ecology*, **7**, 419–426

Doss, S. A. (1979). Effect of host plants on some biological aspects of the boll-worm *Heliothis armigera* (Hubner) (Lepidoptera: Noctuidae). *Zeitschrift für Pflanzenkrankheiten und Pflanzenschutz*, **86**, 143–147

Eckenrode, C. J., Robbins, P. S. and Andaloro, J. T. (1983). Variations in flight patterns of European corn borer (Lepidoptera: Pyralidae) in New York. *Environ. Entomol.*, **12**, 393–396

Edwards, C. A. and Heath, G. W. (1964). *The Principles of Agricultural Entomology*. Chapman and Hall, London

Ellis, P. R., Soni, S. K. and Mayne, H. J. (1986). *National Vegetable Research Station Annual Report* 36, p. 31

Emmett, B. J. (1981). *Tests of Agrochemicals and Cultivars (Annals of Applied Biology*, **100**, Supplement), **3**, 30–31

Emmett, B. J. and Savage, M. J. (1980). *Plant Pathology*, **29**, 159–167

Ertle, E. R. and Davis, C. P. (1975). *Trichogramma nubilale* new species (Hymenoptera: Trichogrammatidae) and egg parasite of *Ostrinia nubilalis* (Hubner). *Ann. Entomol. Soc. Am.*, **68**, 525–528

Evans, D. A. and Medler, J. T. (1967). Flight activity of the corn leaf aphid in Wisconsin as determined by yellow pan trap collections. *J. Econ. Entomol.*, **60**, 1088–1091

Everly, R. T. (1960). Loss in corn yield associated with the abundance of the corn leaf aphid, *Rhopalosiphum maidis*, in Indiana. *J. Econ. Entomol.*, **53**, 924–932

Feng, C. C. and Yang, J. Y. (1987). Influence of temperature on the growth and development of *Schizaphis graminum*. *Insect Knowledge*, **24**, 140–143

Ferrari, R. (1980). *Informatore Fitopatologico*, **10**, 27–28

Ferro, D. N. and Fletcher-Howell, G. (1985). Controlling European corn borer (Lepidoptera: Pyralidae) on successively planted sweet corn in western Massachusetts. *J. Econ. Entomol.*, **78**, 902–907

Finch, S., Cadoux, M. E., Eckenrode, C. J. and Spitler, T. D. (1986a). *Journal of Economic Entomology*, **79**, 736–740

Finch, S., Eckenrode, C. J. and Cadoux, M. E. (1986b). *Journal of Economic Entomology*, **79**, 107–113

Flath, R. A., Forrey, R. R., John, J. O. and Chan, B. G. (1978). Volatile components of corn silk, *Zea mays*, as possible *Heliothis-zea* attractants. *J. Agric. Food Chem.*, **26**, 1290–1293

Forster, R. L., Stoltz, R. L., Fenwick, H. S. and Simpson, W. R. (1980). Maize dwarf mosaic virus in Idaho, USA. *Plant Dis.*, **64**, 410–411

Foster, J. E., Stamenkovic, S. S. and Arayka, J. E. (1988). Life cycle and reproduction of *Rhopalosiphum padi* (L.) (Homoptera: Aphididae) on wheat in the laboratory. *J. Entomol. Sci.*, **23**, 216–222

Foster, R. E. (1989). Strategies for protecting sweet corn ears from damage by fall armyworms (Lepidoptera: Noctuidae) in Southern Florida, USA. *Fla Entomol.*, **72**, 146–151

Frye, R. D. (1972). Evaluation of insect predation on European corn borer in North Dakota. *Environ. Entomol.*, **1**, 535–536

Gage, S. H., Whalon, M. E. and Miller, D. J. (1982). Pest event scheduling system for biological monitoring and pest management. *Environ. Entomol.*, **11**, 1127–1133

Ganguli, R. N. and Raychaudhuri, D. N. (1980). Studies on *Rhopalosiphum maidis* Fitch (Aphididae: Homoptera)—a formidable pest of *Zea mays* in Tripura. *Sci. Cult.*, **46**, 259–261

Gherasim, V. and Lacatusu, M. (1977). *Analele Institului de Cercetari Pentru Protectia Plantelor*, **12**, 229–236

Ghidiu, G. M., Linduska, J. J. and McClung, E. L. (1988). Granular and liquid formulations of insecticides for management of the fall armyworm on midwhorl growth stage of late-planted sweet corn. *Appl. Agric. Res.*, **3**, 209–213

Goix, J. (1986). *Phytoma*, **379**, 37

Gonzalez, D. and Rawlins, W. A. (1969). Relation of aphid populations to field spread of lettuce mosaic virus in New York. *J. Econ. Entomol.*, **62**, 1109–1114

Gordon, D. T., Bradfute, O. E., Gingery, R. E., Knoke, J. K., Louie, R., Nault, L. R. and Scott, G. E. (1981). Introduction: history, geographical distribution, pathogen characteristics, and economic importance. In *Virus and Virus-like Diseases of Maize in the United States* (ed. D. T. Gordon, J. K. Knoke and G. E. Scott). Southern Coop. Ser. Bull. No. 247

Gostick, K. G., Powell, D. F. and Slough, C. (1971). *Plant Pathology*, **20**, 63–65

Goth, G. J., Wellington, W. G. and Contant, H. Y. (1983). *Researches on Population*

Ecology, **25**, 366–386

Gracen, V. E. (1989). Breeding for resistance to European corn borer. In *Proc. Int. Symp. on Methodology for Developing Host Plant Resistance to Insects*. CIMMYT, Londres, Mexico, pp. 203–206

Gross, H. R. Jr. (1988). *Archytas marmoratus* (Diptera: Tachinidae): Field survival and performance of mechanically extracted maggots. *Environ. Entomol.*, **17**, 233–237

Gross, H. R. Jr. and Johnson, R. (1985). *Archytas marmoratus* (Diptera: Tachinidae): advances in large-scale rearing and associated biological studies. *J. Econ. Entomol.*, **78**, 1350–1353

Gross, H. R. Jr., Pair, S. D. and Layton, R. C. (1985). *Archytas marmoratus* (Diptera: Tachinidae): Screened-cage performance of mechanically extracted maggots against larval populations of *Heliothis zea* and *Spodoptera frugiperda* (Lepidoptera: Noctuidae) on whorl and early tassel-stage corn. *J. Econ. Entomol.*, **78**, 1354–1357

Gross, H. R. Jr. and Young, J. R. (1977). Comparative development and fecundity of corn earworm reared on selected wild and cultivated early season hosts common to the southeastern USA. *Ann. Entomol. Soc. Am.*, **70**, 63–65

Hamilton, J. T. and Muirhead, W. A. (1981). Chemical control of *Heliothis armiger* (Hubner) in sweet corn. *Australian J. Exp. Agric. Anim. Husb.*, **21**, 231–235

Hamm, J. J. and Wiseman, B. R. (1986). Plant resistance and nuclear polyhedrosis virus for suppression of the fall armyworm (Lepidoptera: Noctuidae). *Fla Entomol.*, **69**, 541–549

Hanec, W. and Beck, S. D. (1960). Cold hardiness in the European corn borer, *Pyrausta nubilalis* (Hubner). *J. Insect Physiol.*, **5**, 169–180

Harris, M. O. and Miller, J. R. (1983). *Annals of the Entomological Society of America*, **76**, 766–771

Harrison, F. P. (1960). Corn earworm oviposition and the effect of DDT on the egg predator complex in corn silk. *J. Econ. Entomol.*, **53**, 1088–1094

Hassan, S. A., Stein, E., Dannemann, K. and Reichel, W. (1986). Mass production and utilization of *Trichogramma*. 8. Optimizing the use to control the European corn borer, *Ostrinia nubilalis*. *J. Appl. Entomol.*, **101**, 508–515

Heemert, C. Van (1977). *Nature*, **266**, 445–447

Henry, M. and Dedryver, C. A. (1989). Fluctuations in cereal aphid populations on maize, *Zea mays*, in Western France in relation to the epidemiology of BYDV. *J. Appl. Entomol.*, **107**, 401–410

Hill, D. S. (1985). *Agricultural Insect Pests of Temperate Regions and Their Control*. Cambridge University Press, Cambridge

Hill, R. E. and Gary, W. J. (1979). Effects of the microsporidium, *Nosema pyrausta*, on field populations of European corn borers in Nebraska. *Environ. Entomol.*, **8**, 91–95

Hminina, M. (1981). The cycle and economic importance of *Heliothis armigera* Hb. (Noctuidae) on tomato on the Atlantic coast of Morocco. *Awamia, 1979*, **57**, 1–20

Hodgson, B. E. (1928). The host plants of the European corn borer in New England. *US Dep. Agric. Tech. Bull. No. 7*

Hruska, A. J. and Gladstone, S. M. (1988). Effect of period and level of infestation of the fall armyworm *Spodoptera frugiperda* on irrigated maize field. *Fla Entomol.*, **71**, 249–254

Huber, L. L. and Stringfield, G. H. (1942). Aphid infestation of strains of corn as an index of their susceptibility to corn borer attack. *J. Agric. Res.*, **64**, 283–291

Hudon, M. and LeRoux, E. J. (1986a). Biology and population dynamics of the European corn borer (*Ostrinia nubilalis*) with special reference to sweet corn in Quebec. I. Systematics, morphology, geographic distribution, host range, economic importance. *Phytoprotection*, **67**, 39–54

Hudon, M. and LeRoux, E. J. (1986b). Biology and population dynamics of the European corn borer (*Ostrinia nubilalis*) with special reference to sweet corn in Quebec. II. Bionomics. *Phytoprotection*, **67**, 81–92

Hudon, M. and LeRoux, E. J. (1986c). Biology and population dynamics of the European corn borer (*Ostrinia nubilalis*) with special reference to sweet corn in Quebec. III. Population dynamics and spatial distribution. *Phytoprotection*, **67**, 93–115

Hudon, M., LeRoux, E. J. and Harcourt, D. G. (1989). Seventy years of European corn borer (*Ostrinia nubilalis*) research in North America. *Agric. Zool. Rev.*, **3**, 53–96

Isenhour, D. J. (1985). *Campoletis sonorensis* as a parasitoid of *Spodoptera frugiperda*: host range preferences and functional response. *Entomophaga*, **30**, 31–36

Isenhour, D. J. and Wiseman, B. R. (1987). Foliage consumption and development of the fall armyworm (Lepidoptera: Noctuidae) as affected by the interactions of a parasitoid *Campoletis sonorensis* (Hymenoptera: Ichneumonidae), and resistant corn genotypes. *Environ. Entomol.*, **16**, 1181–1184

Jackson, H. B., Coles, L. W., Wood, E. A. Jr. and Eikenbary, R. D. (1970). Parasites reared from the greenbug and corn leaf aphid in Oklahoma in 1968 and 1969. *J. Econ. Entomol.*, **63**, 733–736

Jarvis, J. L. and Guthrie, W. D. (1987). Ecological studies of the European corn borer (Lepidoptera: Pyralidae) in Boone County, Iowa. *Environ. Entomol.*, **16**, 50–58

Jary, S. G. and Rolfe, S. W. (1945). *Agriculture, London*, **52**, 35–37

Johnson, B. (1956). Flight dynamics of Aphididae. *Australian J. Sci.*, **18**, 199–200

Katona, A. and Cziklin, M. (1984). Prediction-based control of European corn borer in sweet corn. *Novenyvedelem*, **20**, 363–366

Kennedy, J. S. and Stroyan, H. L. G. (1959). Biology of aphids. *Ann. Rev. Entomol.*, **4**, 139–160

Kieckhefer, R. W. and Gellner, J. L. (1988). Influence of plant growth stage on cereal aphid reproductions. *Crop Sci.*, **28**, 688–690

Kisha, J. S. A. (1977). *Annals of Applied Biology*, **86**, 219–228

Kisha, J. S. A. (1979). *Pest. Abstr. News Sum.*, **25**, 19–24

Klun, J. A. and co-workers (1973). An evaluation of resistance of 41 corn inbred lines to European corn borer by chemical analyses of the plant tissues. *Rep. Int. Proj. on Ostrinia nubilalis*. Agric. Res. Inst., Hungarian Acad. Sci., Martonva'sa'r, pp. 104–108

Klun, J. A. and co-workers (1975a). Insect sex pheromones: Intraspecific pheromonal variability of *Ostrinia nubilalis* in North America and Europe. *Environ. Entomol.*, **4**, 891–894

Klun, J. A., Chapman, O. L., Mattes, K. C. and Beroza, M. (1975b). European corn borer and redbanded leafroller: disruption of reproduction behavior. *Environ. Entomol.*, **4**, 871–876

Klun, J. A., Tipon, J. L. and Brindley, T. A. (1967). 2,4-Dihydroxy-7-methoxy-1,4-benzoxazin-3-one (DIMBOA) an active agent in the resistance of maize to the European corn borer. *J. Econ. Entomol.*, **60**, 1529–1533

Knoke, J. K. and Louie, R. (1981). Epiphytology of maize virus diseases. In *Virus and Virus-like Diseases of Maize in the United States* (ed. D. T. Gordon, J. K. Knoke and G. E. Scott). Southern Coop. Ser. Bull. No. 247, pp. 92–102

Kochansky, J., Carde, R. T., Liebherr, J. and Roelofs, W. L. (1975). Sex pheromone of the European corn borer, *Ostrinia nubilalis* (Lepidoptera: Pyralidae) in New York. *J. Chem. Ecol.*, **1**, 225–231

Kou, R. and Chow, Y. S. (1987). Emergence time and mating-related behaviour of the cotton bollworm *Heliothis armigera* (Lepidoptera: Noctuidae) in reversed photoperiod. *Bull. Inst. Zool.*, **26**, 179–186

Lamb, K. P. (1959). Composition of the honeydew of the aphid *Brevicoryne brassicae* (L.) feeding on swedes (*Brassica napobrassica* Dc.). *J. Insect Physiol.*, **3**, 1–13

Lambert, J. D., Arnason, J. T., Serratos, A., Philogene, B. J. R. and Faris, M. A. (1987). Role of intercropped red clover in inhibiting European corn borer (Lepidoptera: Pyralidae) damage to corn in eastern Ontario. *J. Econ. Entomol.*, **80**, 1192–1196

Landazabal, A. J., Fernandez, A. F. and Figuerola, P. A. (1973). Biological control of *Spodoptera frugiperda* (J. E. Smith), on maize by the nematode *Neoaplectana carpocapsae*. *Acta Agronomica, Columbia*, **23**, 41–70

Legutowska, H. and Plaskota, E. (1986). *Colloques de l'INRA*, **36**, 61–73

Lewis, L. C., Cossentine, J. E. and Gunnarson, R. D. (1983). Impact of two microsporidia, *Nosema pyrausta* and *Vairimorpha necatrix*, in *Nosema pyrausta* infested European corn borer (*Ostrinia nubilalis*) larvae. *Can. J. Zool.*, **61**, 915–921

Linduska, J. J., Dively, G. P., Harrison, F. P., Hellman, J. L. and Enbrey, M. (1985). Foliar sprays to control corn earworms, fall armyworms and European corn borers in sweet corn, 1984. *Insect. Acar. Tests*, **10**, 105

Linduska, J. J., Harrison, T. S., Patton, T. W., Embry, M. and Dively, G. (1988). Foliar sprays to control corn earworms, dusky sap beetle and European corn borers in sweet corn, 1987. *Insect. Acar. Tests*, **13**, 115–116

Long, B. J., Dunn, G. M., Bowman, J. S. and Routley, D. G. (1977). Relationship of hydroxamic acid content in corn and resistance to the corn leaf aphid. *Crop. Sci.*, **17**, 55–58

Luginbill, P. (1928). The fall army worm. *US Dep. Agric. Tech. Bull. No. 34*

McClanahan, R. J. and Founk, J. (1972). Control of the European corn borer (Lepidoptera: Pyralidae) on sweet corn and peppers in southwestern Ontario. *Can. Entomol.*, **104**, 1573–1579

McLeod, D. G. R. (1976). Geographical variation of diapause termination in the European corn borer *Ostrinia nubilalis* (Lepidoptera: Pyralidae) in southwestern Ontario. *Can. Entomol.*, **108**, 1403–1408

McLeod, P. (1989). Corn earworm control, 1988. *Insect. Acar. Tests*, **14**, 118

McMillian, W. W., Wiseman, B. R. and Widstrom, N. W. (1977). Evaluation of commercial sweet corn hybrids for damage by *Heliothis zea*. *J. Ga Entomol. Soc.*, **12**, 75–79

McMillian, W. W., Wiseman, B. R. and Widstrom, N. W. (1980). Dent and sweet corns: influence of defoliation, plant age, and genotype on yield. *J. Ga Entomol. Soc.*, **15**, 373–377

Madder, D. J. (1981). *Proceedings of 28th Annual Meeting of Canadian Pest Management Society*, p. 113

Malyk, M. R. and Robinson, A. G. (1971). A study of the voracity, fecundity and development rates of some common lady beetle predators of aphids on cereal crops in Manitoba. *Manitoba Entomol.*, **5**, 89–95

Meisner, J., Melamed-Madjar, V., Ascher, K. R. S. and Tam, S. (1985). Effect of an aqueous extract of neem seed kernel on larvae of the European corn borer, *Ostrinia nubilalis*. *Phytoparasitica*, **13**, 173–178

Miles, M. (1953). *Bulletin of Entomological Research*, **44**, 583–588

Milinko, I., Rakk, Z. V. and Kovacs, G. (1983). Population dynamics of aphids, vectors of MDMV and aphid resistance to some maize hybrids. *Acta Phytopathol. Acad. Sci. Hung.*, **18**, 201–208

Miller, J. W. and Coon, B. F. (1964). The effect of barley yellow dwarf virus on the biology of its vector the English grain aphid, *Macrosiphum granarium*. *J. Econ. Entomol.*, **57**, 970–974

Missonnier, J., Arnoux, J., Brunel, E., Portier, S. and Oudinet, R. (1966). *Phytiatrie-Phytopharmacie*, **15**, 193–212

Mitchell, E. R. (1978). Relationships of planting date to damage by earworms in commercial sweet corn in north central Florida. *Fla Entomol.*, **61**, 251–255

Mitchell, E. R., Copeland, W. W., Sparks, A. N. and Sekul, A. A. (1974). Fall armyworm: disruption of pheromone communication with synthetic acetates. *Environ. Entomol.*, **3**, 778–780

Mittler, T. E. (1958). Studies on the feeding and nutrition of *Tuberolachnus salicmus* (Gmelim) (Homoptera: Aphididae). II. Nitrogen and sugar composition of ingested phloem, sap and excreted honeydew. *J. Exp. Biol.*, **35**, 74–84

Moericke, V. (1950). Uber das farbsehen der pfirsichblattlaus (*Myzodes persicae*). *Z. Tierpsychol.*, **7**, 265–267

Moericke, V. (1955). Über die Hebensgewohn heiten der geflügelten Blattläuse (Aphidina) besonderer Berücksichtigung des Verhaltens bien Landen. *Z. Angew. Entomol.*, **37**, 29–91.

Molenaar, N. D. (1984). *Dissertation Abtracts, International, B (Sciences and Engineering)*, **45**, 1075B

Moreau, J. P. (1987). Monitoring aphids, *Rhopalosiphum maidis*, in maize fields protected against the pyralid, *Ostrinia nubilalis* Hbn. In *Aphid Migration and Forecasting 'Euraphid' Systems in European Community Countries*. Comm. Europ. Comm., Luxembourg, pp. 13–19

Mote, U. N. (1981). *Indian Journal of Entomology*, **43**, 236–239

Nault, L. R., Edwards, L. J. and Styer, W. E. (1973). Aphid alarm pheromones: secretion and reception. *Environ. Entomol.*, **2**, 101–105

Neiswander, C. R. and Triplehorn, C. A. (1961). Differential resistance of dent corn strains to the corn leaf aphid, *Rhopalosiphum maidis* (Fitch) in Ohio. *Ohio Agric. Exp. Stn Res. Bull No. 898*

Neuffer, G. (1982). The use of *Trichogramma evanescens* Westw. in sweet corn fields. A contribution to the biological control of European corn borer, *Ostrinia nubilalis* Hubner, in Southwest Germany. *Proc. Int. Symp. on Trichogramma, Antibes, France*, pp. 231–237

Newton, C. and Dixon, A. F. G. (1988). A preliminary study of variation and life-history traits and the occurrence of hybrid vigour in *Sitobion avenae* (F.). *Bull. Entomol. Res.*, **78**, 75–83

North, R. C. and Shelton, A. M. (1986). *Environmental Entomology*, **15**, 695–699

Oatman, E. R. (1966). Parasitization of corn earworm eggs on sweet corn silk in Southern

California, with notes on larval infestations and predators. *J. Econ. Entomol.*, **59**, 830–835

Oatman, E. R., Hall, I. M., Arakawa, K. Y., Platner, G. R., Bascom, L. A. and Beegle, C. C. (1970). Control of the corn earworm on sweet corn in southern California with a nuclear polyhedrosis virus and *Bacillus thuringiensis*. *J. Econ. Entomol.*, **63**, 415–421

Oatman, E. R. and Platner, G. R. (1970). Parasitization of corn earworm larvae infesting sweet corn tassels in southern California. *J. Econ. Entomol.*, **63**, 326–327

Ohnesorge, B. (1988). Investigations on the population dynamics of maize aphids in southwestern Germany. In *International Symposium on Crop Protection*. Inst. Phytomedizin, Univ. Hobenheim, Stuttgart, pp. 1187–1193

Onazi, O. C. and Wilde, G. E. (1974). Factors affecting transmission of MDMV and its control. *Nigerian J. Entomol.*, **1**, 51–56

Ongoren, K., Kayka, N. and Turkmen, S. (1977). Investigations on the morphology, bioecology and control of the tomato fruitworm (*Heliothis armigera* Hb.) injurious to tomatoes in the Aegean region. *Bitki Koruma Bull.*, **17**, 3–28

Orlob, B. D. and Medler, J. J. (1961). Biology of cereal and grass aphids in Wisconsin (Homoptera). *Can. Entomol.*, **93**, 703–714

Palmer, M. A. (1952). *Aphids of the Rocky Mountain Region*. Thomas Say Foundation, A. D. Hirschfeld Press, Denver

Pathak, M. D. and Painter, R. H. (1958). Differential amounts of material taken up by the four biotypes of corn leaf aphid from resistant and susceptible sorghums. *Ann. Entomol. Soc. Am.*, **51**, 250–254

Pearson, M. N. and Robb, S. M. (1984). The occurrence and effects of barley yellow dwarf virus in maize in Southwestern England, UK. *Plant Pathol.*, **33**, 503–512

Phillips, W. J. and Barber, G. W. (1936). Oviposition by *Heliothis obsoleta* Fab. on the silks of corn. *Va Agric. Exp. Stn, Tech. Bull. No. 58*

Pike, K. S. and Allison, D. (1987). Effects on twospotted spider mites of insecticides applied to sweet corn for control of corn earworm. *J. Agric. Entomol.*, **4**, 327–332

Prostak, D. J. (1973). Sweet corn pest management—a better way. *Rutgers University, Ext. Bull. No. 422*

Rabichuk, A. (1985). Resistance of maize to aphids. *Informatsionnyi Bull. po Kururuze*, **4**, 149–167

Rahn, R. (1982a). *Agronomie*, **2** (8), 696–699

Rahn, R. (1982b). *Agronomie*, **2** (10), 957–962

Rainey, R. C. (1973). Airborne pests and the atmospheric environment. *Weather*, **28**, 223–239

Ramos, V. E. and Morallo-Rejesus, B. (1978). Effects of nutrition on larval coloration of *Helicoverpa armigera* (Hubner). *Philippine Entomol.*, **3**, 201–224

Ravensberg, W. J. and Berger, H. K. (1988). Biological control of the European corn borer with *Trichogramma maidis* Pintureau and Voegele in Austria in 1980–1985. In *Trichogramma and Other Egg Parasites*. Colloques de l'INRA, No. 43, pp. 557–564

Redfern, R. E., Warthen, J. D. Jr., Jacobson, M. and Stokes, J. B. (1984). Antifeeding potency of neem formulations. *J. Environ. Sci. Health (A)*, **19**, 477–481

Reid, J. A. K. and McEwen, F. L. (1979). *Canadian Entomologist*, **6**, 749–750

Rice, M. E. and Wilde, G. E. (1988). Experimental evaluation of predators and parasitoids in suppressing greenbugs (Homoptera: Aphididae) in sorghum and wheat. *Environ. Entomol.*, **17**, 836–841

Riedell, W. E. (1989). The influence of nitrogen nutrition on plant response to greenbug infestation. *J. Plant Nutrition*, **12**, 317–325

Risherson, J. V. and DeLoach, C. J. (1973). Seasonal abundance of *Perilitus coccinellae* and its coccinellid hosts and degree of parasitism in central Missouri. *Environ. Entomol.*, **2**, 138–141

Robinson, J. F., Klun, J. A., Guthrie, W. D. and Brindley, T. A. (1982). European corn borer (Lepidoptera: Pyralidae) leaf-feeding resistance: DIMBOA bioassays. *J. Kan. Entomol. Soc.*, **55**, 357–364

Rochow, W. F. (1970). Barley yellow dwarf virus. No. 32. In *Descriptions of Plant Viruses*. Commw. Mycol. Inst. Assoc. Appl. Biol., Surrey, England

Roelofs, W. L., Du, H.-W., Tang, X.-H., Robbins, P. S. and Eckenrode, C. J. (1985). Three European corn borer populations in NY based on sex pheromones and voltinism. *J. Chem. Ecol.*, **11**, 829–836

Rose, A. H., Silversides, R. H. and Lindquist, O. H. (1975). Migration flight by an aphid,

Vegetable Crop Pests

Rhopalosiphum maidis (Hemiptera: Aphididae), and a noctuid, *Spodoptera frugiperda* (Lepidoptera: Noctuidae). *Can. Entomol.*, **107**, 567–576

Saxena, R. C. (1975). *Indian Journal of Agricultural Sciences*, **45**, 434–436

Saynor, M. and Hill, D. S. (1977). *Annals of Applied Biology*, **85**, 113–120

Scaltriti, G. Pellizzari and Rezzadore, G. (1981). *Bollettino del Laboratorio di Entomologia Agraria 'Filippo Silvestri'*, **38**, 213–229

Schalk, J. M. and Radcliffe, R. H. (1977). *FAO Plant Protection Bulletin*, **25**, 9–14

Schepers, A. (1972). Control of aphid vectors in the Netherlands. In *Viruses of Potatoes and Seed-potato Production* (ed. J. A. de Bokx). PUDOC, Wageningen, pp. 167–173

Scott, D. R. (1987). Corn earworm (Lepidoptera: Noctuidae) in southwest Idaho USA: prediction of intensity of infestation. *Environ. Entomol.*, **16**, 821–824

Scott, G. E., Davis, F. M., Beland, G. L., Williams, W. P. and King, S. B. (1977). Host plant resistance is necessary for late-planted corn. *Miss. Agric. Forest. Exp. Stn Res. Rep.*, **3**, 1–4

Sekul, A. A. and Sparks, A. N. (1976). Sex attractant of the fall armyworm moth. *US Dep. Agric. Tech. Bull. No. 1542*

Shanks, C. H. Jr. and Chapman, R. K. (1965). The effects of insecticides on the behavior of the green peach aphid and its transmission of potato virus Y. *J. Econ. Entomol.*, **58**, 79–83

Showers, W. D. (1979). Effect of diapause on the migration of the European corn borer into the southeastern United States. In *Movement of Highly Mobile Insects: Concepts and Methodology in Research* (ed. R. L. Rabb and G. G. Kennedy). University Graphics, Raleigh, N. Carolina, pp. 420–430

Showers, W. D. (1981). Geographic variation of the diapause in the European corn borer. In *Insect Life History Patterns: Habitat and Geographic Variation* (ed. R. F. Denno and H. Dingle). Springer-Verlag, New York, pp. 97–111

Showers, W. D., Berry, E. C. and vonKaster, L. (1980). Management of 2nd-generation European corn borer by controlling moths outside the cornfield. *J. Econ. Entomol.*, **73**, 88–91

Showers, W. D., Reed, G. L. and Oloumi-Sadeghi, H. (1974). European corn borer: attraction of males to synthetic lure and to females of different strains. *Environ. Entomol.*, **3**, 51–58

Singh, H. and Singh, G. (1977). Biological studies on *Heliothis armigera* in the Punjab India. *Indian J. Entomol.*, **37**, 150–164

Sparks, A. N. (1979). A review of the biology of the fall armyworm. *Fla Entomol.*, **62**, 82–87

Starratt, A. N. and McLeod, D. G. R. (1982). Monitoring fall armyworm, *Spodoptera frugiperda*, moth populations in southwestern Ontario with sex pheromone traps. *Can. Entomol.*, **114**, 545–549

Stoeva, R. (1973). The cotton Noctuid *Heliothis armigera* Hb. *Rastitelna Zashchilta*, **21**, 28–31

Story, R. N., Sundstrom, F. J. and Riley, E. G. (1983). Influence of sweet corn cultivar planting date and insecticide on corn earworm (*Heliothis-zea*) damage. *J. Ga Entomol. Soc.*, **18**, 350–353

Straub, R. W. (1968). Host plant resistance in corn to the corn earworm. *Master's Thesis, University of Missouri, Columbia*

Straub, R. W. (1977). European Corn Borer control in early sweet corn: Role of pre-silk applications and leaf feeding resistance. *J. Econ. Entomol.*, **70**, 524–526

Straub, R. W. (1983). Minimization of insecticide treatment for first-generation European corn borer (Lepidoptera: Pyralidae) control in sweet corn. *J. Econ. Entomol.*, **76**, 345–348

Straub, R. W. (1984). Maize dwarf mosaic virus: symptomatology and yield reactions of susceptible and resistant sweet corns. *Environ. Entomol.*, **13**, 318–323

Straub, R. W. (1986). Granular fonofos: efficacy and timing of applications for European corn borer management. *J. Agric. Entomol.*, **3**, 286–291

Straub, R. W. (1989). Red-winged blackbird damage to sweet corn in relation to infestations of European corn borer (Lepidoptera: Pyralidae). *J. Econ. Entomol.*, **82**, 1406–1410

Straub, R. W. and Boothroyd, C. W. (1980). Relationship of corn leaf aphid and maize dwarf mosaic disease to sweet corn yields in southeastern New York. *J. Econ. Entomol.*, **73**, 92–95

Straub, R. W. and Hogan, H. J. (1974). Feasibility of fall armyworm, *Spodoptera frugiperda* (Smith), control on late-planted dent corn. *New York State Agric. Exp. Stn Geneva, Food and Life Sci. Bull. No. 49*

Straub, R. W. and Huth, P. C. (1976). Correlations between phenological events and European corn borer activity. *Environ. Entomol.*, **5**, 1079–1082

Straub, R. W. and VanKirk, J. R. (1987). Insect management guide for New York sweet corn growers. *New York State Agric. Exp. Stn Handbook*

Szatmari-Goodman, G. and Nault, L. R. (1983). Tests of oil sprays for suppression of aphid-borne maize dwarf mosaic virus in Ohio sweet corn. *J. Econ. Entomol.*, **76**, 144–149

Talman, J. H., Harris, C. R. and Tomlin, A. D. (1981). *Proceedings of the Twenty Eighth Annual Meeting, Canadian Pest Management Society*, pp. 109–112

Taylor, L. R. (1958). Temperature relations of teneral development and behaviour in *Aphis fabae* Scop. *J. Exp. Biol.*, **34**, 189–208

Taylor, L. R. (1977). Aphid forecasting and the Rothamsted insect survey. *J. Roth. Agric. Soc.*, **138**, 114–144

Theunissen, J. A. B. M. (1977). *Mededelingen Landbouwhogeschool Wageningen*, 77–12

Thompson, A. R. and Wheatley, G. A. (1973). *Proceedings of the Seventh British Insecticide and Fungicide Conference*, Vol. 2, pp. 535–544

Thompson, M. S. and All, J. N. (1984). The use of oviposition on artificial substrates as a survey tool for the fall armyworm. *Fla Entomol.*, **67**, 349–357

Tomlin, A. D., Miller, J. J., Harris, C. R. and Tolman, J. H. (1985). *Journal of Economic Entomology*, **78**, 975–981

Triplehorn, C. A. (1959). The possible effect of weather on incidence of corn leaf aphid infestation and damage. *Proc. N.C. Branch, Entomol. Soc. Am.*, **14**, 28–29

Tu, J. C. and Ford, R. E. (1971). Factors affecting aphid transmission of maize dwarf mosaic virus to corn. *Phytopathology*, **61**, 1516–1521

Uvah, I. I. I. and Coaker, T. H. (1984). *Entomologia Experimentalis et Applicata*, **36**, 159–167

Vargas, R. and Nishida, T. (1978). Evaluation of corn earworm damage and incidence of blank tip in corn grown in Hawaii and its significance to pest management. *Proc. Hawaiian Entomol. Soc.*, **23**, 455–461

Vargas, R. and Nishida, T. (1982). Parasitization by *Trichogramma chilonis* (Hymenoptera: Trichogrammatidae) of corn earworm eggs on sweet corn in Hawaii, USA. *Proc. Hawaiian Entomol. Soc.*, **24**, 123–126

Vernon, R. S. and Bartel, D. L. (1985). *Environmental Entomology*, **14**, 210–216

Vernon, R. S. and Borden, J. H. (1979). *Journal of the Entomological Society of British Columbia*, **76**, 12–16

Vernon, R. S. and Borden, J. H. (1983). *Environmental Entomology*, **12**, 650–655

Vosselman, L. (1978). *Chromosoma*, **67**, 201–218

Waiss, A. C., Chan, B. G., Eliger, C. A., Wiseman, B. R., McMillian, W. W., Widstrom, N. W., Zuber, M. S. and Keaster, A. J. (1979). Maysin, a flavone glycoside from corn silks with antibiotic activity toward corn earworm. *J. Econ. Entomol.*, **72**, 256–258

Walter, E. V. (1962). Sources of earworm resistance for sweet corn. *Proc. Am. Soc. Hort. Sci.*, **80**, 485–487

Whalen, J. and Breeding, R. (1988). Insect control on sweet corn, 1987. *Insect. Acar. Tests*, **13**, 120–121

Whalen, J. and Breeding, R. (1989). Control of sweet corn, 1988. *Insect. Acar. Tests*, **14**, 121

Whalen, J. and Vanderhoef, H. (1985). Whorl applications to control fall armyworm in sweet corn. *Insect. Acar. Tests*, **10**, 111

Whitfield, G. H., Carruthers, R. I. and Haynes, D. L. (1985). *Agriculture, Ecosystems and Environment*, **12**, 189–200

Whitman, R. J. (1975). Natural control of the European corn borer, *Ostrinia nubilalis* (Hubner) in New York. *Ph.D. Thesis, Cornell University, Ithaca, New York (Diss. Abstr. B, Sci. and Eng., 2611B)*

Widstrom, N. W., McMillian, W. W. and Wiseman, B. R. (1970). Resistance in corn to the corn earworm and the fall armyworm. IV. Earworm injury to corn inbreds related to climatic conditions and plant characteristics. *J. Econ. Entomol.*, **63**, 803–807

Wiesenborn, W. D. and Trumble, J. T. (1988). Optimal oviposition by the corn earworm (Lepidoptera: Noctuidae) on whorl-stage sweet corn. *Environ. Entomol.*, **17**, 722–726

Windels, M. B. and Chiang, H. C. (1975). Distribution of second-brood European corn borer egg masses on field and sweet corn plants. *J. Econ. Entomol.*, **68**, 133

Wiseman, B. R. and Davis, F. M. (1979). Plant resistance to the fall armyworm. *Fla Entomol.*, **62**, 123–130

Wiseman, B. R., Painter, R. H. and Wassom, C. E. (1966). Detecting corn seedling differences in the greenhouse by visual classification of damage by the fall armyworm. *J.*

Econ. Entomol., **59**, 1211–1214

Wiseman, B. R., Wassom, C. E. and Painter, R. H. (1967). An unusual feeding habit to measure differences in damage to 81 Latin American lines of corn by the fall armyworm, *Spodoptera frugiperda* (J. E. Smith). *Agron. J.*, **59**, 279–281

Wiseman, B. R. and Widstrom N. W. (1986). Mechanisms of resistance in 'Zapalote Chico' corn silks to fall armyworm (Lepidoptera: Noctuidae) larvae. *J. Econ. Entomol.*, **79**, 1390–1393

Wiseman, B. R., Williams, W. P. and Davis, F. M. (1981). Fall armyworm: resistance mechanisms in selected corns. *J. Econ. Entomol.*, **74**, 622–624

Wood, J. R., Poe, S. L. and Leppla, N. C. (1979). Winter survival of fall armyworm pupae in Florida. *Environ. Entomol.*, **8**, 249–252

Yin, Y. S. and Chang, J. Y. (1987). A comparative study of the utilization of introduced *Trichogramma* sp. and indigenous species against insect pests. *Nat. Enemies of Insects*, **9**, 45–47

Zitter, T. A. and Simons, J. N. (1980). Management of viruses by alternation of vector efficiency and by cultural practices. *Ann. Rev. Phytopathol.*, **18**, 289–310

Zungia, G. E., Argandona, V. H., Niemeyer, H. M. and Corcuera, L. J. (1983). Hydroxamic acid content in wild and cultivated *Gramineae*. *Phytochemistry*, **22**, 2665–2668

Zurlini, G. and Robinson, A. S. (1978). *Entomologia Experimentalis et Applicata*, **23**, 279–286

CHAPTER 8

Pests of Solanaceous Crops

R. G. McKinlay, A. M. Spaull and R. W. Straub

Potatoes

8.1 INTRODUCTION

The potato, *Solanum tuberosum* L., is one of the world's staple food crops. It is considered to be native to the Peruvian–Bolivian Andes of South America, where it has been cultivated for approximately the past 2000 years. The Spanish introduced potatoes to Europe during the second half of the sixteenth century. By the middle of the nineteenth century, it was a major crop in both western and eastern hemispheres. Potatoes, which are nutritious, supplying some vitamins and amino acids, are used either raw or cooked. They are ground into flour which is used in baking and as a sauce thickener, or they are cooked and served whole or mashed as an easily digestible, starchy food. Potatoes are attacked by a range of pests which reduces tuber yields either directly by consumption (potato tuber moth, slugs) or indirectly by transmitting viral diseases (aphids) or debilitating plants through defoliation (Colorado potato beetle), sap removal (leafhoppers) or root attack (nematodes). Being a vegetatively propagated crop, infection of parent potato plants with systemic virus diseases will reduce the yields of progeny plants grown from parent plant tubers.

8.2 APHIDS

8.2.1 *Myzus persicae* (Sulzer): Green Peach Aphid, Peach-potato Aphid (Hemiptera: Aphididae)

Myzus persicae is one of the most important insect pests of crops in the world because it is such a very efficient vector of plant viruses. It has been shown to transmit more than 100 viruses to plants belonging to various families—e.g. Chenopodiaceae, Leguminosae, Solanaceae, Compositae, Cruciferae and Cucurbitaceae (Kennedy *et al.*, 1962). We shall see shortly why aphids are such efficient plant virus vectors, after outlining the distribution of *M. persicae* in the world, and its description.

Although *M. persicae* probably came from Asia (Blackman and Eastop, 1984), it now has a world-wide distribution: Europe, Middle East, Far East, Africa, Australasia, South and North America. It occurs as far north as southern Scandinavia and northern China and as far south as the southern tip of South America.

Adult apterae of *M. persicae*—the wingless, parthenogenetic females—are small- to medium-sized aphids (1.25–2.5 mm long), with a barrel-shaped body (Figure 8.1). They vary in colour from pale green to pink and are more deeply pigmented green or crimson with blackish hues under cold conditions. Apterae are uniformly coloured, not shiny (Blackman and Eastop, 1984). Adult alatae—the winged, parthenogenetic females—are similar in size and colour to the apterae except for a central black patch on the upper surface of the abdomen.

The main reason aphids are such efficient vectors of plant viruses is that they possess specialised piercing–sucking mouthparts (Forbes, 1977). All insects in the order Hemiptera—to which aphids belong—possess these mouthparts. Most hemipteran insects feed on plants, but some are predators of other insects, and therefore beneficial, and others suck the blood of higher animals. The front of the head of hemipteran insects is drawn out into a snout in which is located the piercing–sucking mouthparts. They lie in a groove on the aphid's snout (or rostrum or proboscis), forming a stylet bundle. The stylets are not all used to suck food. The outer pair of mandibular stylets are solid and are therefore not involved in the process of food ingestion. They give physical support to the inner pair of

Figure 8.1 Adult aptera (wingless, parthenogenetic female) of *Myzus persicae*. © R. McKinlay

maxillary stylets which together form a hypodermic needle-like structure involved directly in ingestion. As well as a food canal, the maxillary stylets also have a salivary duct.

The stylets are used to penetrate and tap the phloem tissue of plants (Pollard, 1977). The mandibular stylets lead the maxillary stylets, with each mandible alternating with the other during the process of penetration. The stylet bundle usually follows an intercellular path on its way to the vascular tissue. The aphid secretes saliva which gels to form a salivary sheath. Pectinase enzymes are also secreted with the saliva to help the intercellular passage of the stylets. Once the stylets have penetrated the sieve tubes of the phloem, the aphid will usually settle down for a prolonged feed. Phloem sap travels along the maxillary stylets as a result of a combination of plant turgor pressure and muscular contractions of the aphid's cibarial pump. The hypodermic needle-like mouthparts of an aphid are then an ideal mechanism for both acquiring and transmitting plant viruses. The effectiveness of *M. persicae* as a virus vector is the result not only of possessing specialised mouthparts, but also of exhibiting a restless behaviour pattern with repetitive probing of leaves (van Emden *et al.*, 1969) when it alights on plants.

Aphids can exhibit two types of life cycle: holocyclic and anholocyclic. Holocyclic life cycles have an annual sexual phase. In contrast, anholocyclic life cycles do not produce sexual forms. Otherwise, aphids with holocyclic and anholocyclic life cycles reproduce parthenogenetically. *M. persicae* has holocycles and anholocycles. We shall consider first the holocyclic and then the anholocyclic life cycles.

In temperate latitudes *M. persicae* with a holocyclic life cycle alternate between a primary woody host plant in the winter and a secondary herbaceous host plant in the summer. The primary hosts are certain *Prunus* L. spp., principally *P. persica* Batsch (the peach). Secondary hosts are to be found in more than 40 different plant families (Blackman and Eastop, 1984), among them being Solanaceae (e.g. potato), Chenopodiaceae (e.g. sugar beet), Compositae (e.g. lettuce), Cruciferae (e.g. brassicas) and Cucurbitaceae (e.g. cucumber). Aphids which alternate annually between primary and secondary host plants are said to be heteroecious.

Eggs of *M. persicae* are laid on the bark of peach trees by sexual adult females, called oviparae, during the autumn. Oviparae mate with males and then lay between 4 and 13 fertilised eggs each, usually in crevices around and in axillary buds (van Emden *et al.*, 1969). The eggs, which are extremely cold-resistant, hatch in the early spring into parthenogenetic females called fundatrices. The fundatrices each give birth viviparously to about 50 or 60 progeny—the fundatrigeniae—which feed on opened buds, flowers, leaves and shoots. Eight wingless (apterous) generations of fundatrigeniae may pass on peach trees before winged (alate) forms are produced. Alatae fly from peach trees to secondary host crop plants, where they may each produce about 20 apterous offspring on 7–10 plants (van

Emden *et al.*, 1969). These apterae reproduce parthenogenetically, to yield large numbers of aphids. Aphid infestations on crops are commonly uniform, but not necessarily similar. Alate aphids are produced periodically throughout the summer. These aphids may colonise other crop plants in the same field or further afield. Towards the end of the summer, the aphid population produces alatae—males and females—which fly back to peach trees. These alatae are produced on the secondary host plants in response to declining day length and temperature. The female alatae—called gynoparae—produce parthenogenetically the sexual females, the oviparae, which mate with the males and subsequently lay eggs. Thus, the holocyclic life cycle of *M. persicae* has come full circle.

M. persicae with an anholocyclic life cycle do not overwinter as eggs on peach trees, but as active adults and nymphs on crops and weeds. Overwintering host plants include cruciferous crops (e.g. brassicas), potatoes, beets, peach nursery stock and various weeds (van Emden *et al.*, 1969). The lower temperature threshold for reproduction and development of *M. persicae* has been estimated by Pozarowska (1987) to be close to 2°C. The lower temperature threshold for survival of *M. persicae* is approximately −10°C to −15°C (Clough *et al.*, 1990). One of the selective advantages of anholocycly over holocycly is that overwintering active stages produce populations earlier in the spring than do overwintering eggs. The earlier populations develop in the spring; the younger and therefore more susceptible will be crop plants to the transmission of aphid-borne viruses. For instance, a succession of mild winters from 1970 to 1976 in Scotland led to early spring migrations of aphids which dramatically increased the incidence of leaf roll virus disease in seed potato stocks (Howell, 1977).

Androcyclic populations of *M. persicae*, which are neither holocyclic nor anholocyclic but produce some males to contribute to the sexual phase, are found in some areas—e.g. Scotland (Pozarowska, 1987) and England (Blackman, 1971). *M. persicae* in warm temperate regions is exclusively anholocyclic (van Emden *et al.*, 1969).

Aphids damage plants directly by sucking sap and indirectly by transmitting viruses.

The direct sap-sucking injury can, if aphid numbers are high at a time when plants are under stress from lack of water, lead to wilting and a loss of yield. More will be said of this when we consider *Macrosiphum euphorbiae* Thomas, the potato aphid (see page 272). Product quality can also be reduced by aphid feeding and the production of honeydew. Aphid saliva is toxic to plants, which commonly respond by forming necrotic spots at feeding sites. These blemishes can spoil lettuces and cucumbers. Because phloem sap is much more rich in sugars than in amino acids which aphids need for growth, much of the sap is excreted from the anus as honeydew. When aphid populations are large, this sugar-rich honeydew can cover leaf surfaces, forming an ideal substrate for the growth of sooty mould fungi. As well as contaminating crop products (e.g. Brussels sprouts and tomatoes), these fungi reduce the photosynthetic efficiency and thereby the final yields of affected plants.

M. persicae transmits two types of plant viruses—persistent and non-persistent. The difference between the two types of virus is one of time available to transmission: once infective, the aphid vector can transmit a persistent virus from then for the rest of its life; on the other hand, a non-persistent virus can be transmitted following acquisition by an aphid vector usually for a period of a few hours only. For aphids to remain vectors of non-persistent viruses, they need to become continuously reinfective by feeding on diseased plants. The persistent viruses appear to be closely associated with the accessory salivary gland of the aphid vectors (Gildow, 1987). The non-persistent viruses have such a tenuous relationship with the aphid vector that they are considered to be simply lying on the stylets and can be readily sloughed off during probing and moulting. Persistent viruses are normally transmitted by a few aphid species only, but non-persistent viruses can be transmitted not uncommonly by any vagrant aphid which stops to probe as it passes through a crop. The suppression of the spread of non-persistent viruses by applying insecticides to control the aphid vectors is a much more difficult objective to achieve, therefore, than controlling the spread of persistent viruses.

The principal persistent and non-persistent viruses spread by *M. persicae* to solanaceous (potatoes and tomatoes) as well as chenopodiaceous (sugar beet), composite (lettuce), cruciferous (leaf and root brassicas) and cucurbit (cucumber, etc.) crops are shown in Table 8.1.

Aphids are controlled naturally by predators, parasites and pathogens. Among the predators of aphids are various beetles (ladybirds, Coccinellidae; ground beetles, Carabidae; rove beetles, Staphylinidae), lacewings (Neuroptera), hoverflies (Syrphidae), cecid flies (Cecidomyiidae), anthocorid bugs (Anthocoridae) and birds (e.g. tits). The parasites (or strictly parasitoids) of aphids are wasps (Hymenoptera) of the families Aphidiidae and Aphelinidae. Aphids can become infected with fungal pathogens (e.g. *Entomophthora* Fresenius spp. (Entomophthorales: Entomophthoraceae).

Biological control is the deliberate manipulation of natural enemies to suppress pest species (McKinlay, 1988). The European ladybird beetle aphid predator *Coccinella septempunctata* L. has been introduced to New York State, USA, where it has successfully contributed to the biological control of aphids on potatoes (Kogan, 1986). The evidence for the role of ground beetles as natural enemies of aphids appears to be conflicting. Dixon (1986) concluded that ground beetles could be important in controlling aphids on potatoes, but Boiteau (1986) considered that they were likely to play a relatively minor role, with aphid-specific, canopy-level natural enemies assuming greater importance.

Cultural control is the deliberate alteration of a crop production system

Table 8.1 Principal viruses spread by *Myzus persicae* to vegetable crops

Plant family	Virus	Persistent (P), non-persistent (NP)	Disease symptoms	Geographical distribution
Solanaceae	Potato leafroll[a]	P	Primary infection: pallor of tip leaves, which become rolled and assume an erect habit. Secondary infection (plants grown from infected tubers): stunting of shoots, upward rolling of leaflets especially on lower leaves. Note: internal net necrosis in tubers of some varieties	World-wide
	Potato virus Y[a]	NP	Depending on variety, causes mild to severe mottle of leaves (Y^N strain), streak or 'leaf-drop streak' with necrosis along the veins of the underside of the leaflets (Y^o strain), or 'stipple-streak' (Y^C strain)	World-wide
	Tomato aspermy[a]	NP	Severe leaf distortion and usually seedless fruits	Western Europe, Russia, USA, Canada, India, Japan, Australia and New Zealand
Chenopodiaceae	Beet mild yellowing	P	Orange-yellow colour on leaves of sugar beet (Smith, 1972)	Europe, USA (Lange, 1987)
	Beet mosaic	NP	Mosaic pattern of light and dark areas on leaves of sugar beet (Lange, 1987)	World-wide[a]
	Beet western yellows[a]	P	Stunting and chlorosis of sugar beet. BWYV also infects lettuce, broccoli, cauliflower, radish and turnip	Europe, North America, Asia; probably common throughout world

Family	Virus		Description	Distribution
	Beet yellows	semi-P	Partial to complete yellowing of sugar beet leaves (Lange, 1987), followed by development of necrotic tissue and thickening. Leaves feel leathery and are brittle to touch	World-wide[a]
Compositae	Lettuce mosaic[a]	NP	Various mosaic and mottle symptoms in almost all types of lettuce	World-wide
Cruciferae	Cauliflower mosaic[a]	NP	Induces mosaic and mottle diseases of many cruciferous crop plants. Often found in mixed infections with turnip mosaic virus	World-wide
	Turnip mosaic[a]	NP	Causes mottling and black necrotic spots and ringspots in cabbage, cauliflower and Brussels sprouts; mosaic with leaf distortion and stunting in turnip, swede, radish, rape, mustard, Chinese cabbage, watercress and horseradish	World-wide
Cucurbitaceae	Cucumber mosaic[a]	NP	Causes mosaic of cucumber and many other cucurbits, blight of spinach, fern leaf of tomato, and mosaics of many other species of dicotyledonous and monocotyledonous crops. The virus also often causes plant dwarfing and flower breaking	World-wide
	Watermelon mosaic virus 2[a]	NP	Causes mosaic and mottle diseases of cantaloupe, cucumber, pumpkin, squash and watermelon. Reduces fruit production and quality in squash and other cucurbits	Widely distributed throughout the world

[a]Commonwealth Agricultural Bureaux/Association of Applied Biologists, 1970, 1971, 1972, 1979, 1981, 1984. Commonwealth Mycological Institute, Ferry Lane, Kew, Surrey, England.

to the disadvantage of pests (McKinlay, 1988). Among cultural measures recommended to reduce aphid-borne transmission of viruses to potatoes in the Netherlands, Harrewijn (1983) recommends applying less nitrogen fertiliser. Mixed plant stands can help to reduce aphid populations. In Scotland McKinlay (1985) found that numbers of peach-potato aphids were generally smaller on potatoes undersown with grass than on potatoes not undersown with grass. However, mixed cropping practice may incur a yield penalty.

Aphids are routinely controlled on solanaceous, chenopodiaceous, composite, cruciferous and cucurbit crops by insecticides. Insecticides are likely to be used regularly to prevent virus transmission and less regularly, even only occasionally, to prevent direct feeding losses. Granules of systemic organophosphorus or carbamate insecticides (e.g. phorate and thiofanox, respectively) incorporated into the soil at sowing or planting protect young plants from early aphid colonisation. Subsequent control is given by foliar sprays of organophosphorus or carbamate insecticides (e.g. demeton-*S*-methyl and pirimicarb, respectively). Aphid spray warning schemes based on monitoring are available for potato and sugar beet growers in Europe (Robert, 1987).

Insecticidal granules have been found to give a consistent and substantial reduction in the spread of aphid-borne virus disease in potatoes (McKinlay and Franklin, 1983). By contrast, sprays have not. McKinlay and Franklin (1983) suggested that the occasionally poor suppression of potato leaf roll virus spread by foliar sprays was previously poor control of the peach-potato aphid population on the lower leaves of the potato crop.

Principally because of the difference in the times required by *M. persicae* to acquire and transmit persistent and non-persistent plant viruses, the spread of persistent viruses is much more easily suppressed by insecticides. Pyrethroids, mineral oils, and mixtures of pyrethroids and mineral oils as well as pyrethroids and organophosphates are being tested for aphid control and suppression of (usually non-persistent) virus spread. Mineral oils which limit virus spread by aphid vectors in an as yet unknown way are used in Switzerland to suppress potato virus spread (Cornu and Gehriger, 1981). The repellency/antifeedant activities of the pyrethroids have been combined with the systemic toxicity of the organophosphates in a commercial product for aphid virus control on potatoes and sugar beet in the UK. This product is a mixture of the pyrethroid deltamethrin and the organophosphate heptenophos.

Insecticide resistance is an increasingly important problem. By 1984 multiple resistance to organochlorine, organophosphorus, carbamate and pyrethroid insecticides was widely distributed in peach-potato aphids (Metcalf, 1986). In southern and eastern England Ffrench-Constant and Devonshire (1986) reported an increased frequency of very resistant peach-potato aphids, so-called R_2 and R_3 variants. Two other variants exist: R_1, moderately resistant, and S, susceptible. An increasing frequency of peach-potato aphids very resistant to organophosphorus, carbamate and pyrethroid insecticides may lead to serious problems of controlling virus spread in susceptible crops, particularly if *M. persicae* continues to be selected for resistance as a result of the application of insecticidal sprays on other crops. Dewar *et al.* (1988) have drawn attention to the possible influence of the selection of resistant *M. persicae* colonising vegetable brassica crops on virus yellows spread in sugar beet in England. However, the integration of resistance into the aphid genome may result in a selective disadvantage to the R allele in the absence of insecticide. McKinlay (1988) suggested that the 'cost' of selecting for insecticide resistance in peach-potato aphid populations may be diminished cold-hardiness and, consequently, reduced overwintering success. Roush and McKenzie (1987), noting that the mechanism of resistance in peach-potato aphid populations—gene duplications leading to an increase in the quantity of a carboxylesterase (E4)—comprises a relatively large amount (about 3%) of total body protein, observed that 'the most serious and consistent [fitness] disadvantages often seem to be associated with resistance due to generalized esterases'. The best strategy to limit the evolution of resistance would seem to be to use a high dose of a rapidly degraded insecticide in a circumscribed area (to allow immigration of non-selected insects) (Taylor, 1986). Resistance evolves faster when the opposite conditions prevail.

Integrated control (or integrated pest management) is a package of mutually compatible control measures, preferably involving minimal use of insecticides, designed to suppress pest species (McKinlay, 1988). Few examples exist of successfully integrated pest control programmes on potatoes. Radcliffe (1982) suggested that one of the main reasons for this is a lack of experimentally determined economic thresholds. McKinlay (1988) suggested another reason: the demanding attention to detail that must be paid by growers to monitoring aphid numbers, often at the expense of other, perhaps commercially more important, farm operations at the time. Aphid monitoring was perceived by potato growers in Scotland to be the responsibility of remote specialists (McKinlay and Kerr, 1985).

Integrated control procedures have been developed in many countries to combat yellows disease of sugar beets: England, France, Germany and the Netherlands in Europe; and California, Oregon, Washington and the Central Plains states in the USA (Lange, 1987). In California the integrated control programme included planting resistant seed; avoiding virus sources and following the 'beet-free' recommended planting and harvesting dates for each district; avoiding peaks in aphid flights; practising good cultural methods such as crop clean-up following harvest, weed control, and proper irrigation, fertilisation and spacing; using pesticides judiciously; protecting natural enemies, when possible, by using systemic granular insecticides; and watching for resurgences of minor pests while monitoring insect populations on beets during the season and particularly during the early developmental period.

Plant resistance—the ability of a plant species or variety to grow and ultimately produce a harvestable yield despite the presence of a pest species (McKinlay, 1988)—is pivotal to the success of integrated control programmes. Some wild potato species possess a trichome-mediated resistance mechanism to the peach-potato aphid (Gregory *et al.*, 1986). Mndolwa *et al.* (1984) concluded that breeding for resistance to peach-potato aphids may not be as promising for potato leaf roll virus control as developing leaf roll-resistant varieties.

8.2.2 *Macrosiphum euphorbiae* (Thomas): Potato Aphid (Hemiptera: Aphididae)
Aulacorthum solani (Kaltenbach): Glasshouse Potato Aphid (Hemiptera: Aphididae)

Adult aphids of both of these species are larger than adult *M. persicae* (see page 263): *M. euphorbiae* 2–3.5 mm (Figure 8.2), *A. solani* 2–3 mm. *M.*

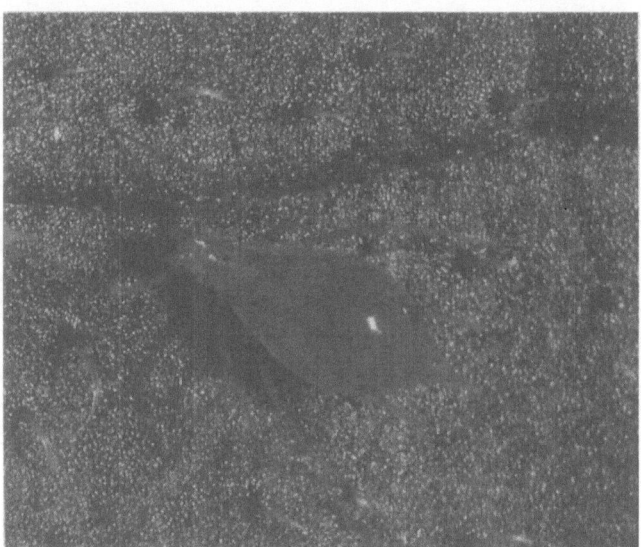

Figure 8.2 Adult female potato aphid, *Macrosiphum euphorbiae*. © Crown copyright

euphorbiae is either green or pink in colour and *A. solani* varies in colour from yellow-green to almost brown. Both species have an almost world-wide distribution, with *M. euphorbiae* originating in North America and *A. solani* originating probably in Europe (Blackman and Eastop, 1984). Like *M. persicae*, they exhibit both holocycly and anholocycly. *M. euphorbiae* has a heteroecious holocyclic life cycle in north-eastern USA, with wild or cultivated roses as primary host plants (Blackman and Eastop, 1984). It is probably mainly anholocyclic in Europe (Möller, 1971), overwintering as apterae on herbaceous hosts or sprouting potatoes in store. The egg stage of the heteroecious holocyclic life cycle of *A. solani* can overwinter on many different herbaceous host plants, with the common foxglove (*Digitalis purpurea* L.) being a principal primary host. Apterae are capable of overwintering on sprouting potatoes. As suggested by its common name, *A. solani* can breed continuously in glasshouses. The secondary host plants of *M. euphorbiae* and *A. solani* in the field are principally solanaceous (potato, tomato), but Compositae (lettuce) are also colonised. Like *M. persicae*, *M. euphorbiae* and *A. solani* damage host plants directly by sucking sap and indirectly by transmitting viruses. They are not as efficient as *M. persicae* at transmitting viruses but are nevertheless each capable of transmitting 35–45 plant viruses. Both aphid species transmit potato leaf roll virus, potato virus Y and tomato aspermy virus: in contrast to *A. solani*, *M. euphorbiae* is capable of transmitting lettuce mosaic virus (Kennedy *et al.*, 1962; see Table 8.1 for descriptions of viruses). While individual *M. euphorbiae* and *A. solani* may not be as efficient virus vectors as individual *M. persicae*, numerically dominant populations of *M. euphorbiae* and *A. solani* may compensate partially for poorer transmission—for example, although Tamada and Harrison (1981) found that potato leaf roll virus was transmitted much less efficiently by *M. euphorbiae* than by *M. persicae*, *M. euphorbiae* is numerically dominant to *M. persicae* every year on Scottish potato crops and may, therefore, be responsible for more of the spread of the virus. Large numbers of *M. euphorbiae* on potato plants can lead to a disorder called top-roll. This disorder of the upper leaves of plants—typified by light green spots on young leaves, followed soon after by an upward rolling of the leaf margins around the midrib, with the margins becoming violet and necrotic (Veen, 1985)—is caused directly by the sap-sucking activity of many *M. euphorbiae*. Veen (1985) suggested that photosynthesis in leaves showing top-roll is inhibited by impaired phloem transport, with subsequent accumulation of carbohydrates. Once plants are showing top-roll symptoms, tuber yields have been substantially reduced. Large infestations of *M. euphorbiae* can reduce potato tuber yields by more than 60% in the field (Gibson, 1974). To prevent yield loss in the UK, a single spray of insecticide is recommended to be applied when there are 3–5 aphids to a complete leaf. Control of *M. euphorbiae* and *A. solani* is essentially the same as for *M. persicae*.

8.3 LEAFHOPPERS

8.3.1 *Empoasca fabae* Harris: Potato Leafhopper (Hemiptera: Cicadellidae)

The potato leafhopper is an important pest of potatoes in north-central and north-eastern USA and eastern Canada. The ecology of *Empoasca fabae* has recently been reviewed by Hower (1987).

The adult potato leafhopper is about 3 mm long and pale green in colour. It has a narrow body which bears long somewhat pointed wings and three pairs of walking legs. The hindlegs, which are longer and larger than the others, have tibiae armed with spines. Adults are very active and walk with great agility. The five nymphal stages are essentially smaller versions of the adult insect, with a similar shape and colour. Eggs are white in colour, banana-shaped and not longer than 1 mm. They are inserted into the leaf veins or petioles of the host plant.

In the USA the potato leafhopper migrates northward each spring from its overwintering sites in southern Louisiana and northern Florida. It is unable to overwinter successfully in northern USA and Canada, because of low temperatures. Adult leafhoppers overwinter on leguminous host plants. The annual northward migration of *E. fabae* in the spring is associated with warm north winds and air masses which move up the Mississippi valley. The migrant insects are not usually found before mid-May on alfalfa in the north-central and north-eastern states. Soybeans and potatoes are infested between 2 and 4 weeks later. Potato leafhoppers infest Canadian potato crops during June.

Following crop colonisation, breeding is continuous, with 3–4 generations developing each year. The migrant leafhoppers are predominantly female. Each female lays between 33 and 200 eggs at a rate of 2 or 3 eggs a day (Delong, 1971). On potatoes, eggs are laid more frequently in the terminal leaflets and on leaves between apical and basal leaves than in either of these two extremes. Eggs and nymphs will not develop at a temperature less than 8.4°C (Hogg, 1985). The upper developmental threshold temperature appears not to be as marked as the lower threshold, with development continuing to occur at temperatures in excess of 30°C. Nymphal development takes approximately 12–15 days (Delong, 1971). On potatoes, population growth is continuous during the life of the crop.

Flying by *E. fabae* is dependent on temperature and light. In late summer temperature and light conditions change as cold fronts approach from the north. Taylor and Reling (1986) suggested that adult potato leafhoppers are stimulated to fly in response to these changing environmental circumstances. The insects are then transported by the northerly airflows behind the fronts in a south to south-westerly direction back to their overwintering sites in the southerly regions of the states bordering the Gulf of Mexico.

The potato leafhopper causes the foliage of infested potatoes to become curled and yellow in colour. The characteristic yellowing or browning of affected foliage is known as 'hopperburn'. Hower (1987) reports that injury to the plant has not been precisely defined, although a salivary sheath appears to be produced at the feeding site and the vascular system, primarily the phloem, becomes plugged, restricting food transport within the plant. Injured plants are typically stunted and yields are reduced. Johnson *et al.* (1986) found that the potato leafhopper caused an average maximum yield reduction of 54% in potato cultivars Russet Burbank, Norland and Red Pontiac during 1983 and 1984 in Minnesota, USA. Potato leafhoppers cause greatest yield reductions in potatoes when the tubers are rapidly bulking. Not surprisingly, perhaps, potato leafhoppers reduced photosynthesis and respiration of infested potato plants (Ladd and Rawlins, 1965). The feeding of *E. fabae* causes a localised increase in the concentration of carbohydrates, particularly sucrose, which seems to stimulate further feeding by the insect. The potato leafhopper is not known to transmit diseases, although other closely related leafhoppers are known to be disease vectors.

Hower (1987) states that little information exists on biotic agents affecting potato leafhopper population growth. Among the natural enemies of *E. fabae* are tetragnathid, oxyopid, salticid and thomisid spiders; nabid, mirid and anthocorid bugs; chrysopid lacewings; coccinellid beetles; dolichopodid flies; ants; sphecid wasps; and an *Erynia* Nowakowski sp. (Entomophthorales) fungus. None has yet been manipulated to exert a degree of biological control. Although potato leafhopper populations on alfalfa are reduced by the presence of grass weeds (Lamp *et al.*, 1984), mixed cropping as a cultural technique to control leafhopper on potatoes has still to be tested. In contrast to glabrous plants, pubescent soybeans are virtually immune from *E. fabae* infestation (Lee *et al.*, 1986). Similarly, *E. fabae* may not colonise potato cultivars with trichomes.

Potato leafhoppers are routinely controlled by insecticidal sprays (Radcliffe, 1982) of, for instance, azinphos-methyl, carbaryl or fenvalerate. The sprays are applied to the foliage every 2 weeks from first appearance of adult leafhoppers. Granular systemic insecticides can be applied at planting to give early season control and a substantial reduction of leafhopper-borne disease (Radcliffe, 1982).

8.4 BEETLES

8.4.1 *Leptinotarsa decemlineata* (Say): Colorado Potato Beetle (Coleoptera: Chrysomelidae)

The Colorado potato beetle is a serious pest of potatoes. As its name suggests, it originated and spread from the eastern slopes of the Rocky Mountains in western USA. Casagrande (1985) considers that the beetle should be more properly called the 'Iowa potato beetle', because it was first collected in what is now the border of Iowa and Nebraska. Whether Colorado or Iowa, the beetle began about the middle of the nineteenth century to move from its natural wild solanaceous host plants (e.g.

buckthorn (*Rhamnus* L. spp.)) to colonise the newly introduced, cultivated solanaceous crop, the potato. It spread across the USA and into Canada and was first reported in Europe (Germany) during 1877 (Bartlett, 1982). The Colorado potato beetle is at present distributed throughout the USA, southern Canada, Mexico, Costa Rica and most of Europe. It has to date been kept out of the UK.

The adult Colorado potato beetle is an attractive-looking insect. It has a yellow-coloured oval-shaped body, about a centimetre long, with irregular black markings on the thorax and five black stripes along each elytron (Figure 8.3). The larva is also attractively coloured (Figure 8.4): the young larva is brown; the older larva is bright pink to red. Each larva has a distinct head with biting mouthparts, three pairs of walking legs and a humped-back body with two rows of black dots along either side.

Adult *L. decemlineata* overwinter about 30 cm deep in the soil. They become active during May, crawl up to the soil surface and fly if the air temperature is high enough, usually more than 10°C. Males and females mate, with females each laying several hundred eggs on the lower surfaces of leaves of host plants. Larvae hatch from these eggs in approximately 7 days and begin feeding voraciously on the leaves. The larvae take about 2–3 weeks to develop fully, during which time they pass through four larval instars. When fully grown, the larvae fall to the soil surface, burrow into the soil and pupate for 2–3 weeks. Adult beetles will either emerge from these pupae to give rise to another generation before winter (e.g. USA) or not emerge from these pupae and overwinter in the soil (e.g. mainland Europe).

Both adult and larval Colorado potato beetles damage plants by feeding on their leaves (see Figures 8.3 and 8.4). As a result of the gregarious behaviour of the beetles, infested plants can become quickly defoliated, leaving only stems and leaf midribs. Crop yields are, of course, substantially reduced, a 50% loss being common (Hill, 1987). A less serious form of damage is the contamination of infested plants with insect excreta (or 'frass'). The black deposits of frass will lower the physiological efficiency of photosynthesis and therefore help to lower crop yields.

Crop rotation can be a useful cultural control practice with a pest which has an overwintering stage in the soil under affected crops (McKinlay, 1988). In New Jersey, USA, Lashomb and Ng (1984) found that oviposition and first appearance of larvae of the Colorado potato beetle were substantially delayed by rotating potatoes with winter wheat, compared with unrotated potatoes. This delay was attributed to physical and environmental barriers that retard emigration from wheat by overwintering adults. However, developments in fertiliser and herbicide technologies have diminished the need for crop rotation, originally introduced partly as a weed control measure, in the developed countries, where continuous monoculture of, for example, cereals is common. Monocultural cropping of potatoes is not generally practised, principally to prevent a build-up in

Figure 8.3 Adult Colorado potato beetle, *Leptinotarsa decemlineata*, on damaged leaves. © Crown copyright

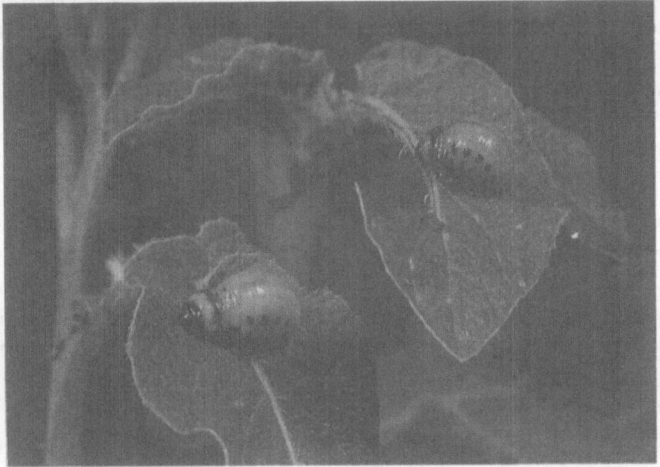

Figure 8.4 Two larvae of the Colorado potato beetle, *Leptinotarsa decemlineata*, and damaged leaves. © Crown copyright

the numbers of potato cyst nematodes (see p. 282).

Mixed cropping may have potential to suppress Colorado potato beetle populations by masking potato plant odours (Thiery and Visser, 1986). Jansson and Smilowitz (1985) found that the Colorado potato beetle, unlike potato aphids (Jansson and Smilowitz, 1986), did not respond to varying nitrogen fertiliser levels. Veverka and Oliberius (1987) found that nitrogen fertiliser was toxic to the beetles, particularly the very young larvae.

Biological control—the deliberate manipulation of natural enemies—of the Colorado potato beetle has involved predators, parasites and pathogens. Among the recorded predators are ants (Godzińska, 1986), pentatomid bugs (Izhevskaya, 1983; Ruberson *et al.*, 1986) and ground beetles (Sorokin, 1981); among the recorded parasites are tachinid flies (Izhevskaya, 1983), eulophid wasps (Schroder and Athanus, 1985), podapolipid mites (Eickwort and Eickwort, 1986) and nematodes (MacVean *et al.*, 1982); and among the recorded pathogens are bacteria (*Bacillus thuringiensis* Berliner sub-spp. *thuringiensis* Heimpel and Angus (Bacillaceae)—Sikura and Sikura, 1983—and *tenebrionis*—Langenbruch *et al.*, 1985) and fungi (*Beauveria bassiana* Vuill. (Hyphomycetes)—Sikura and Sikura, 1983).

McKinlay (1988) has recently briefly reviewed chemical control of the Colorado potato beetle. Contact insecticides are used to control the beetle, mostly organophosphates and carbamates, but increasingly pyrethroids as well. The Colorado potato beetle is controlled by low-volume sprays of, for example, the organophosphorus insecticide azinphos-methyl or the carbamate insecticide carbaryl. In some areas of the USA (e.g. Long Island, New York), the beetle has been controlled by a combination of insecticides formulated as granules (e.g. aldicarb) applied to the open furrow at potato planting time with follow-up sprays (e.g. fenvalerate, as necessary). Aldicarb was withdrawn from use in Long Island because of its appearance as a groundwater contaminant (Radcliffe, 1982). Colorado potato beetles became resistant to the pyrethroid insecticide fenvalerate within 2 years of its introduction to the Long Island area.

Apart from groundwater contamination, one other problem associated with the use of carbamates as soil-applied insecticides is enhanced degradation (Kogan, 1986). Previous treatment of soils with carbofuran has led in some cases to 'problem soils' with an ability to degrade not only carbofuran, but also a variety of other carbamate insecticides, including aldicarb, carbaryl and propoxur. These 'problem soils' result apparently from an active microbial population which is able to use carbamates as a source of carbon and nitrogen. This problem poses a real threat to the future protection of potato crops by soil-applied insecticides.

Like the peach-potato aphid, the Colorado potato beetle—particularly in the USA—had developed multiple resistance to organochlorine, organophosphorus, carbamate and pyrethroid insecticides by 1984 (Metcalf, 1986). The toxicity of fenvalerate, aldicarb and oxamyl to resistant beetles was significantly increased by the addition of the synergist piperonyl butoxide (Forgash, 1985). Silcox and Ghidiu (1986) found that piperonyl butoxide without insecticide was as effective as fenvalerate without synergist at protecting common eggplants (*Solanum melongena* L. var. *esculentum* Nees) in the field from Colorado potato beetle attack. The synergist was not toxic to the beetle larvae, but appeared to be repellent. Among new materials being tested against resistant Colorado potato beetles is a derivative of nereistoxin (Forgash, 1985).

The Colorado potato beetle has so far been prevented from becoming established in Britain by legislative control. The Plant Health (Great Britain) Order 1987 (made under the Plant Health Act 1967) controls the importation of potato plants, tubers and other vegetables from anywhere outside the UK and the compulsory destruction of any crop in the UK found to be infested with Colorado potato beetles. The Order requires also that occupiers of land report to the agricultural authorities any suspected outbreaks and prohibits the keeping of any live beetles and the spraying or other treatment of any affected crop without the express permission of the agricultural authorities. By employing such drastic measures—and breeding colonies of beetles have been found on many occasions since 1877, when the first beetle was found in the UK at Liverpool docks on a ship carrying produce from Texas, USA—Britain remains free of Colorado potato beetles.

8.5 MOTHS

8.5.1 *Phthorimaea operculella* Zeller: Potato Tuber Moth (Lepidoptera: Gelechiidae)

The potato tuber moth is a serious pest of potatoes in the warmer countries of the world: Spain, Portugal, southern France, Italy, Greece, north and south Africa, Indian subcontinent, Australasia, southern USA, central America and most of South America. The damaging stage in the life cycle of this almost cosmopolitan pest is the caterpillar.

The adult moth has a 15 mm wingspan and a body 10–12 mm long. The wings are narrow and pointed at the tips. The colour of the forewings is light grey, mottled with black dots and darkened at the tips. The hindwings are fringed with long hairs. The caterpillar is 10–15 mm long, with a purplish-white body colour and a dark head, legs, prothoracic plate and last abdominal tergite (Figure 8.5). Each abdominal segment has a row of spots, each bearing a seta.

The adult female moth lays elliptical, 0.5-mm-long, creamy-coloured

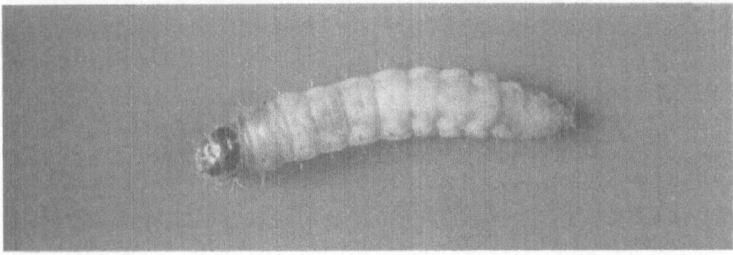

Figure 8.5 Caterpillar of the potato tuber moth, *Phthorimaea operculella.* © Crown copyright

eggs in eyes, cracks or folds of the potato tuber, both in store and in the field. Eggs can also be laid on the leaves and in the soil near the tuber. The eggs hatch in about 18 days at 15–16°C, in 9 days at 20°C and in 3 days at 31–35°C (Rivnay, 1962). The temperature threshold of egg development is about 10°C (Hovey, 1943). Humidity has no effect on the length of development of the eggs (Hovey, 1943). On hatching from the eggs, the caterpillars bore into the flesh of the tubers. If the eggs had been laid on leaves, the caterpillars bore into the main veins, petioles or young shoots. The caterpillar can grow and develop quickly under favourable conditions. Larval development can be completed in 7 days. Pupation occurs in the field within tubers, between leaves or under soil surface litter. In store, pupation occurs within tubers, structural cracks or folds of bags. Larval and pupal development require about 80 days at 18°C and 15–17 days at 31–35°C (Rivnay, 1962). Development from egg to adult requires about 54 days at 18°C and 19–20 days at 31–35°C (Rivnay, 1962). Female moths begin laying eggs 3–4 days after emerging from the cocoons. During their short lives of a few days only, female moths can lay up to 87–88 eggs each (Rivnay, 1962). The potato tuber moth has more generations a year in warmer than cooler areas: for instance, nine in Egypt and six in France (Rivnay, 1962).

Potato tuber moth caterpillars consume tuber tissue (Figure 8.6). As a result, they destroy the feeding value as well as the sprouting potential of affected tubers. Caterpillar damage can also lead to secondary rotting. Caterpillar mining of the main veins, petioles and young shoots of plants in

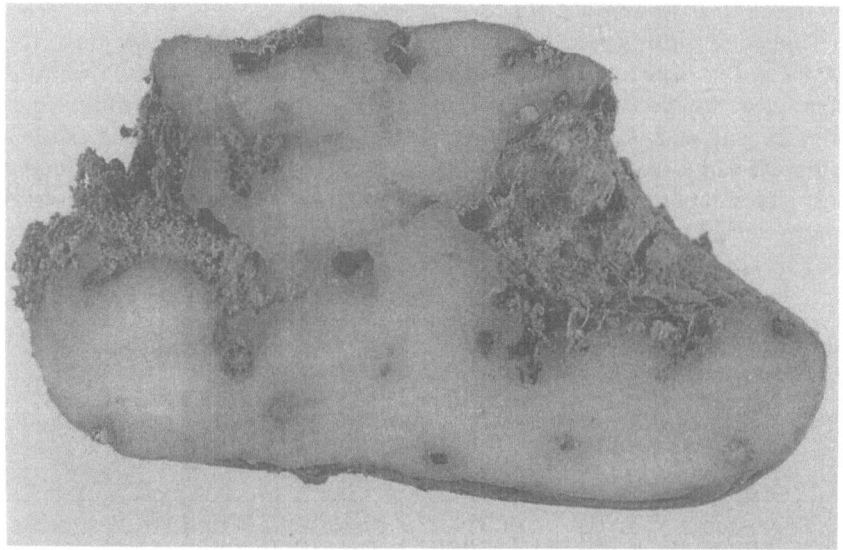

Figure 8.6 Tuber tissue damaged by caterpillars of the potato tuber moth, *Phthorimaea operculella*. © Crown copyright

the field leads ultimately to wilting and a resultant loss in crop yield.

Mixed cropping as a cultural control practice can help to reduce potato tuber moth outbreaks. Tuber infestation by the moth in Peru was significantly higher on potatoes alone than on potatoes growing with onions, tomatoes, green beans, soybeans or maize (Anon., 1982). Irrigation is another cultural control practice which may be used. It was shown not only to reduce tuber infestation by the potato tuber moth (Temerak, 1983), but also to increase the numbers of predatory natural enemies (Sorokin, 1982). Cultural practices such as deeper planting of seed tubers, repeated hilling, preharvest haulm destruction, prompt lifting of harvested tubers and elimination of groundkeepers were largely used to control the moth before the advent of insecticides (Radcliffe, 1982).

The natural enemies—principally parasitoids—of the potato tuber moth have been studied to determine whether they can be deliberately manipulated to exert a degree of biological control. The larval parasitoids of the moth which have been evaluated as potential biological control agents are the braconids *Apanteles* Foerster spp., *Chelonus* Jurine spp., *Orgilus* Haliday spp. and *Agathis* Latreille sp.; the ichneumonids *Diadegma* Foerster sp. and *Temelucha* Foerster sp.; and the encyrtid *Copidosoma koehleri* Blanchard (Khandge *et al.*, 1979; Kfir, 1981; Choudhary *et al.*, 1983; Simchuk, 1985; Flanders and Oatman, 1987). Of these, *C. koehleri* released continuously has given effective control of the potato tuber moth in India (Khandge *et al.*, 1979). Parasitoids have a reduced impact in the colder areas of the moth's range (Briese, 1986).

Insecticides can have an adverse influence on the effectiveness of the natural enemies of the potato tuber moth. Shelton *et al.* (1981) found that organophosphorus and carbamate insecticides applied to potato foliage in California to control the moth population markedly reduced larval parasitism by braconid wasps (*Agathis gibbosa* Say; *Chelonus phthorimaea* Gaham; *Apanteles scutellaris* Muesebeck) in treated areas. Predation of eggs and pupae (principally by Anthicidae, Araneida and Staphylinidae), which was high in untreated potatoes, was greatly reduced by applications of the organophosphorus insecticide azinphos-methyl. Following applications of azinphos-methyl, the moth population resurged rapidly, adults requiring less time to resurge than larvae. However, different types of pest control—in the example just given, chemical and natural controls—need not interact adversely with each other. For instance, natural control of the potato tuber moth may be positively integrated with a cultural practice: parasitoid effectiveness is increased by ridging of soil against potato stems (Kogan, 1986).

Potato moths are controlled chemically by sprays of the organochlorine insecticide DDT; the organophosphorus insecticide chlorfenvinphos; or the carbamate insecticide methomyl (McKinlay, 1988). Pyrethroid insecticides have not been found generally to give adequate control in the field. Protectant sprays of insecticide are applied to the foliage every 2 weeks after the first leaf mine has been seen. Stored tubers can be protected from the potato tuber moth by dipping or fumigation.

Raman and Palacios (1983) reported that breeding for plant resistance to potato tuber moth is promising. Nabi (1984) found that the number of larvae and the degree of damage on the leaves were reduced in lines of *Solanum acaule* Bitt.

Traps baited with the sex pheromones of the potato tuber moth can be used effectively to protect both field crops and stored tubers (Raman and Palacios, 1983). Using a funnel trap, Raman (1983) reported that a 9:1 mixture of (4*E*,7*Z*)-4,7-tridecadienyl acetate and (4*E*,7*Z*,10*Z*)-4,7,10-tridecatrienyl acetate, the two components of the moth sex pheromone, captured the highest numbers of male moths. Attempts have been made in New Zealand to sterilise male potato tuber moths. In field cage trials Nabi (1983) found that male moths chemically sterilised with thiotepa (tris(1-aziridinyl)phosphine sulphide) were effective at suppressing the population in the cages.

8.6 CYST NEMATODES

8.6.1 *Globodera pallida* (Stone) Behrens: White Potato Cyst Nematode (Tylenchida: Heteroderidae)
Globodera rostochiensis (Wollenweber) Behrens: Yellow Potato Cyst Nematode (Tylenchida: Heteroderidae)

Both *G. pallida* and *G. rostochiensis* are indigenous to South America but are predominant in different areas of the continent: *G. pallida* in Peru and *G. rostochiensis* in Bolivia. They were introduced accidentally into Europe during the nineteenth century and subsequently to many other parts of the world as an unfortunate consequence of potato trading. The distributions of potatoes and *G. pallida* and *G. rostochiensis* now largely coincide.

Potato cyst nematodes are sophisticated plant parasites. They are well adapted to potatoes and have few hosts apart from *Solanum* L. spp. The most important hosts other than potatoes are tomatoes and aubergines. *G. pallida* and *G. rostochiensis* have similar life cycles and cause similar damage, but may be distinguished with the naked eye only by the colour of the adult females on the roots. The females, which are about 0.5 mm in diameter and white when they first appear on the roots, turn a golden-yellow colour in the case of *G. rostochiensis* but remain white in the case of *G. pallida*. On their death, the females' integument, in both species, tans, to form a brown cyst, each of which can contain several hundred eggs. In this form the nematodes can survive for many years (at least 10 and frequently longer) in the absence of a host crop.

When a host crop is grown, the roots release specific exudates (collectively known as 'hatching factor') that stimulate hatching of most of the dormant juveniles in the cysts. The hatched juveniles are eel-like (hence their original common name, 'eelworms') and are attracted to the root tips of the host plants by the exudates. The juveniles must then penetrate the roots to continue their development. The growing region immediately behind the root tip is the usual point of entry and, once within the root, the juveniles move through the cortex, disrupting cells, until they reach the stele. Feeding by the juveniles induces the formation of specialised cells (syncytia). The syncytia are formed mainly from phloem parenchyma cells by the breakdown of cell walls and cell enlargement (Jones and Northcote, 1972). The juveniles become sedentary once they have started to feed from the syncytia. They pass through three moults within the roots, with the sexes being distinguishable after the first moult. After the third moult, the nematodes are adult and the vermiform males escape from the roots into the soil. The males do not feed and are therefore fairly short-lived. The females increase so much in size during their development that they rupture the root cortex around them and leave only their heads and necks embedded in the root tissue. The females release a sex attractant and, after fertilisation, remain attached to the root for some time while eggs are produced. The first juvenile moult occurs in the egg, which, when it hatches, releases a second-stage juvenile.

Damage and Economic Importance

It has been estimated that some 12% of the world's potential annual potato production is lost as a result of nematode attack (Sasser and Freckman, 1987) but the effects of specific nematodes such as *G. pallida* and *G. rostochiensis* were not assessed. In the UK the annual yield loss caused by potato cyst nematodes is estimated to be between 2% and 9%, worth £54m approximately. The annual cost of potato cyst nematode control in the UK has been estimated at £15m (Trudgill *et al.*, 1987), with some 40000 ha being treated each year. In the Netherlands even larger areas are treated annually, principally in the starch-producing regions. The cost of potato cyst nematode control is even greater in countries producing both ware (table) and seed potatoes, or other crops for transplanting. The rapid spread of potato cyst nematodes around the world as a contaminant of seed potato lots and the potential of these nematodes to devastate potato yields have induced governments to impose statutory measures, which generates costs to the exporting nation by requiring soil and crop testing. Consignment-testing costs may be borne by exporter and importer alike. There can also be additional costs to an individual seed producer if soil is found to be infested, as further seed production may not be permitted. In the European Economic Community such land is said to be 'scheduled' and restrictions are placed on future cropping, preventing seed potato production while the schedule is in force.

In the ware crop, damage occurs to the growing roots as juvenile nematodes invade and is usually in direct proportion to nematode numbers (Trudgill *et al.*, 1985). Affected plants grow more slowly and become noticeable in the field as patches of poor, often yellowed growth (see Figure 8.7). These plants are more prone to drought and nutritional

Figure 8.7   Damage caused by the potato cyst nematode, *Globodera* spp. Note the poor, yellowed growth. © A. M. Spaull

stresses and will wilt more rapidly than unaffected plants. It is also common for senescence to occur earlier in infested plants and, at harvest, for yields to be reduced. These effects can be related to changes in nutrient and water uptakes and to haulm growth (see Evans and Trudgill, 1978).

Damage, in terms of lost yield, increases in proportion to the initial density of potato cyst nematodes in the soil. Little damage is thought to result from infestations of less than 5 eggs/g soil but would be expected in any soil type with an infestation of 10 eggs/g or more, assuming a susceptible variety is grown. Yield loss is thought also to be proportional to the yield potential of the crop—i.e. yield loss will be more in absolute terms where the potential is for a 60 t/ha than a 40 t/ha crop (Seinhorst, 1982).

The relationship between the nematode and its host is complicated by interactions with other biotic factors. Some potato varieties are better able to withstand similar levels of nematode attack than others and are described as tolerant. Evans (1982) showed that yields were from 8% to 71% less, depending on variety, in soil heavily infested with *G. rostochiensis* than in very lightly infested soil. Tolerance can be influenced considerably by environmental conditions. Trudgill (1986) showed that fertiliser use could increase the tolerance of potatoes growing in infested soil. Tolerance can also be influenced by concomitant pathogen attack—e.g. *Verticillium dahliae* Kleb. (Hyphomycetes) (Evans *et al.*, 1984) infection, causing Verticillium wilt, is often facilitated by nematode invasion. Combined attack does not alter the reaction in some varieties (e.g. Maris Anchor) but others (e.g. Maris Peer) suffer more damage than from either nematode or pathogen alone. It seems probable that tolerance to potato cyst nematode attack will be affected by other micro-organisms and, indeed, by other nematodes.

Soil type and differences in yield potential between sites complicate further the host–parasite relationship. Trudgill (1986) concluded that, in contrast to differences in yield potential between sites, soil type accounted for a large part of the variation observed when relating yield to numbers of potato cyst nematodes. In common with many other nematode problems, potato cyst nematodes are more damaging in lighter, sandy soils than in heavier, clay soils.

Population Dynamics

The host–parasite relationship—potato plant–potato cyst nematodes—is dynamic, with both the nematodes and the host plant influencing and reacting to changes in each other. Jones and Perry (1978) have developed a model that attempts to describe the effects on cyst nematodes of factors such as host–plant resistance, nematicide use, competition between nematodes, etc. The population dynamics differ according to whether the variety of potato is susceptible or resistant to the nematode. The rate of multiplication of potato cyst nematodes on susceptible potatoes is density-dependent. In lightly infested soils (up to 3–5 eggs/g), populations can increase fiftyfold and occasionally more. In heavily infested soils competition for feeding sites and root damage may be so severe that few new females are formed and populations then decrease. Tolerant or vigorous varieties produce large root systems that tend to be able to support larger potato cyst nematode populations than can intolerant or less vigorous varieties.

Plant resistance is measured in terms of a variety's ability to reduce the rate of, or even prevent, potato cyst nematode multiplication. Although the roots of resistant varieties may sometimes be invaded by fewer juveniles than susceptible varieties (Forrest and Phillips, 1984), most of the effects of resistant plants on female production result from a failure to produce multinucleate syncytia, effectively preventing further nematode development. Resistance and tolerance are independent of each other: although a tolerant variety may yield more than a resistant one in the field, it may fail to exert the same degree of nematode control.

Jones and Perry (1978) assumed in their model that nematodes will behave similarly in a particular potato variety—known now not to be the case. Internationally, several pathotypes of both *G. pallida* and *G. rostochiensis* are recognised by their ability or failure to reproduce on different varieties. It is now well known that resistance to the most widespread pathotypes of *G. rostochiensis* (Ro1 and Ro4) is conferred by a single dominant gene, designated *H1*, which is present in UK varieties such as Maris Piper. These pathotypes can therefore be distinguished readily from others. However, pathotypes of *G. pallida* are less easily distinguishable, as resistance to this species is conferred by several genes acting together and, to date, only partial resistance is available against *G. pallida*. Currently, there is much debate in Europe about the basis for distinction between *G. pallida* pathotypes and it has been suggested that pathotypes should be separated only by clearly identified genes in the host plant (Anderson and Anderson, 1982; Trudgill, 1985).

Populations recognised as being of the same pathotype may differ in their virulence—i.e. they multiply unexpectedly on varieties normally considered resistant, or relatively resistant (see Spaull *et al.*, 1987). This result has obvious implications for nematode management, especially when only partial resistance is currently available against *G. pallida*.

Control

Potato cyst nematodes may be controlled by cultural, varietal, chemical and integrated methods. The discovery that *Heterodera avenae* Wollenweber (cereal cyst nematode) could be controlled by fungal parasites (Kerry and Crump, 1977) made it apparent that biological control of nematodes was a possibility. Some fungi have been tested for their pathogenicity to potato cyst nematodes *in vitro* (Jatala *et al.*, 1979) but a pathogen suitable for field control has not yet been found. This topic has been reviewed by Kerry (1987).

Cultural Control   The rotational interval between potato crops was used historically to control potato cyst nematodes, when potatoes were grown usually only one year in six. Potato cyst nematode populations decline gradually in the absence of a host crop. This natural decline reduces encysted egg numbers by an annual average of about 30% (Cooper, 1954). It is assumed that the rates of decline are similar for both species of potato cyst nematode, although there is evidence from New Zealand that *G. pallida* declines more slowly than *G. rostochiensis*. It is known, however, that this annual mortality is affected by soil type, soil temperature and possibly by the variety last grown (Anderson, 1986).

Potato varieties that may be harvested before the nematodes have completed their life cycles can exert some control. Nevertheless, some females will produce eggs successfully and some encysted eggs in the soil will not have hatched, leaving a residue to survive and attack a future crop. Trap-cropping—growing a susceptible host crop for a few weeks and then destroying it—fails for the same reasons as early harvesting. Although some control is achieved, timing of both planting and removal of the trap crop is critical (Whitehead, 1977; Balandras, 1985).

The use of cultural control methods is declining. Trends in potato production are now changing in Europe as other crops become less profitable. There is a tendency to grow potatoes on lighter soils for easier, cleaner harvesting and to intensify production. Some growers are even attempting potato monoculture.

Varietal Control The process of intensification of potato production was assisted by the introduction in the 1960s of varieties with the *H1* gene that conferred full resistance to *G. rostochiensis*. As a result, potatoes could be grown for the first time in infested soil without the risk of greatly increasing the infestation by harvest. These fully resistant varieties give excellent, cost-effective control of *G. rostochiensis* only. Many populations contain, however, at least a trace of *G. pallida* which usually remains undetectable until *G. rostochiensis*-resistant varieties are grown. The varieties first introduced containing the *H1* gene were fully susceptible to *G. pallida*, which was able to multiply without competition from *G. rostochiensis* and could predominate after just three such crops.

Sources of resistance to *G. pallida* are currently very limited. The most useful material in European breeding programmes derives its resistance from *Solanum andigena* L. and/or *S. vernei* Bitt. and Wittm. As resistance is conferred polygenically, developing agronomically suitable varieties is a slow and difficult process. There are, however, some marketed varieties able to restrict *G. pallida* multiplication to about twofold (Spaull, unpublished). These varieties need to be treated with nematicide to achieve a degree of control similar to the *G. rostochiensis*-resistant varieties.

Chemical Control Two types of nematicide are available: fumigants (usually formulated as liquids) and non-fumigants (formulated as granules). The fumigants are based on either halogenated aliphatic hydrocarbons (e.g. dichloropropene) or methyl isothiocyanate precursor compounds (e.g. dazomet). These chemicals tend to be expensive, usually need specialist application and are unpleasant to use. Most of the nematicide used in Britain for control of potato cyst nematodes in the field is in the form of non-fumigant granules, containing either an organophosphate or an oximecarbamate, which must be applied to the soil before potatoes are planted. In mainland Europe the fumigants are used more widely for field control of potato cyst nematodes. As fumigant materials are phytotoxic, they must be used and have dispersed before planting time. All nematicides currently available are non-specific, very toxic materials and there is an urgent need to find more environmentally acceptable compounds. Although fumigants can kill nematodes (i.e. they are truly nematicidal), to give worth-while control, they must be at least 94% effective (Whitehead, 1986). Fumigants diffuse through soil pores and small amounts dissolve in the water films surrounding soil particles and can penetrate the cuticle of all stages of the nematode. There are two commonly used non-fumigant, carbamate nematicides in Europe: aldicarb and oxamyl. In the USA other chemicals such as carbofuran are also used. Aldicarb breaks down to products that are toxic to nematodes, but the breakdown products of oxamyl are not effective nematicides. The parent materials dissolve in the soil water and may be toxic to all soil-inhabiting stages, although eggs are not usually affected at normal application rates. These compounds have a reversible blocking action on acetylcholinesterase, and nematodes are not directly killed at the rates in commercial use. Repeated applications of certain nematicides have allowed a soil microflora to develop in some areas, such as parts of the USA, which is capable of enhancing the breakdown of these compounds (see page 278). Although this phenomenon is not yet widespread for most products, enhanced microbial breakdown of 1,3-dichloropropene (a fumigant) and several carbamates is known. There is no evidence yet of resistance to any nematicide in *Globodera* Skarbilovich species, despite the period over which they have been in use.

Integrated Control   The useful life of both nematicides and resistant plant varieties will be prolonged if both are used judiciously and in combination with other methods (e.g. rotation) in an integrated control system. At the moment, economic pressures can make rotational control unpopular, but *G. pallida* problems can become so pronounced that production is threatened unless potatoes are grown less intensively. Using integrated methods, it is possible to achieve control of *G. pallida* similar to the control of *G. rostochiensis* given by varieties containing the *H1* gene for resistance (Spaull *et al.*, 1987). As resistant plant material exerts selective pressure on the nematode, there is value in retaining susceptible varieties grown with nematicidal protection in the rotation. Thus, the aim becomes one of potato cyst nematode management, rather than futile attempts at extermination.

8.7 SLUGS

8.7.1 *Deroceras reticulatum* (Müller): Field Slug (Gastropoda: Limacidae)
Milax budapestensis (Hazay), *Milax sowerbyi* (Férussac): Keeled Slugs (Gastropoda: Limacidae)
Arion hortensis Férussac, *Arion distinctus* (Mabille): Garden Slugs (Gastropoda: Arionidae)

Slugs occur in many parts of the world, but are important agricultural pests anywhere there is a moist, temperate climate (see Godan, 1983). A variety of vegetable crops can be attacked indiscriminately, but the largest economic loss probably occurs to potatoes. In most vegetable crops other than potatoes, injury is possible from the seedling stage onwards. In addition to direct losses resulting from slugs feeding, much 'loss' also results from contamination by the slugs themselves or their faeces or slime. Most of the injury to potatoes is done to the tubers, which are hollowed out. Several species of slug have been associated with injury to potatoes. Barnes and Weil (1944) considered *A. hortensis*, *M. budapestensis* and *M. sowerbyi* to be the primary pests of potato tubers in the UK and *D. reticulatum* (Figure 8.8) to be a secondary pest, only increasing any original damage. However, Warley (1970) and Airey (1987) considered *D. reticulatum* to be capable of direct injury to tubers. Airey (1987) also found *A. fasciatus* to be injurious. In other parts of Europe *Limax maximus* L., the great slug, can also be a damaging species.

The slug pest species have been spread in recent times by trade and travel, so that their distribution is no longer limited to their continent of origin: *D. reticulatum* has spread from Europe to Australia, New Zealand and South America (Godan, 1983); and *M. budapestensis*, which originated south of the Alps and Carpathians, is now found in northern Europe. *M. budapestensis* has also spread, apparently within the last 40 years, to the USA, Canada, Hawaii and Australia (Godan, 1983).

Life Cycle and Biology

Slug life cycles vary in length. Reproduction may occur over a prolonged period, with mating and egg laying occurring when conditions are generally favourable for activity. Although there are recognised peaks of egg laying in most species, field populations of single species usually have a mixed age structure. The egg-laying peaks vary with year and location, but the information below on *M. budapestensis* and *D. reticulatum* is based on UK

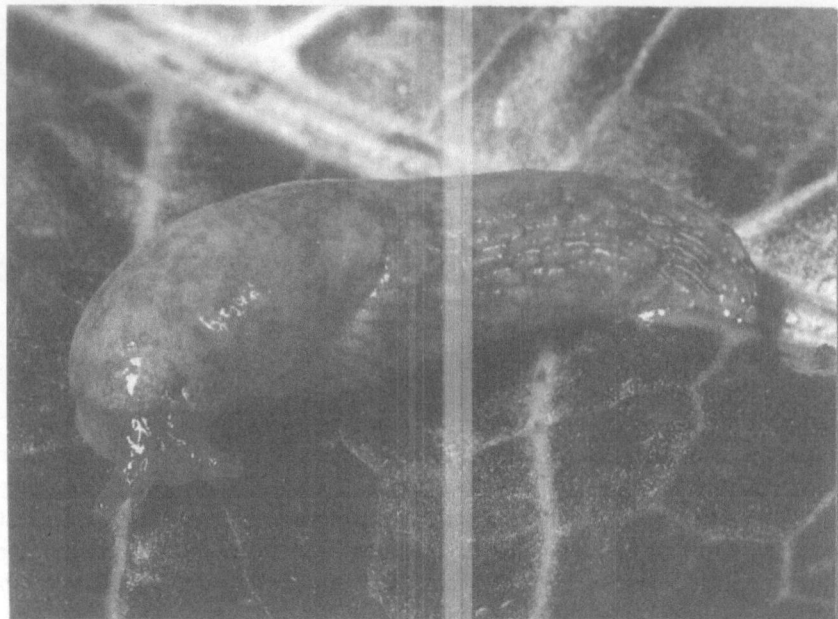

Figure 8.8 The field slug, *Deroceras reticulatum*, a secondary pest of potato. © A. M. Spaull

studies by Hunter (1968, 1978) and on the *A. hortensis* aggregate from Port and Port (1986).

M. budapestensis will breed from autumn to spring, with a peak in late autumn. Eggs can be found between December and April, and new hatchlings may be numerous in June. Low temperatures during the winter delay hatching, so that eggs laid over several months may all hatch within a few weeks of each other. The young slugs grow during the autumn, spring and summer, and are able to lay eggs by the end of that year.

D. reticulatum has a more variable life cycle than has *M. budapestensis* and was originally thought to have two generations per year. The life cycle is probably more complex, as the intervals between generations are approximately 9 months (Hunter, 1978), and the pattern is probably very susceptible to changes in temperature.

A. hortensis agg. comprises several species that are difficult to distinguish without dissection (Davies, 1977, 1979). *A. hortensis* s.s. breeds throughout the winter until April (approximately). *A. distinctus* is a slightly later species, breeding mostly in spring and summer, although it may continue to breed the year round. There is a peak of egg hatching for both species in late spring and the hatchlings can develop rapidly, so that eggs of the next generation can be laid within a year.

Feeding and mating are both dependent on weather conditions, as slugs are not wholly adapted to life on land and are unable to survive very dry conditions. Slugs' activity is greatly influenced by temperature, light, moisture and soil structure. Slugs show a circadian rhythm of activity, being generally more active at night. According to Dainton (1954), this rhythm is triggered by changes in temperature, with increases in activity being less marked at higher temperatures. Certain species of slug differ in their ability to remain active as temperatures fall. *D. reticulatum* is not completely immobilised even at 0°C, whereas *M. budapestensis* is inactive and *A. hortensis* barely mobile at 5°C. Light is generally held to be the factor responsible for entraining the circadian rhythm but there are conflicting views on the relative importance of light and temperature (see Port and Port, 1986). Field assessments of activity have always found temperature to be of significance, although *A. hortensis* does not appear to be as affected by meteorological factors as are other species. Moisture, either rain or dew, also influences activity, as the mucus necessary for locomotion is 98% water. The effect of rain varies with its intensity: a sudden downpour inhibits activity, whereas a light shower or persistent mist will increase it. Slug activity can be markedly increased after a potato crop has been irrigated. Barnes (1944) found that *D. reticulatum* was most active in September, but *M. budapestensis*, *M. sowerbyi* and *A. hortensis* were most active in October/November.

Slugs are far more numerous on heavy, wet, little-cultivated soil with large clods than on soil that is loose and frequently tilled. Extremes of weather on the soil surface, such as heavy rain or strong winds, discourage activity above ground.

Damage and Economic Importance

Slugs do no economic damage to seed tubers and do not affect plant stand. However, Godan (1983) reported that in Germany the foliage of potato plants could be extensively eaten by slugs. The greatest economic injury is done between late summer and harvest by slugs feeding on the maincrop tubers. By comparison with maincrops, seed crops and early varieties of table crops (harvested by the end of August) are not usually at risk. Damage is always exacerbated in wet autumns and may also be aggravated by irrigation. Port and Port (1986), considering the UK alone, put a value of £0.3–7 million on the losses attributable to slugs. However, such estimates are bedevilled by the influence of weather on the amount of slug damage, of season on potato prices and, in the UK, of varieties that are particularly prone to slug injury being grown on an increasing area.

Actual yield (tuber weight) is reduced little by slug feeding but the resulting holes render tubers unacceptable for human consumption. Slug holes are superficially similar to cutworm holes, but slug damage tends to be deeper and more extensive, with tunnels being created. Damaged tubers need to be separated from the remainder, involving the cost of extra labour, and can be sold only for stock feed or industrial use (usually achieving only 20–30% of the value of undamaged tubers sold for human consumption). Slug damage can account for over 2.5% of rejects but may vary widely between years. Port and Port (1986) quote data from S. P. Kerr for damage to CV Maris Piper of 0.2% in one year but 55% in the following year. This variety is particularly prone to slug damage.

Damage is usually not uniform across a crop but is evident as scattered foci that may not be obviously related to soil or other site variations. The random nature of slug damage is presumed to reflect apparently random differences in slug abundance (Airey, 1984). Interestingly, Airey (1987) reports that interactions between species in a mixed population evidently reduce the overall degree of damage.

Damage to potatoes occurs throughout western Europe and other temperate parts of the world, such as New Zealand, but is probably greatest in countries such as Britain, where the climate, with its high relative humidity, moderate temperatures and mild winters, provides an ideal environment for slugs. Under these conditions and with plant cover, slugs may multiply rapidly and survive well, to cause extensive crop damage.

To direct potato tuber damage must be added the often unknown cost of indirect damage. Dawkins *et al.* (1986) implicated slug injury in the transmission of the bacterial soft rot *Erwinia carotovora* var. *atroseptica* (van Hall) Dye (Eubacteriales: Enterobacteriaceae), which causes the disease blackleg. In some varieties in some seasons, considerable yield loss could result.

Control

Sadly, much of slug control remains more of an art than a science. No precise relationship may be derived between slug numbers and tuber injury, as the influence of the weather is so great. Furthermore, assessing true slug numbers is time-consuming and the sampling process would be disruptive to a well-developed potato crop. Thus, the need for control measures is usually judged on the basis of past experience and/or assessments of slug activity on the soil surface—less than ideal when the damage will occur underground. Control may be achieved by cultural, varietal and chemical means, although, to date, none is particularly satisfactory. There is no obvious candidate as a biological control agent of slugs, although parasites and predators are known (see Stephenson and Bardner, 1977).

Cultural Control   Slug problems are fewer on lighter, well-tilled soils, and, with potato production across Europe being intensified on these soils, some reduction in problems might be predicted. However, lighter soils are often irrigated, and if this practice is continued until close to harvesting the crop, little protection from slug damage will be afforded.

Most of the holing and tunnelling of tubers occurs from late summer onwards. Early varieties and seed crops, harvested before the end of August in the UK, usually escape significant slug damage. Tuber damage in maincrop varieties harvested from September onwards will be reduced, especially in a wet autumn, if harvested without delay: damage will increase with every extra week that the tubers remain in the soil.

Many pests may be partially controlled by rotation, but slugs are so polyphagous and can cause such extensive damage to crops in rotation (e.g. winter wheat) that this approach only offers a practical control method of doubtful value. There is, however, a school of thought that argues that because a weedy crop offers alternative sources of food, attacks on potato tubers are reduced. While this approach may serve to reduce injury in practice, it is unlikely to be adopted as a deliberate method of control!

Varietal Control It has been recognised for at least 40 years that some varieties of potatoes are more prone to attack by slugs than others (Thomas, 1947). A scale of susceptibilities has been drawn up for some UK varieties:

Very susceptible—Cara, Kerrs Pink, Kingston, Maris Peer, Maris Piper
Moderately susceptible—Desiree, King Edward, Pentland Crown, Record, Romano
Slightly susceptible—Pentland Dell, Pentland Squire
Low susceptibility—Pentland Ivory

However, it is likely that even the less susceptible varieties could be fed upon if the tuber skin has been already damaged (Airey, 1987). It is probable that slugs will feed on the flesh rather than on the skin, but the tuber skin of some varieties may contain feeding deterrents (Stephenson, 1965).

Chemical Control Most molluscicide is used in the form of pellets that may be applied by a number of farm machines, often designed primarily for another purpose—e.g. fertiliser spreaders. The chemicals used are usually formulated with attractants, such as bran. However, it is known that other products (e.g. sugar or sugary substances) would improve the attractiveness of pellets to some species (Godan, 1983). A balance must be struck when formulating pellets such that the consistency is not so hard (or soft) that slugs will not feed, yet the pellets must survive normal weathering for at least a week. Furthermore, for prolonged control, the toxic constituent must not be very repellent at the concentration used. It is probable that control could be increased by improved formulations, even of existing compounds.

The two most commonly used chemical compounds are metaldehyde and methiocarb. These chemicals have different modes of action and one may be more effective on some species than the other. Airey (1986) showed, for instance, that at 5°C *A. hortensis* was more susceptible to methiocarb than was *D. reticulatum*, but there was no difference between the species in their susceptibility to metaldehyde. However, at temperatures of 14–17°C—likely ambient temperatures during the period of tuber attack—*A. hortensis* was less susceptible than *D. reticulatum* to both chemicals (Crawford-Sidebotham, 1970). Several authors have quoted increasing toxicity of metaldehyde at higher temperatures (see Airey, 1986).

The presence of green plant material can compete with the degree of attractiveness exerted by molluscicide pellets to *D. reticulatum* (Airey, 1986). In practice, methiocarb and metaldehyde may achieve similar control in the field.

Metaldehyde poisons slugs either by contact action or by ingestion: the primary mechanism of action is unknown. The poisoned slug becomes immobile and is stimulated to produce copious mucus, with dehydration frequently being the eventual cause of death. If the slug is able to reach shelter, however, recovery is possible. The slugs may also learn to avoid similarly poisoned baits in the future. Methiocarb is primarily a stomach poison with some secondary contact activity. Poisoned slugs move normally for a short time, but then become immobilised as muscle tone is lost. A toxic dose of methiocarb is usually consumed in one feed, whereas several

Figure 8.9 A slug trap, consisting of a small quantity of methiocarb pellets on the soil surface and a cover. © A. M. Spaull

feeds of metaldehyde pellets are needed before a sufficiently toxic dose of the molluscicide has been consumed. There is no evidence of slugs evolving resistance to either chemical, nor is it likely, since efficacy of control is so variable and, therefore, selection pressure is weak (Port and Port, 1986).

Neither compound is specific to molluscs: methiocarb is also insecticidal and both are toxic to vertebrates. Domestic pets are the usual vertebrate victims with metaldehyde being more commonly consumed than methiocarb.

As slug pellets cost up to *c*. £26/ha to apply at current UK prices, adding directly to the variable costs of potato production, the advisory services recommend that the potential benefit from applying pellets should be judged before use, by assessing slug activity. Traps, consisting of a small quantity of methiocarb pellets covered with a supported tile, a piece of hardboard or a similar structure (Figure 8.9), are set across the field. Slug numbers under the traps are counted after three nights and an average of one slug/trap justifies treatment. This method does not assess the likelihood of damage but gives a crude guide to surface activity and, hence, the chances of achieving some control. Current recommendations suggest a timing for application of the second half of July or early August. However, Newell (personal communication) argues that applying pellets before potatoes are planted could do much to prevent adults then present from breeding subsequently, and thus limit the extent of damage towards the end of the season. This method deserves to be more widely tested.

Tomatoes

8.8 INTRODUCTION

The tomato, *Lycopersicon esculentum* Mill., is grown world-wide, both outdoors and in greenhouses, for fresh-market consumption (green or red fruit) and processing (canned, soups, condiments). The roots, stems, foliage and fruits of tomatoes are attacked by a variety of invertebrate pests. The nature and extent of damage vary greatly, but general symptoms include abnormal growth or death of the plant, and reduction in fruit quality manifest as scarring, tissue destruction, or aberrations in shape and colour. Tomato quality standards may differ relative to the ultimate market—e.g. fresh market or processing. Because it affects customer appeal, superficial or 'cosmetic' injury to the fruit is cause for rejection or downgrading of a fresh-market product. A greater degree of damage may be allowed, however, for processing fruit because some types of superficial injury can be eliminated at the processing plant. Invertebrate pests also cause indirect or 'secondary' effects through the introduction of decay organisms into tomato products or the production of 'honeydew' which provides a substrate for development of black sooty mould fungi on fruits and foliage. Insects also vector virus diseases that affect yield through growth disorders or plant mortality.

8.9 WHITEFLIES

8.9.1 *Bemisia tabaci* (Gennadius): Cotton Whitefly (Hemiptera: Aleyrodidae)

The cotton whitefly (also sweet potato whitefly, tobacco whitefly) occurs throughout most tropical and subtropical regions of the world. It occurs naturally in a band around the world approximately 30–35 degrees latitude, north and south of the equator. Distribution out of this band is limited basically by low winter temperatures. *B. tabaci* adapts well, however, to greenhouse conditions and has become a sporadic outdoor pest in Europe and northern regions of the USA, probably originating as escapes from greenhouses or as transients on greenhouse-grown vegetable transplants and ornamentals. In temperate regions, as the common names imply, the primary hosts are cotton, sweet potato and tobacco. Alternative hosts include a very wide range of wild and cultivated plant families.

Insects in the family Aleyrodidae display simple metamorphosis. Typically, insects with simple metamorphosis have immature stages that closely resemble the adult. However, aleyrodids are unusual (resembling complete metamorphosis), in that the nymphs appear 'scale-like' and the last nymphal instar is quiescent and 'pupa-like'. The four-winged adult is 1–3 mm long and the entire insect is covered by a white-coloured, waxy bloom. Adults can fly for only short distances, but may be dispersed over large areas by wind. Females usually lay their first eggs on the lower surface of the leaf on which they emerged, but soon move upwards to young leaves, generally on the same plant. Pear-shaped eggs, which hatch in about 7 days, are about 0.2 mm long and are inserted vertically into the leaf tissue. They are anchored at the larger end by a stalk, which penetrates the leaf epidermis through a puncture made by the ovipositor, and passes into the spongy parenchyma. Water can pass from the plant tissue into the egg, and when there are high numbers of eggs, the plant may become water-stressed (Gameel, 1974). Early in the season, eggs are laid singly but later are laid in groups (Gerling and Or, 1984). The average daily egg-laying capacity is 2 eggs/female, and total capacity may be 160 eggs/female (Gameel, 1974). Upon hatching, nymphs move only a very short distance before settling down to feed. Once feeding begins, nymphs do not move again. All nymphal instars are greenish-white in colour, oval in outline, scale-like and somewhat shiny. The last instar is about 0.7 mm long, and the red-coloured eyes of the adult can often be seen through the larval integument. Nymphs complete three moults before pupation and emergence as adults.

The undersides of tomato leaves are usually infested with adults and nymphs that extract fluids from plant tissue through sucking mouthparts. The stylets penetrate between the epidermal and parenchyma cells of a leaf to the phloem (path of sugar and nutrient movement). Infested plants are of low vigour, and may wilt, turn yellow in colour and die when whitefly infestations are severe or of long duration. Damage may be accentuated when plants are under water stress. Mor and Marani (1984), comparing various degrees of water stress to cotton plants, found dramatic whitefly increases in treatments that suffered from lack of water, resulting in plant defoliation and yield reductions. Whiteflies excrete honeydew, a clear fluid containing unabsorbed organic matter and sugars, which often completely covers foliage. As a result of sooty mould fungi growing on the honeydew, leaves may turn a black colour, thus reducing the efficiency of respiration and photosynthesis.

Perhaps the greatest damage to tomato caused by *B. tabaci* is through the transmission of virus diseases. Among the more serious viruses vectored are tomato yellow leaf curl virus (TYLCV), tomato leaf curl virus (TLCV), tomato yellow net virus (TYNV) and tomato yellow top virus (TYTV). In temperate climes TYLCV is probably the most destructive. Affected plants are severely stunted, shoots are erect, and leaflets are reduced in size, curled upwards and chlorotic. When very young plants are attacked, they show poor vigour and produce very few marketable fruits. Although there is no definitive evidence for the multiplication of TYLCV in whiteflies, there are indications that the virus pathway in the insects is not a passive one—i.e. the acquisition of TYLCV is cyclic, owing to an anti-virus mechanism triggered by the virus (Cohen *et al.*, 1984).

Control of *B. tabaci* infesting tomato and other crops throughout the world is primarily achieved by the use of insecticides. Since *B. tabaci* efficiently vectors a number of plant viruses, management schemes are usually directed towards reducing the vector populations to a point where transmission is minimised, a practice that has yielded mixed degrees of success (Sharaf and Allawi, 1980). Foliar sprays of the pyrethroids cypermethrin and fenpropathrin appear to be generally more effective than similar sprays of other classes of insecticides in controlling *B. tabaci* and reducing TYLCV infection (Berlinger *et al.*, 1986; Mishra, 1986). Some systemic granular insecticides (aldicarb, carbofuran, disulfoton), applied at planting, effectively reduce TLCV incidence (Rataul and Butter, 1975). Mineral oil sprays alone and in combination with the insecticides per-methrin, pirimiphos-methyl or monocrotophos have been shown to reduce significantly the incidence of TLCV (Butter and Rataul, 1973; Mote, 1979) and TYLCV (Sharaf and Allawi, 1981).

The integration of cultural practices and insecticidal use to prevent the spread of TYLCV has met with some success. Straw mulch placed around germinating tomato plants is more attractive than tomato leaves to *B. tabaci* (Cohen *et al.*, 1974), but whiteflies are more attracted to yellow polyethylene mulch than to straw, and a combined treatment of yellow polyethylene and azinphos-methyl sprays effectively prevented the spread of TYLCV (Cohen and Melamed-Madjar, 1978).

Cultural control methods alone may also be effective. Drip irrigation appears to be instrumental in reducing adult whitefly populations and TYLCV incidence because, when compared with other methods, it promotes the growth of shorter plants with fewer leaves (Sharaf *et al.*, 1984). The intercropping of cucumber (Al-Musa, 1982) and sorghum (Gravena *et al.*, 1985) in tomato plantations can increase beneficial arthropod populations, reduce *B. tabaci* infestations and delay the onset of TYLCV.

Biological control of the greenhouse whitefly (*Trialeurodes vaporariorum* Westwood: see page 300), another whitefly pest of tomato, has been effectively employed in greenhouses because optimal parasitoid/prey ratios can be achieved by controlling the environment and the releases of parasitoids. To date, biological control of *B. tabaci* either by means of parasitoid releases or through management of natural parasitoid populations, has not been fully exploited under field conditions. In many instances, standard pest control methods utilising insecticides are incompatible with efforts to establish or maintain an equilibrium between pest and parasitoid. In the south-western USA, where *B. tabaci* has occurred since the late 1920s without achieving major pest status, populations have recently risen sharply, causing severe economic damage to cotton, sugar beet and various vegetable crops (Duffus and Flock, 1982). The explosion of whitefly populations in California, USA, and similar phenomena in Thailand and Sudan, were related to reductions in parasitoid numbers (primarily the wasp parasitoid *Encarsia formosa* Gahan (Hymenoptera: Eulophidae)) caused by applications of pyrethroid-based insecticides that were directed against other insect pests (Johnson *et al.*, 1982).

The defensive role of foliar trichomes of certain members of the Solanaceae against herbivorous insects has become an area of intense research (see Duffey, 1986). Some insects fail to colonise tomato plants effectively because the tarsi become stuck in a 'gum-like' exudate from the foliage, or because leaf hairs broken or disturbed by insect contact secrete a 'lethal exudate' (Gentile and Stoner, 1968). In tomato species there are four major types of glandular trichomes. The type with a short multicellular stalk and with a 2–4-celled glandular head appears to be the most important in resistance to whiteflies (Gentile *et al.*, 1968; Gentile and Stoner, 1968). A study of the role of glandular leaf hairs of four tomato cultivars in trapping *B. tabaci* adults showed that 86% of trapped whiteflies were female, and many became trapped during oviposition (Kisha, 1981). Berlinger *et al.* (1983) reported that resistance in selected accessions appeared to be polygenic in nature and comprised of various mechanisms such as: differences in pH; the content of secondary plant substances; and sticky exudations that were probably toxic. The incorporation of trichome-type resistance from related species into commercial lines of tomato is possible and promising. Future control strategies for the reduction of direct plant damage and disease transmission may include the integration of whitefly-resistant tomato cultivars, controlled releases of parasitoids such as *E. formosa*, the use of selective insecticides that are not disruptive to biological control and the utilisation of attractive mulches and trap crops.

8.9.2 *Trialeurodes vaporariorum* (Westwood): Greenhouse Whitefly (Hemiptera: Aleyrodidae)

The greenhouse whitefly is a common and persistent pest of tomatoes and other vegetables as well as ornamental crops grown throughout the temperate climates of the world. Introductions into many non-temperate regions have resulted as whiteflies spread from controlled-environment greenhouses and from crops that are transplanted from infested greenhouses. As an aleyrodid, it displays simple metamorphosis (see *B. tabaci*, page 296). The four-winged adult is 2–4 mm in length and covered with a white-coloured pubescence. The wings are held 'roof-like' over the body, at approximately 45° angles when at rest. Adults can move only short distances between host plants, but may be dispersed over large areas by wind. Most adults leave the plant on which they originated and the rate of within-plant movement and migration onto neighbouring plants is dependent upon daytime temperatures (Noldus *et al.*, 1985, 1986a). They appear to prefer to feed on young leaves and this apparent preference may have adaptive value, for young leaves are relatively higher in both protein and soluble nitrogen than older leaves (Noldus *et al.*, 1986b). Preparatory to copulation, females attract males by emitting a sex pheromone (Li and Maschwitz, 1983) and appear to assess males according to tactile signals for their courtship stamina and intensity (Las, 1980; Ahman and Ekbom, 1981). Females oviposit small (0.2 mm), yellowish-coloured eggs in clusters on leaf undersides, each egg being attached to the leaf by a short stalk. The average daily egg-laying capacity is 3 eggs/female (Castresana-Estrada *et al.*, 1981) and the total egg-laying capacity may range from 92 to 250 eggs/female, depending upon tomato variety (Curry and Pimentel, 1971). Eggs hatch in 8 days at 23 °C (Hernandez-Roque, 1973) and in 11 days at 20 °C (Laska *et al.*, 1980). On hatching, the flat and nearly transparent nymphs settle upon the leaf near the point of hatching and remain there until adulthood. The first three nymphal instars are greenish-white in colour, oval in outline and somewhat spiny, while the fourth and last instar (pupa) is 'disc'- or 'cake'-shaped. The pupa is about 0.8 mm long. The period of development from egg to adult takes 21 days at 26 °C (Collmann and All, 1980) and 34 days at 20 °C (Laska *et al.*, 1980). *T. vaporariorum* may display asexual reproduction. There are parthenogenetic races which produce females from fertilised eggs and males from unfertilised eggs and a race consisting almost exclusively of females (see Wigglesworth, 1965).

The undersides of tomato leaves are usually infested with adults and nymphs, which extract fluids from plant tissues through sucking mouth-parts. The stylets penetrate between the epidermal and parenchyma cells of the leaf to the phloem (path of sugar and nutrient movement). Infested plants have low vigour, and may wilt, turn yellow in colour, shed leaves and display reduced growth rates if infestations are severe or of long duration. During feeding, honeydew is excreted. Sooty mould fungi grow on the honeydew, turning the foliage and fruit black in colour and thus reducing the efficiency of respiration and photosynthesis. Honeydew deposition by each adult and pupa may average 10 to 125 droplets/h, and is a principal factor causing injury to tomato plants and fruits (Hussey *et al.*, 1959). Unlike *Bemisia tabaci*, *T. vaporariorum* transmits no diseases that seriously damage tomatoes.

Much of the research involving management of *T. vaporariorum* on tomatoes has been directed towards greenhouse rather than field culture. However, much of the information is directly or indirectly applicable to the field situation. Insecticides remain an integral tool of most control prog-rammes, although continual usage of the same type of insecticide may lead to problems from resistance. Eggs, nymphs and pupae of *T. vaporariorum* are at least partially resistant to many currently used insecticides, with 6- to 4000-fold resistance to permethrin having been reported (Lindquist, 1972; Wardlow, 1985). Because of the toxicity of many insecticides to non-target organisms, control materials are often chosen with the preservation of natural enemies in mind. Some systemic granular insecticides such as oxamyl, when applied to the soil, have shown good activity against adults and nymphs while having no effect on parasitic insects (Martin *et al.*, 1984).

The corpora allata of larval insects produce juvenile hormone that acts upon epidermal cells to ensure they retain their larval characteristics. The synthetic insect growth regulators (IGRs) methoprene, kinoprene and triprene, which are juvenile hormone analogues, have been successfully employed against all whitefly stages (Lindquist, 1972; van de Veire *et al.*, 1974). Because of its selective activity against *T. vaporariorum* and its safety to introduced parasitoids, buprofezin IGR is commonly utilised in greenhouses but its effectiveness in the field has yet to be determined (Martin and Workman, 1986). IGRs tend to work more slowly than conventional insecticides. However, the speed of activity can be acceler-ated by the addition of compatible insecticides such as bioresmethrin (Giustina, 1976).

The parasitoid *E. formosa* (see page 298) has been widely studied and used for management of *T. vaporariorum* in the controlled environments of greenhouses, but it has not been consistently effective in tomato fields. Foster (1980) computed that high numbers of *E. formosa* (30 000 parasitised scale insects/ha) were necessary for effective control when the initial density of whiteflies did not exceed 0.1 adult/upper leaf, but such introductions were ineffective at higher pest densities. Because this *Encarsia* Förster species is sensitive to low temperatures, its success outdoors will depend upon the development of strains better adapted to cool temperatures. *E. pergandiella* How., introduced from the USA into Italy, has proven able to overwinter, and parasitises whiteflies at a success rate of 40–80% (Viggiani and Mazzone, 1980; Mazzone and Viggiani, 1985). Other biological control agents are potentially effective in reducing whitefly populations. Studies in Japan have shown that the hemipteran bugs *Campylomma* Reuter sp. (Miridae) and *Orius sauteri* Popp. (Anthocoridae) are successful whitefly predators (Kajita, 1982, 1984). Spore suspensions of the entomopathogenic fungus *Aschersonia* Mont. spp. (Deuteromycotina: Coelomycetes), when sprayed on tomatoes, provided 90–94% infection of eggs and early nymphal instars (Fransen *et al.*, 1987), but the fungus appears to be influenced significantly by temperature and humidity under field conditions (Solovei and Kol'tsov, 1976). Effectiveness of another entomopathogenic fungus, *Verticillium lecanii* (Zimmenn) Viegas (Hyphomycetes), may be similarly limited by humidity requirements (Ekbom, 1981).

Although little host resistance to *T. vaporariorum* exists in currently cultivated tomato varieties, marked resistance is present in related wild species. Relative resistance to whiteflies has been shown in *L. hirsutum* f. *glabrum* C. H. Mull (also *L. hirsutum* Humb. and Bonpl.) and *Solanum pennellii* Correll (Gentile *et al.*, 1968). It is manifest in less frequent whitefly visits, smaller numbers of hatched eggs and lower percentage of larvae/unit leaf area (Georgiev and Sotirova, 1978). This type of resistance may be attributable to a number of leaf lamellar factors, acting nonpreferentially, antibiotically, or both (Duffey, 1986). The sticky exudate of glandular hairs or leaf trichomes on stems and leaves is a striking mechanism of resistance in *L. hirsutum* and *S. pennellii*. Tomato trichomes have become the subject of much investigation since McKinney (1938) noticed that aphids were unable to colonise leaves because the tarsi became stuck in a 'gum-like exudate from the foliage', and similar effects involving trichomes were reported for whiteflies (Gentile *et al.*, 1968). Glandular trichomes play a critical role in resistance to whiteflies and other small insects, not only by encumbering the insect with an exudate that impedes or immobilises, but also by secreting, in some cases, toxic exudates (Duffey, 1986). These traits may be undesirable where parasitoids such as *E. formosa* are used, because parasitoids as well as whiteflies become entrapped (de Ponti *et al.*, 1983). Future management strategies against *T. vaporariorum* could integrate the use of selective insecticides that are safe to non-target arthropods with the introduction and/or protection of adapted parasitoids, predators and pathogens and with the utilisation of cultivars that incorporate the physical and chemical properties of trichome exudates found in wild tomato species.

8.10 BUGS

8.10.1 *Nezara viridula* L.: Southern Green Stinkbug (Hemiptera: Pentatomidae)

The southern green stinkbug (also cotton greenbug, green pumpkin bug) is a cosmopolitan species, occurring throughout the tropics and subtropics, and some temperate regions of the world (Europe, Asia, Africa, Australia and the Americas). It feeds on a wide range of cultivated and wild host plants within 18 plant families (Todd and Herzog, 1980). It is an important pest of soybeans and other legumes, cotton, and various vegetables, including tomatoes.

Members of the Hemiptera undergo simple or incomplete metamorphosis. Southern green stinkbugs, like other pentatomids, produce a strong scent that is discharged from dorsal, metathoracic glands as a defence reaction against predators and also acts as an alarm pheromone causing individuals of the same species to disperse (Ishiwatari, 1974). *N. viridula* has a complex population profile, being polyandrous, polygynous, multivoltine and long-lived as adults (Harris and Todd, 1980). Adults are relatively large (15 × 8 mm) and may occur in three colour variations, ranging from apple-green to reddish-brown. They emerge from overwintering sites in early spring, and begin feeding and mating. Mate attraction involves a complex repertoire of behaviourally and acoustically distinct songs that are inaudible to the human ear: females have three songs and males produce seven, which are used during the singing of rivalry duets (Harris *et al.*, 1982). Both sexes remate approximately three times, and males may guard females by way of prolonged copulation to protect the precedence of their sperm (Azmy, 1980; McLain, 1981). Barrel-shaped eggs (1.2 × 0.75 mm) are laid in clusters on the undersides of leaves. At the time of deposition, eggs are yellowish-white in colour, turning pinkish with progressive development. In the laboratory, females may lay from 25 to 700 eggs each in from 1 to 10 separate batches (Singh, 1973; Rodriquez-Velez, 1974; Harris and Todd, 1980). Eggs hatch into nymphs in 4–7 days. Nymphs pass through five instars over a period ranging from 35 to 40 days. Instar durations depend on photoperiod, temperature and humidity, and on the quality and quantity of nutritional resources. Nymphs may remain gregarious during the first three instars, dispersing during the fourth and fifth instars. They feed on plant sap, developing seeds or fruits. The entire egg-to-adult life cycle ranges from 28 to 48 days on tomatoes and from 22 to 56 days on other various crops (Harris and Todd, 1980). Generations can overlap.

Almost invariably, *N. viridula* attacks the fruiting structures of plants. Punctures of immature and mature fruits caused by insertion of the sucking mouthparts of the insect cause local necrosis, cloudy blotches, surface depressions, fruit malformation and, in severe cases, fruit drop. Some symptoms may be the result of enzymes present in the saliva. Pest density and feeding duration have significant negative effects upon fruit circumference and weight and can reduce yields by inducing early maturity (Lye *et al.*, 1988). In the USA fresh-market tomato yields have been reduced by up to 31% (Chalfant, 1973). Because of the manner in which it feeds, *N. viridula* mechanically transmits tomato bacterial spot and, perhaps, a yeast that causes fruit decay. Feeding wounds permit infection by secondary plant pathogens.

General success in the management of *N. viridula* has been achieved through spray applications of organophosphate and carbamate insecticides during the fruiting period (Chalfant, 1973; Hoffmann *et al.*, 1987). In the USA some Hawaiian populations are resistant to chlorinated hydrocarbon insecticides, and in California a pyrethroid (fenvalerate) was ineffective in reducing green stinkbug populations on tomatoes (Miyasaki and Sherman, 1966; Hoffmann *et al.*, 1987). Investigations in Australia have shown that adults and nymphs move to the upper plant canopy during early morning and remain there for a period of time (Waite, 1980). Insecticidal effectiveness can be improved if applications are timed to coincide with this basking behaviour. Although insecticidal sprays may be advisable under high infestation conditions, the use of broad-spectrum insecticides has the potential of causing resurgences of secondary tomato pests due to harmful effects on natural enemies (see Shepard *et al.*, 1977).

Studies on the biological control of *N. viridula* have been performed mostly on soybeans (see Panzzi and Slansky, 1985) but such information is directly or indirectly applicable to tomato culture. Pentatomid eggs are parasitised by numerous hymenopterous species belonging to the Scelionidae, Encyrtidae and Eupelimidae. The most important wasp parasitoid of *N. viridula* eggs in Europe, Africa and North America appears to be *Trissolcus basalis* Wallaston, a scelionid species that has been successfully introduced into many countries world-wide. Other egg parasitoids include *Ooencyrtus* Ashmead spp. (Encyrtidae). Tachinid flies are common parasitoids of adult green stinkbugs, with *Trichopoda pennipes* L. being the most extensively studied. Harris and Todd (1982) reported that, although attraction may be mediated by host plant odour, *T. pennipes* caused on average a 49% reduction in the lifespan of *N. viridula* adults. A number of predators are also active against this tomato pest. Ragsdale *et al.* (1981) tested 6 orders and 15 families of insects and spiders for predation on *N. viridula* eggs and nymphs: *Geocoris punctipes* Say (Lygaeidae) was found to be the most efficient predator. They concluded, however, that the predator complex is apparently unable to maintain *N. viridula* below levels that cause economic damage to soybeans. The pentatomid *Euthyrhynchus floridanus* L. and the red imported fire ant, *Solenopsis invicta* Buren (Formicidae), prey upon *N. viridula* in the USA (Mead, 1976; Krispyn and Todd, 1982).

Husbandry techniques can be useful in green stinkbug management. Early planting of soybeans has been used in the USA to attract *N. viridula* (McPherson and Newsom, 1984). A high proportion (70–72%) of the total pest population remains on the trap crop until mid-summer, at which time insecticidal sprays can be used to suppress spread into the main host crop. Trap crops of legumes or highly attractive vegetables could be utilised prior to tomato fruit set to concentrate pest populations of *N. viridula* and then control them by insecticides. In many regions the coordination of early-maturing trap crops and late-maturing tomato cultivars could serve to regulate infestation levels.

8.11 FLIES

8.11.1 *Liriomyza bryoniae* (Kaltenbach): Tomato Leafminer (Diptera: Agromyzidae)
Liriomyza sativae Blanchard: Vegetable Leafminer (Diptera: Agromyzidae)
Liriomyza trifolii (Burgess): American Serpentine Leafminer (Diptera: Agromyzidae)

Leafminers are pests of several vegetable and ornamental crops grown throughout the temperate and subtropical climates of the world. Of the numerous species that infest tomatoes, three are most important. *L. sativae* and *L. trifolii* are distributed throughout the Nearctic (most of the North American continent). They have overlapping host ranges and are morphologically similar, characteristics which have contributed to taxonomic confusion (see Parrella and Keil, 1984). *L. trifolii* has been recently introduced into Europe, probably on ornamentals shipped from the USA, spreading first to greenhouse tomatoes and then to cultivated and wild hosts outdoors (Powell, 1982). *L. bryoniae* is at present limited to the Palaearctic (Europe, Africa and Asia). It was once thought, however, to be synonymous with *L. langei* Frick in North America (Spencer, 1965). Management problems are often compounded by the close proximity of vegetables and ornamentals. Leafminer establishment in many areas has occurred through the introduction of infested chrysanthemum plants (Parrella and Robb, 1982). All three species of leafminer have a wide range of weed hosts.

Of *L. bryoniae*, *L. sativae* and *L. trifolii*, the most complete published information on description, life cycle and damage appears on *L. sativae* (see Musgrave *et al.*, 1975). Therefore, much of the information presented here is likely to pertain more precisely to *L. sativae* and more generally to the other species. Agromyzid leafminers display complete metamorphosis. Adults are very small (wing length, 1.5 mm), with females being slightly larger than males. They may be observed on leaves, where the females repeatedly puncture the leaves and feed on exuding sap. Males apparently lack structures to puncture, but may feed at wounds produced by females. Mating occurs within 24 h of emergence from puparia, and eggs are inserted into leaves after a preoviposition period of about 2 days. A single female can lay approximately 250 eggs (Audemard and D'Aguilar, 1969; Parrella *et al.*, 1981), which hatch in about 3 days, depending on temperature. Adult flies are weak fliers and movement into fields is often from the direction of prevailing winds (Tyron *et al.*, 1980). Zehnder and Trumble (1984) reported that flight activity in California, USA, peaked from 0700 hours to 1100 hours.

The larva is a laterally compressed, legless, headless maggot with darkly coloured intestinal contents. A serpentine mine appears in the upper leaf surface 3–5 days after oviposition, within which frass produced by the larva can be seen through the epidermis 'window'. A larva typically feeds and rests lying on its side and, as it turns inside the mine, faeces are deposited along the opposite side. It feeds (by extruding sickle-like mandibles, rasping away the leaf mesophyll) for 1–3 weeks, depending on temperature, after which it usually cuts a semicircular hole in the end of the mine and emerges to pupate. The shiny, golden-brown-coloured puparia adhere to the foliage, or fall to the soil surface, where they may be found in the upper soil zone. Puparia may overwinter in this fashion or give rise to another generation of adults in 7–14 days. The entire life cycle of *L. sativae* and *L. trifolii* can be completed in 21–28 days in Florida and California, USA, facilitating many generations per year. Although *L. bryoniae* may complete the life cycle in 14–32 days (30–18°C) (Nedstam, 1985) in the laboratory, it may take 2–3 months outdoors in England (Esbjerg, 1977).

Leafminers directly damage tomatoes by stippling and mining the leaves. Leaf wounding creates habitats for infection by bacterial and fungal pathogens. Heavily mined leaves may have nearly all of their mesophyll removed. Photosynthesis rates in mined tissues can be reduced by 62% and even low incidences of mining activity greatly affect leaflet photosynthesis (Johnson *et al.*, 1983). Mining of leaves adjacent to fruit during its early stages of development has a great effect on ultimate yield (Ledieu and Helyer, 1985). Plant stress, moisture loss or sun scald of fruit owing to the absence of shading by foliage may occur as secondary problems.

Pesticides used for leafminer management include all the major classes of insecticides and the development of resistance has been relatively rapid. Average insecticidal longevity against *L. sativae* in the USA has been brief since 1975, and no Californian populations are at present regarded as completely susceptible. In Florida the projected field lives of an organophosphate (methamidophos), a carbamate (oxamyl) and a pyrethroid (permethrin) have ranged from 2 to 4 years (Parrella and Keil, 1984). In Europe, where chemical control has mostly been performed in greenhouses, tolerance to insecticides has not yet become a major problem. Candidate materials have to be selected with care, however, lest they disrupt the biological control of the greenhouse whitefly by the parasitoid *E. formosa*.

Regular usage of conventional insecticides against leafminers and other insect species that infest tomatoes may intensify leafminer damage owing to insecticidal reduction of natural enemies (e.g. Hayslip, 1961; Poe *et al.*, 1978; Johnson *et al.*, 1980a, b). 'Selective' insecticides such as the microbe *Bacillus thuringiensis* Berliner var. *kurstaki* Kurstak (Bacillaceae), the insect growth regulator 'avermectin' and extracts of neem tree leaves have efficacy against leafminers and are safe on associated parasites (Johnson *et al.*, 1980b; Fagoonee and Toory, 1984; Trumble, 1985). Insects are, perhaps, less likely to develop resistance to these natural or naturally derived insecticides. However, resistance by Indian meal moth (*Plodia interpunctella* (Hübner)) to *Bacillus thurigiensis* is known to occur.

In many field situations, a complex of hymenopterous parasitoids may maintain leafminers below economically damaging levels. In the USA at least 14 parasitic species (from Braconidae, Pteramalidae, Eulophidae and Cynipidae) have been reared from the larvae and pupae of *L. sativae* (Musgrave *et al.*, 1975; Oatman and Kennedy, 1976b). Native or introduced parasitoids that have shown varying degrees of effectiveness against leafminers in Europe and America include: *Dacnusa sibirica* Telenga (Braconidae); *Diglyphus isaia* Ashmead, *D. intermedius* Ashmead, *D. begini* Ashmead, *D. pulchripes* Crawford (Eulophidae); *Opius pallipes* Wesmeal, *O. bruneipes* Goh. (Braconidae); and *Cryptogaster vulgaris* Walker (Braconidae). There is evidence that *L. trifolii* can encapsulate the eggs of certain parasitoids, making it less amenable to biological control: in the Netherlands parasitoids that provide good control of *L. bryoniae* are much less successful against *L. trifolii* (Parrella and Keil, 1984). Clearly, the potential for biological control in the management of leafminers on tomatoes and other crops is real, but must be approached, for the most part, on an individual leafminer species basis.

Host resistance to leafminers has not yet been fully exploited in the wild relatives and cultivars of tomatoes. Intensified research is advisable, for relatively low levels of resistance in tomatoes are sufficient to aid the management of leafminers, because high levels of damage are needed to effect reductions in yield (Wolfenbarger and Wolfenbarger, 1966). Webb *et al.* (1971) reported that several breeding lines of tomato have genes contributing to larval antibiosis, adult non-preference, or both. All accessions of *L. hirsutum* and *L. hirsutum* f. *glabrum* (see page 301) were virtually immune to feeding and egg deposition in both greenhouse and field tests. Webb *et al.* (1971) concluded that *L. hirsutum* has potential for use in breeding programmes, but the incorporation of resistant genes into lines, while concurrently maintaining horticulturally desirable qualities, is likely to be difficult. Cultivars having resistance to leafminers are at present available. VF-145, a commonly grown, mechanically harvested tomato, is resistant to *L. sativae*. The apparent mechanism of resistance is non-preference: the curled leaves seem to be unacceptable to females, which feed and lay eggs on flat leaves (Lange and Bronsin, 1981). Bethke *et al.* (1987) evaluated seven cultivars and found that *L. trifolii* pupal survival ranged from 24% to 60%, with VF-7718 having the lowest. They found, as suggested by Webb *et al.* (1971), that agronomic conditions affecting host-plant vigour can influence resistance: high fertiliser rates applied to two cultivars, Patio and Beefsteak, increased larval survival and size of pupae.

Strategies for management of leafminer species on tomatoes should incorporate cultural, biological and chemical methods. Destruction of alternative hosts, particularly broad-leaved weeds, near tomato fields at least 1 month prior to establishment may delay leafminer appearance. Removal or turning-under of host-crop residues prior to planting can prevent emergence of nearly 100% of the viable larvae and pupae (Musgrave *et al.*, 1975). Prior to insecticidal applications, tomato fields should be quantitatively sampled for pest abundance, by utilising either yellow 'sticky boards' to trap adults (Musgrave *et al.*, 1975) or Styrofoam trays to trap prepupae falling from foliage (Johnson *et al.*, 1980c). Information on pest abundance can be used to schedule pesticidal treatments in accordance with economic thresholds, where available. Pesticides that are detrimental to naturally occurring or introduced parasitoids should not be used. The consistent usage of a single class of insecticides should be avoided, to delay or prevent the development of resistance by leafminer species.

8.12 MITES

8.12.1 *Aculops lycopersici* (Massee): Tomato Russet Mite (Prostigmata: Eriophyidae)

A thorough review of the historical perspective and world status of *A. lycopersici*, from which portions of the following information have been compiled, is presented by Perring and Farrar (1986). The tomato russet mite occurs world-wide in a band approximately 60 degrees latitude north and south of the equator. This eriophyid mite is a damaging pest of cultivated tomatoes in all areas where the crop is extensively grown. It also attacks wild tomatoes and, to a lesser degree, other Solanaceae. Alternative hosts are solanaceous weeds (*Solanum* L. spp. and *Datura* L. spp.), field bindweed (*Convolvulus arvensis* L.), cultivated and wild gooseberries (*Ribes* L. spp.) and tobacco (*Nicotiana* L. spp.).

A. lycopersici was first described in 1916 as a pest of tomatoes in Queensland, Australia (Tyron, 1917). Morphological descriptions of tomato russet mite stages are relatively incomplete (Bailey and Keifer, 1943; Lamb, 1953; Anderson, 1954). The typical developmental cycle is adult, egg, larva, larval chrysalis, nymph and nymphal chrysalis.

Members of the subclass Acari (mites) typically possess four pairs of legs. Mites in the family Eriophyidae are exceptional, however, in that larvae, nymphs and adults have two pairs of legs only. Adults are yellow-orange in colour, wedge-shaped, and microscopic (approximately 190 μm long and 60 μm wide), with males being slightly smaller than females. Eggs are minute (45–55 μm in diameter), being opalescent-white in colour when laid, but turning cloudy-yellowish as they mature. The translucent-white-coloured larvae are 90–100 μm long. The yellowish-coloured nymphs are 140–160 μm long. Neither the larva nor the nymph has genital organs. Reproduction may be either sexual, with fertilised eggs producing individuals of both sexes, or asexual, with unfertilised females producing only males (arrhenotoky).

In warmer climes, tomato russet mites probably overwinter on crop debris in unheated greenhouses and on wild solanaceous hosts in protected areas such as ditch banks. Small initial infestations can become acute within a few weeks as colonies develop and individuals disperse to early-planted tomatoes or available wild hosts. Some dispersal may be accomplished by crawling to nearest-neighbour plants, but the primary mode of dispersal is probably by means of spinning a silk thread on which the mites are carried by the wind. They can also be easily spread from one plant to another on the clothes of workers during cultivation, pruning and picking (Sekeroglu and Ozgur, 1984). Developmental rates are rapid, facilitating many generations per season. At 21°C, there is a 2-day pre-oviposition period, after which females may each lay from 10 to 53 eggs that hatch in 2 days: the larval and nymphal periods are 1 day and 2 days, respectively (Bailey and Keifer, 1943). More recently, Abou-Awad (1979) studied the development of tomato russet mites at 25°C and reported specific developmental times for all life stages: egg, 2.3 days; larva, 0.38 day; larval chrysalis, 0.53 day; nymph, 0.96 day; and nymphal chrysalis, 0.72 day. Males developed slightly faster than females (4.6 and 5.2 days, respectively) and females lived longer than males (22.1 and 16.5 days, respectively). Population increases are directly and positively correlated with temperature increases (Taylor, 1978). Conditions near 27°C and 30% relative humidity are the most favourable for development, and under high temperature conditions high humidity is detrimental to mite survival (Rice and Strong, 1962).

Injury due to feeding by tomato russet mites starts on the lower stems and leaves and spreads upwards. The initial indication of infestation is a silvering of the lower leaves, which then become bronze-coloured (russet), wither and drop. As more mite colonies are produced, individuals move upwards, with the greatest concentrations of mites being on the green undamaged tissue adjacent to obvious damage. The loss of foliage exposes fruit to sunscald. Both green and mature fruits may have yellow-to-white, halo-shaped blotches over much of the skin surface. With severe infestations, the skin may become discoloured, with numerous small cracks in the region close to the peduncle (DeOliveira *et al.*, 1982; Kay, 1986; Zalom *et al.*, 1986a). Infestations are often more widespread and severe during dry periods.

Tomato russet mite infestations and resulting yield reductions are a function of both plant phenology and the environmental parameters at a particular time of season. Various reports from South America suggest that infestations occurring soon after seedling emergence or soon after transplanting are most detrimental to tomato production. Eschiapati *et al.* (1975) reported that when mite infestations occurred 25 days after seedling emergence, tomato yields were reduced by 65.5%. Infestations at 45 and 65 days after emergence effected 48.5% and 7.5% yield reductions, respectively. Similarly, DeOliveira *et al.* (1982) reported that mite abundance increased progressively until 82 days after transplanting, at which time mite numbers decreased. They concluded that infestations coinciding with early plant development most severely reduced tomato yields. Studies in South Africa and the USA, however, showed that infestations occurring at later plant developmental stages were most detrimental to yield. Daiber (1985) reported up to 50% yield losses due to sunscald, when infestations occurred shortly before or during harvest maturity. These results were corroborated by Zalom *et al.* (1986b), who found that the greatest damage occurred when most tomato plants had 'mature green' or pink fruit.

The potential for tomato plants that are resistant to tomato russet mites is unknown, for the literature is lacking in reports on this method of management. Research in this area is needed and possibly forthcoming, since resistance to tetranychid mites has been documented for various *Lycopersicon* Mill. species (Stoner and Stringfellow, 1967; Rodriquez *et al.*, 1972; Cantelo *et al.*, 1974).

Traditional programmes for control of tomato russet mites have utilised numerous acaricides, insecticides and fungicides with varying degrees of success (a comprehensive list of materials used between 1940 and 1985 may be found in Perring and Farrar, 1986). Recent tests have shown that the acaricides dicofol and cyhexatin (cyhexatin is no longer available) and the insecticides avermectin B_1 and cypermethrin are efficacious against tomato russet mites (Abou-Awad and El-Banhawy, 1985; Royalty and Perring, 1987). Sulphur or sulphur-containing compounds have been widely used for control of mites since 1934, when the effectiveness of this element was first discovered in Australia. Sulphur dust, which was first recommended in 1943, remains the primary acaricide used in California, USA (Zalom *et al.*, 1986a).

A number of predatory mites have been investigated for control of tomato russet mites. Generally, predators found to be promising in controlled laboratory studies have not been effective in field situations (see Perring and Farrar, 1986). Recently Hessein and Perring (1986) documented control of tomato russet mites by the tydeid mite *Homeopronematus anconai* Baker under laboratory conditions. *H. anconai* is common in many natural and agricultural ecosystems including tomatoes, but is seldom encountered in commercial plantings. *H. anconai* may be susceptible to the pesticides that are commonly used against tomato russet mites and various other tomato arthropod pests. Sulphur, the long-time industry standard for tomato russet mite control, is fivefold more toxic to *H. anconai* than to russet mites. Management programmes designed to control tomato russet mites while maintaining *H. anconai* populations might utilise low concentrations of avermectin B_1 which are relatively selective (Knop and Hoy, 1983; Royalty and Perring, 1987). Pyrethroid insecticides, because of their renowned residual properties and toxicity to predatory mites, should be used only in situations where the pest complex is difficult to manage otherwise (R. W. Weires, personal communication). The feasibility of using predatory mites for the control of tomato russet mites remains to be determined. Integrated methods for the management of mites could include the destruction of crop residues, the elimination of weed hosts in and around tomato fields, the monitoring of mite occurrence and tomato phenology, and the selection of pesticides that are not detrimental to the natural predators of the russet mites.

8.12.2 *Polyphagotarsonemus latus* (Banks): Broad Mite (Prostigmata: Tarsonemidae)

The broad mite (also yellow tea mite and white citrus mite) is a polyphagous pest that attacks a wide range of outdoor crops throughout the tropical and temperate regions of the world. Primarily a pest of tea and citrus, other recorded hosts include: coffee, cotton, solanaceous vegetables, cucumbers, beans and many annual and perennial ornamentals. Broad mites are important pests of vegetables grown in greenhouses. Damage to tomatoes was first recorded in the UK in 1935 (Wilson, 1950). In many temperate regions broad mites have since spread from greenhouses to outdoor crops, presumably by migration or as transients on vegetable and ornamental transplants.

Morphological and behavioural characteristics of the broad mite are not well documented. Reproduction can be either sexual, with fertilised eggs producing individuals of both sexes, or asexual, with unfertilised females producing males only (arrhenotoky). Diapausing adult females overwinter in bud scales and axils on perennial plants (Yang and Chen, 1982) and probably in leaf debris in annual crop situations. The typical developmental cycle is adult, egg, larva and nymph. Adults are yellow or white in colour, are about 1.5 mm long and have eight legs. Eggs (approximately 0.7 mm long) are laid on the undersides of leaves. The eggs are oval in outline, with the upper surface covered in five or six rows of white tubercles. The six-legged larvae which hatch from the eggs are minute, white and pear-shaped. The larva turns into a quiescent pupal stage (nymph) which is apparently 'stuck' to the underside of the leaf. A female pupa is usually picked up and carried about by an adult male, thus ensuring the male of a partner for mating (Costilla, 1955). A female mates only once. At 25°C, the approximate durations of life stages are: male, 4 days; female, 4 days with a 1 day preoviposition period; egg, 2 days; larva, 1 day; and nymph, 1 day. A female lays 8–24 eggs at a rate of 1–6 eggs/day (Sombatsiri, 1978). The optimum environmental parameters for development are 18–25°C at 80–90% relative humidity (Yang and Chen, 1982).

All active stages of broad mite may be found on the undersides of young tomato leaves and on immature tomato fruits. General symptoms of infestation include curling and reduction in size of leaves, stunting of stems and a silver-coloured cast on leaf undersurfaces (Sombatsiri, 1978). When infestations are severe, corky formations occur on leaves, flower buds, growing shoots and fruits (Iacob, 1978). Feeding by broad mites on tomato fruits may cause them to become cracked and mis-shaped (Cross and Bassett, 1982).

Broad mites can be controlled by dicofol or chlorbenzilate sprayed two to three times at 5–7 day intervals (Iacob, 1978; Sombatsiri, 1978; Cross and Bassett, 1982). These chlorinated hydrocarbon acaricides have long residual effects and are generally harmless to beneficial insects, but may be toxic to predacious mites (van de Veire, 1985). Sombatsiri (1978) reported that broad mite adults are highly susceptible to most acaricides, but eggs and nymphs are much more difficult to control. During the presence of early developmental mite stages, emulsions of petroleum oils could be used.

Reports of biological control efforts against broad mites on tomatoes are essentially non-existent. On citrus fruits in the USA, predacious phytoseid mites (*Amblyseius* Berlese spp.) feed on broad mites and may contribute to control (Brown and Jones, 1983). Brown and Jones (1983) found that dicofol, an effective acaricide against broad mites, was toxic to *Amblyseius* spp.

8.12.3 *Tetranychus urticae* Koch: Two-spotted Spider Mite (Prostigmata: Tetranychidae)

Tetranychid mites are found throughout the world on virtually every major food crop and type of ornamental plant. The subfamily Tetranychinae includes a number of economically significant forms, of which *T. urticae* is the most important on tomatoes, beans and cucurbit crops. Owing to the number of cultivated plants on which it feeds, taxonomic confusion existed within the *Tetranychus* Dufour genus for many years prior to 1955, which resulted in an accumulation of more than 50 different common names for this pest (see Gutierrez, 1985).

Tetranychid mites may reproduce sexually, with eggs of mated females producing individuals of both sexes, or asexually, with males only hatching from eggs produced by virgin females (arrhenotoky). The life cycle, which may take 6–8 days at 30°C (Ho and Lo, 1979; Carey and Bradley, 1982), includes five stages: egg, larva (first instar); protonymph (second instar); deutonymph (third instar); and adult. The three feeding stages are each followed by intervening periods of quiescence (chrysalis).

Eight-legged adults overwinter as diapausing females beneath fallen leaves and other vegetable debris. They are 0.4–0.5 mm long. Males are slightly smaller than females. As described by Cone (1985), *T. urticae* performs extensive mating rituals involving male precopulatory behaviour, including aggression, attraction, guarding, web-spinning and territoriality. After the female deutonymph exoskeleton splits and is partially shed, copulation ensues. Males are sexually long-lived and may mate repeatedly. Under crowded conditions, a low percentage of females may be doubly inseminated, but second matings are ineffective. Females pass through a brief preoviposition period, after which they each lay from 60 to 115 spherical, whitish-coloured eggs (0.1 mm diameter) at a rate of 5 or 6 per day (Ho and Lo, 1979; Carey and Bradley, 1982). Eggs may be covered with webs, the presumed purpose of which is to regulate humidity and to protect them from mite predators (Gerson, 1985). The larva is a pinkish colour, is slightly larger than the egg and has three pairs of legs. Protonymphs and deutonymphs have a green or red colour and four pairs of legs. The influence of varying temperatures and humidities on the development of the egg and immature stages is difficult to assess. Fluctuating temperatures and humidities may accelerate or decelerate development, or development may remain unaffected, depending on the magnitude and duration of the fluctuations. The chemical composition of the host plant may influence fecundity, mortality and development of the immature stages (see Crooker, 1985). Childs *et al.* (1976) found that spider mite oviposition, for instance, ranged from 2 to 8 eggs per mite per day among eight cultivars of common bean (*Phaseolus vulgaris* L.). Suski and Badowska (1975) studied mite reproduction on *P. vulgaris* plants treated with varying rates of nitrogen and found that high doses caused the highest innate capacity for increase.

Spider mites have well-developed dispersal mechanisms, enabling them to spread over large areas and to colonise widely separated host plants. *T. urticae* is able to crawl over the soil surface to infest neighbouring plants. Spider mites are also carried by the wind after first assuming a dispersal posture which involves raising the forelegs upright. All active stages except adult males display dispersal posturing (see Kennedy and Smitley, 1985).

In the USA *T. urticae* commonly overwinters on forage legumes, and in late spring becomes wind-blown to fields of beans and other susceptible crops. Bean fields adjacent to perennial crops are usually infested first (Hagel and Landis, 1972). Overwintering sites other than forage legumes include weeds and other vegetable debris (Atanasov, 1971).

Damage by spider mites to tomatoes, beans and cucurbits is usually first evidenced as clusters of yellow spots on the upper surface of leaves, which may also appear chlorotic. On the leaf undersides varying degrees of webbing containing active mite stages will be present. Heavily infested leaves may 'cup' inward, or each entire leaf may turn yellow in colour and die.

Tomato plant yields may be reduced by a mite-induced physiological shock, by reduced size and numbers of fruit, and by sunscalded fruit arising from loss of leaves. When tomato plants at different stages of growth were subjected to *T. urticae* damage (Stacey *et al.*, 1985), it was found that of all the leaves on a plant the top 12 contribute most to yield. Damage to leaves near a truss of setting fruit is likely to cause yield loss at a rate approximately equal to the proportion of leaf area affected: when 10% of the leaf area was damaged, a 9% loss in yield resulted. There was a delay of 5–6 weeks between the onset of damage and a diminished-yield response. The economic injury level, particularly with fresh-market tomatoes, is extremely low because aesthetic or 'cosmetic' elements are important factors in marketability.

Feeding by *T. urticae* on bean plants may cause reductions in plant height, flowering, pod number and length, number of seeds per pod and mean seed weight (Papaioannou-Souliotis, 1979). Kropczynska and Tomczyk (1986) concluded that damage to beans is most severe when mite feeding occurs early in the vegetative period: when mite infestations increased after flowering, no influence on yield resulted, even at densities of 3 mites/cm^2 of leaf, but low-density infestations (0.3–0.5/cm^2) early in the vegetative period caused an approximate 40% reduction in flowering and pod formation. As spider mites feed on bean plants, they inject saliva which physiologically interferes with growth-promoting substances present in the plants (Storms, 1971), thus accounting for the inhibitory or stimulatory effects of mite feeding on plant growth.

All cultivars of Cucurbitaceae are susceptible to injury from spider mites, with cultivars lowest in cucurbitacins suffering greatest injury. Cucumbers growing in greenhouses are particularly susceptible.

Problems with the control of *T. urticae* have been greatly aggravated by this organism's marked propensity to develop resistance to a variety of chemical groups (see Cranham and Helle, 1985). Petroleum oil emulsions continue to be widely used, because no resistance to their primarily physical mode of action has occurred. However, oils are effective only on eggs and must be applied frequently, often causing phytotoxic reactions. Pyrethroid acaricides control mites in much the same fashion as conventional acaricides, but the predominant mode of action appears to be repellency. *T. urticae* is highly sensitive to fenvalerate and this sensitivity is seen as a discrimination by the mites against sites with fenvalerate residues (Penman *et al.*, 1981). On organophosphate (phosmet)-treated apple leaves, two-spotted spider mites spend 70–80% of their time feeding, but only 5–40% of their time is spent feeding on pyrethroid (permethrin or fenvalerate)-treated leaves (Iftner *et al.*, 1986). Iftner *et al.* (1986) reported that mites on pyrethroid-treated leaves exhibit hyperactivity, longer search periods, reduced oviposition and an apparent preference for areas with lower residues. The presence of fenvalerate residues on bean leaves induces *T. urticae* to vacate plants by forming silk threads on which the mites drop to the ground (Penman and Chapman, 1983). Penman and Chapman (1983) also reported that no colonisation occurs on fenvalerate-treated leaves. Other pyrethroid acaricides (e.g. deltamethrin, fluvalinate, fenpropathrin, cypermethrin, flucythrinate) display even greater repellent action than fenvalerate (Penman *et al.*, 1986). Such behavioural responses to pyrethroids have created concern that avoidance of treated tissues promotes secondary outbreaks, as mites migrate or disperse to untreated hosts (Penman and Chapman, 1983; Iftner *et al.*, 1986).

The biological control of tetranychid mites in greenhouses is well developed, and remarkable progress has been made towards integrated control of mites on perennial outdoor crops such as almonds and apples. However, the use of biological agents to control *T. urticae* on annual outdoor crops has been limited. The predatory mite *Phytoseiulus persimilis* Athias-Henriot (Mesostigmata: Phytoseiidae) has been used successfully to control *T. urticae* on greenhouse crops. Although inundative releases of *P. persimilis* within outdoor crops have been somewhat successful (Cochereau, 1976; Pickett and Gilstrap, 1986), this predator has generally failed to effect consistent economic control, because it has limited capacity for dispersal, is solely dependent on tetranychid mites for survival, has a number of natural enemies and is susceptible to high temperature and humidity (Wysoki, 1985). These factors limit its ability to maintain an adequate predator: prey ratio in the field. The predatory mite *Amblyseius fallacis* (Garmen) (Mesostigmata: Phytoseiidae) may be more efficient than *P. persimilis* in field situations, because it does not feed only on tetranychid mites, is relatively mobile, has good searching ability, can overcome the webbings of its prey and has a high reproductive potential in a wide range of environments (Smith and Newsom, 1970). The maintenance of predatory mites in tomatoes and other vegetable crops will be rendered null and void by applications of broad-spectrum pesticides (Brown and Shanks, 1976; Hassan, 1982). Non-specific pesticides, particularly pyrethroids, should be avoided in situations where mite predator: prey balances are to be maintained.

The development of tetranychid mite-resistant tomato cultivars has received considerable emphasis since McKinney (1938) noticed that aphids became stuck in a 'gummy exudate' on tomato foliage, and that the exudates were derived from foliar trichomes (Johnson, 1956). It has been clearly demonstrated that a type of trichome having a short multicellular stalk and a 2–4-celled glandular head is involved in the resistance of a number of tomato species to tetranychid mites (Aina *et al.*, 1972; Rodriquez *et al.*, 1972). Aside from the entrapping nature of the exudates from trichomes, repellent volatiles extracted from trichomes and foliage have been implicated as major resistance factors (Rodriquez *et al.*, 1972; Cantelo *et al.*, 1974). Although glandular trichomes occur in all species of *Lycopersicon*, Snyder and Carter (1984) found that at equivalent trichome densities, *L. hirsutum* was more resistant than *L. esculentum* to *T. urticae*. F_2 hybrids of *L. hirsutum* (PI251303)×*L. esculentum* (CV Ace) provided mite mortalities that were apparently due to the association of glandular trichomes with some toxic factor (Carter and Snyder, 1985). Although it has not been definitively linked with resistance to mites, the toxic chemical (2-*n*-tridecanone) extracted from trichomes of wild tomatoes (PI134417) may be involved. This chemical and the natural volatiles from trichomes are potentially lethal to four other major tomato pests: *Manduca sexta* L. (tobacco hornworm); *Heliothis zea* Boddie (tomato fruitworm) (see page 220); *Leptinotarsa decemlineata* Say (Colorado potato beetle) (see page 275); and *Aphis gossypii* Glover (melon aphid) (see page 140) (Duffey, 1986). The incorporation of *T. urticae* resistance as found in wild tomato types into commercially acceptable cultivars is possible, but made difficult because undesirable tomato quality traits are transferred along with resistance. However, owing to the propensity of *T. urticae* to develop resistance to a variety of acaricides, and the general failure of biological control efforts in the field, the breeding of commercially acceptable resistant cultivars would be an economically important achievement.

Bean cultivars show varying degrees of susceptibility to *T. urticae* feeding. Childs *et al.* (1976) reported differences in intrinsic rates of spider mite growth among a number of *Phaseolus* L. cultivars. Kropczynska and Tomczyk (1986) alluded to conspicuous varietal differences in mite susceptibility and suggested that such differences should be considered in the establishment of economic thresholds. Resistance characteristics, even in small degrees, would be useful in attempts to integrate control methods.

REFERENCES

Abou-Awad, B. A. (1979). The tomato russet mite, *Aculops lycopersici* (Massee) (Acari: Eriophyidae) in Egypt. *Anz. Schadlingskde Pflanzenschutz.*, **52**, 153–156

Abou-Awad, B. A. and El-Banhawy, E. M. (1985). Susceptibility of the tomato russet mite, *Aculops lycopersici*, in Egypt to methamidophos, pyridaphenthion, cypermethrin, dicofol, and fenarinol. *Exp. Appl. Acarol.*, **1**, 11–16

Ahman, I. and Ekbom, B. S. (1981). Sexual behavior of the greenhouse whitefly (*Trialeurodes vaporariorum*): orientation and courtship. *Entomol. Exp. Appl.*, **29**, 330–338

Aina, O. J., Rodriquez, J. G. and Knaval, D. E. (1972). Characterizing resistance to *Tetranychus urticae* in tomato. *J. Econ. Entomol.*, **65**, 641–643

Airey, W. J. (1984). The distribution of slug damage in a potato crop. *J. Mollusc. Studies*, **50**, 239–240

Airey, W. J. (1986). The influence of an alternative food on the effectiveness of proprietary molluscicidal pellets against two species of slugs. *J. Mollusc. Studies*, **52**, 206–213

Airey, W. J. (1987). Laboratory studies on damage to potato tubers by slugs. *J. Mollusc. Studies*, **53**, 97–104

Al-Musa, A. (1982). Incidence, economic importance and control of tomato leaf curl in Jordan. *Plant Dis.*, **66**, 561–563

Anderson, J. A. D. (1986). The effect of resistant potato cultivars on a field population of *Globodera pallida* Pa2 over three years of cropping. *Ann. Appl. Biol.*, **108**, 81–87

Anderson, L. D. (1954). The tomato russet mite in the United States. *J. Econ. Entomol.*, **47**, 1001–1005

Anderson, S. and Anderson, K. (1982). Suggestions for determination and terminology of pathotypes and genes for resistance in cyst-forming nematodes, especially *Heterodera avenae*. *EPPO Bull.*, **12**, 379–386

Anon. (1982). *Report* Int. Potato Center, Lima, Peru

Atanasov, N. (1971). The overwintering of the Atlantic mite. *Rastitelna Zashchita*, **19**, 14–17

Audemard, H. and D'Aguilar, J. (1969). A dipterous pest of vegetable crops under glass: *Liriomyza bryoniae* Kalt. *Compt. Rend. Herb. des Seances de l'Acad. d'Agric. de France*, **55**, 896–901

Azmy, N. M. (1980). Sexual activity, fecundity and longevity of *Nezara viridula* (L.). *Bull. Entomol. Soc. Egypt*, No. 60, 323–330

Bailey, S. F. and Keifer, H. H. (1943). The tomato russet mite, *Phyllocoptes destructor* Keifer: Its present status. *J. Econ. Entomol.*, **36**, 706–712

Balandras, C. M. (1985). Limits to methods for management of potato cyst nematodes. In *Cyst Nematodes* (ed. F. Lamberti and C. E. Taylor). Plenum Press, London, pp. 413–432

Barnes, H. F. (1944). Discussion on slugs. 1. Introduction. Seasonal activity of slugs. *Ann. Appl. Biol.*, **31**, 160–163

Barnes, H. F. and Weil, J. W. (1944). Slugs in gardens: their numbers, activities and distribution Part 1. *J. Anim. Ecol.*, **13**, 140–175

Bartlett, P. W. (1982). *Colorado Beetle (a Crop Pest not yet Established in Britain)*. Ministry of Agriculture, Fisheries and Food Leaflet No. 71

Berlinger, M. J., Dahan, R. and Mordechi, S. (1986). Prevention of tomato leaf curl virus by controlling its vector, *Bemisia tabaci. Hasadeh*, **66**, 686–689

Berlinger, M. J., Dahan, R. and Shevach-Urkin, E. (1983). Breeding for resistance to whiteflies in tomatoes—in relation to integrated pest control in greenhouses. *Bull. SROP*, **6**, 172–176

Bethke, J. A., Parella, M. P., Trumble, J. T. and Toscano, N. C. (1987). Effect of tomato cultivar and fertilizer regime on the survival of *Liriomyza trifolii* (Diptera: Agromyzidae). *J. Econ. Entomol.*, **80**, 200–203

Blackman, R. L. (1971). Variation in the photoperiodic response within natural populations of *Myzus persicae* (Sulz.). *Bull. Ent. Res.*, **60**, 533–545

Blackman, R. L. and Eastop, V. F. (1984). *Aphids on the World's Crops: An Identification and Information Guide*. Wiley, Chichester

Boiteau, G. (1986). Native predators and the control of potato aphids. *Can. Ent.*, **118**, 1177–1183

Briese, D. T. (1986). Geographic variability in demographic performance of the potato moth, *Phthorimaea operculella* (Zeller) (Lepidoptera: Gelechiidae), in Australia. *Bull. Ent. Res.*, **76**, 719–726

Brown, G. C. and Shanks, C. H. (1976). Mortality of twospotted spider mite predators caused by the systemic insecticide, carbofuran. *Environ. Entomol.*, **5**, 1155–1159

Brown, R. D. and Jones, V. P. (1983). The broad mite on lemons in Southern California. *Calif. Agric.*, **37**, 21–22

Butter, N. S. and Rataul, H. S. (1973). Control of tomato leaf curl virus (TLCV) in tomatoes by controlling the vector whitefly, *Bemisia tabaci* Gen., by mineral-oil sprays. *Current Sci.*, **42**, 864–865

Cantelo, W. W., Boswell, A. L. and Argauer, R. J. (1974). *Tetranychus* mite repellent in tomato. *Environ. Entomol.*, **3**, 128–130

Carey, J. R. and Bradley, J. W. (1982). Development rates, vital schedules, sex ratios and life tables for *T. urticae*, *T. turkestani* and *T. pacificus* on cotton. *Acarologia*, **23**, 333–345

Carter, C. and Snyder, J. (1985). Mite responses in relation to trichomes of *Lycopersicon esculentum* × *L. hirsutum* F₂ hybrids. *Euphytica*, **34**, 177–185

Casagrande, R. A. (1985). The 'Iowa' potato beetle, its discovery and spread to potatoes. *Bull. Ent. Soc. Am.*, **31**, 27–29

Castresana-Estrada, L., Notario-Gomez, A. and Brizuela, G. C. (1981). Fecundity of *Trialeurodes vaporariorum* (Westw.) on tomato at 22°C. *An. Inst. Nacional Invest. Agrarias, Agric.*, **17**, 127–132

Chalfant, R. B. (1973). Chemical control of the southern green stinkbug, tomato fruitworm and potato aphid on vining tomatoes in Southern Georgia. *J. Ga Entomol. Soc.*, **8**, 279–283

Childs, G., Poe, S. L. and Bassett, M. J. (1976). Response of two-spotted spider mite to *Phaseolus* cultivars. *Proc. Fla State Hort. Soc.*, **89**, 149–150

Choudhary, R., Prasad, T. and Raj, B. T. (1983). Field evaluation of some exotic parasitoids of potato tuber moth, *Phthorimaea operculella* (Zell.). *Indian J. Entomol.*, **45**, 504–506

Clough, M. S., Bale, J. S. and Harrington, R. (1990). Differential cold hardiness in adults and nymphs of the peach-potato aphid *Myzus persicae*. *Ann. Appl. Biol.*, **116**, 1–9

Cochereau, P. (1976). Biological control in New Caledonia of *Tetranychus urticae* by means of *Phytoseiulus persimilis*. *Entomophaga*, **21**, 151–156

Cohen, S., Antignus, Y. and Ben-Joseph, R. (1984). Whitefly-borne viruses in Israel. *Phytoparasitica*, **12**, 140

Cohen, S. and Melamed-Madjar, V. (1978). Prevention by soil mulching of the spread of tomato leaf curl virus transmitted by *Bemisia tabaci* in Israel. *Bull. Entomol. Res.*, **68**, 465–470

Cohen, S., Melamed-Madjar, V. and Hameiri, J. (1974). Prevention of the spread of tomato yellow leaf curl virus transmitted by *Bemisia tabaci* in Israel. *Bull. Entomol. Res.*, **64**, 193–197

Collmann, G. L. and All, J. N. (1980). Quantification of the greenhouse whitefly life cycle in a controlled environment. *J. Ga Entomol. Soc.*, **15**, 432–438

Cone, W. W. (1985). The Tetranychidae—mating and chemical communication. In *Spider Mites. Their Biology, Natural Enemies and Control*, Vol. 1A (ed. W. Helle and M. W. Sabelis). Elsevier, Amsterdam, Oxford, New York, Tokyo, pp. 243–250

Cooper, B. A. (1954). Eelworm problems in north Fenland with special reference to crop rotation. In Horticultural Educational Association, *Report for 1953*, pp. 106–115

Cornu, P. and Gehriger, W. (1981). [The protection of potatoes against infection by virus Y by means of mineral oil treatments.] *Rev. Suisse Agric.*, **13**, 97–102

Costilla, M. A. (1955). Bioecological aspects of the white citrus mite, *Polyphagotarsonemus latus* (Banks). *Revista Ind. Agric. Tucum'an*, **57**, 15–21

Cranham, J. E. and Helle, W. (1985). Pesticide resistance in Tetranychidae. In *Spider Mites. Their Biology, Natural Enemies and Control*, Vol. 1B (ed. W. Helle and M. W. Sabelis). Elsevier, Amsterdam, Oxford, New York, Tokyo, pp. 405–419

Crawford-Sidebotham, T. J. (1970). Differential susceptibility of slugs to metaldehyde/bran and to methiocarb baits. *Oecologia*, **5**, 303–324

Crooker, A. (1985). The Tetranychidae—embryonic and juvenile development. In *Spider Mites. Their Biology, Natural Enemies and Control*, Vol. 1A (ed. W. Helle and M. W. Sabelis). Elsevier, Amsterdam, Oxford, New York, Tokyo, pp. 149–160

Cross, J. V. and Bassett, P. (1982). Damage to tomato and aubergine by broad mite, *Polyphagotarsonemus latus*. *Plant Pathol.*, **31**, 391–393

Curry, J. P. and Pimentel, D. (1971). Life cycle of the greenhouse whitefly, *Trialeurodes vaporariorum*, and population trends in the whitefly, and its parasite, *Encarsia formosa*, on two tomato varieties. *Ann. Entomol. Soc. Am.*, **64**, 1188–1190

Daiber, K. C. (1985). Tomato pests in South Africa, their potential damage and their control. *Hort. Sci./Tuinbouwetenskap.*, **2**, 1–5

Dainton, B. H. (1954). The activity of slugs. I. The induction of activity by changing temperatures. *J. Exp. Biol.*, **31**, 165–187

Davies, S. M. (1977). The *Arion hortensis* complex, with notes on *A. intermedius* Normand (Pulmonata: Arionidae). *J. Conchol.*, **29**, 173–187

Davies, S. M. (1979). Segregates of the *Arion hortensis* complex (Pulmonata: Arionidae), with the description of a new species *Arion owenii*. *J. Conchol.*, **30**, 123–127

Dawkins, G., Hislop, J., Luxton, M. and Bishop, C. (1986). Transmission of bacterial soft rot of potatoes by slugs. *J. Mollusc. Studies*, **52**, 25–29

Delong, D. M. (1971). The bionomics of leafhoppers. *Ann. Rev. Entomol.*, **16**, 179–210

DeOliveira, C. A. L., Eschidpapti, D., Velho, D. and Sponchiado, O. J. (1982). Quantitative losses caused by the tomato russet mite, *Aculops lycopersici* (Massee) in field tomato crop. *Ecossistema*, **7**, 14–18

de Ponti, O. M. B., Steenhuis, M. M. and Elzinga, P. (1983). Partial resistance of tomato to the greenhouse whitefly (*Trialeurodes vaporariorum* Westw.) to promote its biological control. *Med. Fac. Landbouw. Rijksuniv. Gent*, **39**, 1482–1489

Dewar, A., Devonshire, A. and Ffrench-Constant, R. H. (1988). The rise and rise of the resistant aphid. *Br. Sugar Beet Rev.*, **56**, 40–43

Dixon, P. L. (1986). Pesticides and natural enemies (particularly ground beetles) of aphids on potato. *PhD Thesis, University of Edinburgh*

Duffey, S. S. (1986). Plant glandular trichomes: their partial role in defence against insects. In *Insects and the Plant Surface* (ed. B. Juniper and R. Southwood). Edward Arnold, London, pp. 151–172

Duffus, J. E. and Flock, R. A. (1982). Whitefly-transmitted disease complex of the Desert Southwest. *Calif. Agric.*, Nov–Dec, 4–6

Eickwort, R. C. and Eickwort, G. C. (1986). Effects of parasitism by the mite *Chrysomelobia labidomerae* (Acari: Podapolipidae) on the longevity and fecundity of its host beetle, *Labidomera clivicollis* (Coleoptera: Chrysomelidae). *Int. J. Acarol.*, **12**, 223–227

Ekbom, B. S. (1981). Humidity requirements and storage of the entomopathogenic fungus, *Verticillium lecanii*, for use in greenhouses. *Ann. Entomol. Fennici*, **47**, 61–62

Esbjerg, P. (1977). The tomato mining fly, *Liriomyza bryoniae*. *Manedsoversigt over Plantesygdomme*, No. 503, 122–123

Eschiapati, D., DeOliveira, C. A. L., Velho, D. and Sponchiado, O. J. (1975). Efieto da epoca de infestacao do microacaro, *Aculops* sp., na cultura de tomateiro. *Cien. Cultura*, **27**, 1336–1337

Evans, K. (1982). Effects of infestation with *Globodera rostochiensis* (Wollenweber) Behrens Ro1 on the growth of four potato cultivars. *Crop Protection*, **1**, 169–179

Evans, K., Greet, D. N. and Inge, N. (1984). Tolerance of cyst nematode attack by potatoes. Rothamsted Experimental Station, *Report for 1983*, p. 115

Evans, K. and Trudgill, D. L. (1978). Pest aspects of potato production. Part 1. Nematode pests of potatoes. In *The Potato Crop* (ed. P. M. Harris). Chapman and Hall, London, pp. 440–469

Fagoonee, I. and Toory, V. (1984). The biology and ecology of the leaf miner and its control by neem, *Azadirachta indica*. *Insect. Sci. Appl.*, **5**, 23–30

Ffrench-Constant, R. H. and Devonshire, A. L. (1986). The effect of aphid immigration on the rate of selection of insecticide resistance in *Myzus persicae* by different classes of insecticides. *Aspects Appl. Biol.*, **13**, 115–125

Flanders, R. V. and Oatman, E. R. (1987). Competitive interactions among endophagous parasitoids of potato tuberworm larvae in southern California. *Hilgardia*, **55**

Forbes, A. R. (1977). The mouthparts and feeding mechanism of aphids. In *Aphids as Virus Vectors* (ed. K. F. Harris and K. Maramorosch). Academic Press, New York

Forgash, A. J. (1985). Insecticide resistance in the Colorado potato beetle. *Research Bulletin, Massachusetts Agricultural Experiment Station* No. 704, 33–52

Forrest, J. M. S. and Phillips, M. S. (1984). The effect of *Solanum tuberosum* × *S. vernei* hybrids on hatching of the potato cyst nematode *Globodera pallida*. *Ann. Appl. Biol.*, **104**, 521–526

Foster, G. N. (1980). Biological control of whitefly—initial pest density as the most important factor governing success. *Bull. SROP*, **3**, 41–43

Fransen, J. J., Winkelman, K. and Van Lenteren, J. C. (1987). Control of greenhouse whitefly, *Trialeurodes vaporariorum*, by entomopathogenic fungus, *Aschersonia aleyrodis*. *J. Invert. Pathol.*, **50**, 158–165

Gameel, O. I. (1974). Some aspects of the mating and oviposition behavior of the cotton whitefly, *Bemisia tabaci* (Gen.). *Rev. Zool. Afric.*, **88**, 784–788

Gentile, A. G. and Stoner, A. K. (1968). Resistance in *Lycopersicon* and *Solanum* species to the potato aphid. *J. Econ. Entomol.*, **61**, 1152–1154

Gentile, A. G., Webb, R. E. and Stoner, A. K. (1968). Resistance in *Lycopersicon* and

Solanum to greenhouse whiteflies. *J. Econ. Entomol.*, **61**, 1355–1357

Georgiev, K. and Sotirova, V. (1978). A study of resistance of wild tomato species to the greenhouse whitefly, *Trialeurodes vaporariorum*. *Genet. Sel.*, **11**, 214–217

Gerling, D. and Or, R. (1984). Influence of host plant on the oviposition strategy of *Bemisia tabaci*. *Phytoparasitica*, **12**, 142

Gerson, U. (1985). The Tetranychidae—webbing. In *Spider Mites. Their Biology, Natural Enemies and Control*, Vol. 1A (ed. W. Helle and M. W. Sabelis). Elsevier, Amsterdam, Oxford, New York, Tokyo, pp. 223–230

Gibson, R. W. (1974). The induction of top-roll symptoms on potato plants by the aphid *Macrosiphum euphorbiae*. *Ann. Appl. Biol.*, **76**, 19–26

Gildow, F. E. (1987). Virus membrane interactions involved in circulative transmission of luteoviruses by aphids. In *Current Topics in Vector Research*, Vol. 4 (ed. K. F. Harris). Springer-Verlag, New York, pp. 93–120

Giustina, W. D. (1976). Study of the effectiveness of two juvenile hormone analogues and of bioresmethrin against the whitefly (*Trialeurodes vaporariorum* West) and the green peach aphid *Myzus persicae* Sulz.) in the greenhouse. *Phytiatrie-Phytopharm.*, **24**, 255–264

Godan, D. (1983). *Pest Slugs and Snails. Biology and Control*. Springer-Verlag, Berlin

Godzińska, E. J. (1986). Ant predation on Colorado beetle (*Leptinotarsa decemlineata* Say). *J. Appl. Entomol.*, **102**, 1–10

Gravena, S., Churata-Masca, M. G. C., Arai, J. and Raga, A. (1985). Control of the whitefly, *Bemisia tabaci*, in tomato cultivars of determined growth to avoid virus diseases. *Ann. Soc. Entomol. Bras.*, **13**, 35–46

Gregory, P., Ave, D. A., Bouthyette, P. Y. and Tingey, W. M. (1986). Insect-defensive chemistry of potato glandular trichomes. In *Insects and the Plant Surface* (ed. B. Juniper and R. Southwood). Arnold, London, pp. 173–183

Gutierrez, J. (1985). The Tetranychidae—systematics. In *Spider Mites. Their Biology, Natural Enemies and Control*, Vol. 1A (ed. W. Helle and M. W. Sabelis). Elsevier, Amsterdam, Oxford, New York, Tokyo, pp. 75–89

Hagel, G. T. and Landis, B. J. (1972). Chemical control of the twospotted spider mite on field beans. *J. Econ. Entomol.*, **65**, 775–778

Harrewijn, P. (1983). The effect of cultural measures on behaviour and population development of potato aphids and transmission of viruses. *Med. Fac. Landbouw. Rijksuniv. Gent*, **48**, 791–799

Harris, V. E. and Todd, J. T. (1980). Duration of immature stages of the southern green stink bug, *Nezara viridula* (L.), with a comparative review of previous studies. *J. Ga Entomol. Soc.*, **15**, 114–124

Harris, V. E. and Todd, J. T. (1982). Longevity and reproduction of the southern green stink bug, *Nezara viridula*, as affected by parasitization by *Trichopoda pennipes*. *Entomol. Exp. Appl.*, **31**, 409–412

Harris, V. E., Todd, J. W., Webb, J. C. and Benner, J. C. (1982). Acoustical and behavioral analysis of the songs of the southern green stink bug, *Nezara viridula*. *Ann. Entomol. Soc. Am.*, **75**, 234–249

Hassan, S. A. (1982). Relative tolerance of 3 different strains of the predatory mite *Phytoseiulus persimilis* to 11 pesticides used on glasshouse crops. *Z. Angew. Entomol.*, **93**, 55–63

Hayslip, N. C. (1961). Leafminer control on tomatoes in the Indian River area. *Proc. Fla State Hort. Soc.*, **74**, 134–137

Hernandez-Roque, F. (1973). Studies on the whitefly, *Trialeurodes vaporariorum* (Westwood) in the state of Morelos. *Agric. Tec. Mexico*, **3**, 165–172

Hessein, N. A. and Perring, T. M. (1986). Feeding habits of the Tydeidae with evidence of *Homeopronematus anconai* (Acari: Tydeidae) predation on *Aculops lycopersici* (Acari: Eriophyidae). *Int. J. Acarol.*, **12**, 215–221

Hill, D. S. (1987). *Agricultural Insect Pests of Temperate Regions and Their Control*. Cambridge University Press, Cambridge

Ho, C. C. and Lo, K. C. (1979). Influence of temperature on life history and population parameters of *Tetranychus urticae*. *J. Agric. Res. China*, **28**, 261–271

Hoffmann, M. P., Wilson, L. T. and Zalom, F. G. (1987). Control of stinkbugs in tomato. *Calif. Agric.*, May–June, 4–6

Hogg, D. B. (1985). Potato leafhopper *Empoasca fabae* (Homoptera: Cicadellidae) immature development life tables and population dynamics under fluctuating temperature regimes.

Environ. Entomol., **14**, 349–355

Hovey, C. L. (1943). Effects of temperature and humidity on certain developmental stages of the potato tuber moth. *J. Econ. Ent.*, **36**, 627–628

Howell, P. J. (1977). Recent trends in the incidence of aphid-borne viruses in Scotland. *Proc. Symposium on Problems of Pest and Disease Control in Northern Britain* (ed. R. A. Fox). Scottish Crop Research Institute, Dundee, pp. 26–28

Hower, A. A. (1987). Ecology of the potato leafhopper on alfalfa and other crops. *Proc. 2nd Int. Workshop on Leafhoppers and Planthoppers of Economic Importance* (ed. M. R. Wilson and L. R. Nault). Commonwealth Institute of Entomology, London, pp. 257–266

Hunter, P. J. (1968). Studies on slugs of arable ground. II. Life cycles. *Malacologia*, **6**, 379–389

Hunter, P. J. (1978). Slugs—a study in applied ecology. In *Pulmonates*, Vol. 2a, *Systematics, Evolution and Ecology* (ed. V. Fretter and J. Peake). Academic Press, London, pp. 271–286

Hussey, N. W., Parr, W. J. and Gurney, J. (1959). The effect of whitefly populations on the cropping of tomatoes. *Ann. Rep. Glasshouse Crops Res. Inst.*, pp. 79–86

Iacob, N. (1978). New mite pests on greenhouse crops and on grapevine. *Anal. Inst. Cercetari Pentru Protect. Plantelor*, **14**, 115–120

Iftner, D. C., Hall, F. R. and Sturm, M. M. (1986). Effects of residues of fenvalerate and permethrin on the feeding behavior of *Tetranychus urticae*. *Pestic. Sci.*, **17**, 242–248

Ishiwatari, T. (1974). Studies on the scent of stinkbugs (Hemiptera: Pentatomidae). I. Alarm pheromone activity. *Appl. Entomol. Zool.*, **9**, 153–158

Izhevskaya, S. S. (1983). [The role of insect natural enemies.] *Zashchita Rastenii*, No. 5, 37–38

Jansson, R. K. and Smilowitz, Z. (1985). Influence of nitrogen on population parameters of potato insects: abundance, development and damage of the Colorado potato beetle, *Leptinotarsa decemlineata* (Coleoptera: Chrysomelidae). *Environ. Entomol.*, **14**, 500–506

Jansson, R. K. and Smilowitz, Z. (1986). Influence of nitrogen on population parameters of potato insects: abundance, population growth, and within-plant distribution of the green peach aphid, *Myzus persicae* (Homoptera: Aphididae). *Environ. Entomol.*, **15**, 49–55

Jatala, P., Kaltenbach, R. and Bocangel, M. (1979). Biological control of *Meloidogyne incognita acrita* and *Globodera pallida* on potatoes. *J. Nematol.*, **11**, 303 (Abstr.)

Johnson, B. (1956). The influence of aphids on the glandular hairs on tomato plants. *Plant Pathol.*, **5**, 131–133

Johnson, K. B., Radcliffe, E. B. and Teng, P. S. (1986). Effects of interacting populations of *Alternaria solani*, *Verticillium dahliae*, and the potato leafhopper (*Empoasca fabae*) on potato yield. *Phytopathology*, **76**, 1046–1052

Johnson, M. W., Oatman, E. R. and Wyman, J. A. (1980a). Effects of insecticides on populations of the vegetable leafminer and associated parasites on summer pole tomatoes. *J. Econ. Entomol.*, **73**, 61–66

Johnson, M. W., Oatman, E. R. and Wyman, J. A. (1980b). Effects of insecticides on populations of the vegetable leafminer and associated parasites on fall pole tomatoes. *J. Econ. Entomol.*, **73**, 67–71

Johnson, M. W., Oatman, E. R., Wyman, J. A. and Van Steenwyk, R. A. (1980c). A technique for monitoring *Liriomyza sativae* in fresh market tomatoes. *J. Econ. Entomol.*, **73**, 552–555

Johnson, M. W., Toscano, N. C., Reynolds, H. T., Sylvester, E. S., Kido, K. and Natwick, E. T. (1982). Whiteflies cause problems for Southern California growers. *Calif. Agric.*, Sept.–Oct., 24–26

Johnson, M. W., Welter, S. C., Toscano, N. C., Ting, I. P. and Trumble, J. T. (1983). Reduction of tomato leaflet photosynthesis rates by mining activity of *Liriomyza sativae* (Diptera: Agromyzidae). *J. Econ. Entomol.*, **76**, 1061–1063

Jones, F. G. W. and Perry, J. N. (1978). Modelling populations of cyst-nematodes (Nematoda: Heteroderidae). *J. Appl. Ecol.*, **15**, 349–371

Jones, M. G. K. and Northcote, D. J. (1972). Nematode-induced syncytium—a multinucleate transfer cell. *J. Cell Sci.*, **10**, 789–809

Kajita, H. (1982). Predation by adult *Orius sauteri* Poppius (Hemiptera: Anthocoridae) on the greenhouse whitefly, *Trialeurodes vaporariorum* (Westwood) (Homoptera: Aleyrodidae). *Appl. Entomol. Zool.*, **17**, 424–425

Kajita, H. (1984). Predation of the greenhouse whitefly, *Trialeurodes vaporariorum* (Westwood) (Homoptera: Aleyrodidae), by *Campylomma* sp. (Hemiptera: Miridae). *Appl. Entomol. Zool.*, **19**, 67–74

Kay, I. R. (1986). Tomato russet mite: a serious pest in tomatoes. *Queensland Agric. J.*, Sept.–Oct., 231–232

Kennedy, G. G. and Smitley, D. R. (1985). The Tetranychidae—dispersal. In *Spider Mites. Their Biology, Natural Enemies and Control*, Vol. 1A (ed. W. Helle and M. W. Sabelis). Elsevier, Amsterdam, Oxford, New York, Tokyo, pp. 233–240

Kennedy, J. S., Day, M. F. and Eastop, V. F. (1962). *A Conspectus of Aphids as Vectors of Plant Viruses*. Commonwealth Institute of Entomology, London

Kerry, B. R. (1987). Biological control. In *Principles and Practice of Nematode Control in Crops* (ed. R. H. Brown and B. R. Kerry). Academic Press, London, pp. 233–263

Kerry, B. R. and Crump, D. H. (1977). Observations on fungal parasites of females and eggs of the cereal cyst nematode, *Heterodera avenae*, and other cyst nematodes. *Nematologica*, **23**, 193–201

Kfir, R. (1981). Fertility of the polyembryonic parasite *Copidosoma koehleri*, effect of humidities on life length and relative abundance as compared with that of *Apanteles subandinus* in potato tuber moth. *Ann. Appl. Biol.*, **99**, 225–230

Khandge, S. V., Parlekar, G. Y. and Naik, L. M. (1979). Inundative releases of *Copidosoma koehleri* Blanchard (Hymenoptera: Encyrtidae) for control of the potato tuber worm *Phthorimaea operculella* Zeller. *J. Maharashtra Agricultural Universities*, **4**, 165–169

Kisha, J. S. A. (1981). Observations on the trapping of the whitefly *Bemisia tabaci* by glandular hairs on tomato leaves. *Ann. Appl. Biol.*, **97**, 123–127

Knop, N. F. and Hoy, M. A. (1983). Biology of a tydeid mite, *Homeopronematus anconai* (n. comb.) (Acari: Tydeidae), important in San Joaquin Valley vineyards. *Hilgardia*, **51**, 1–30

Kogan, M. (ed.) (1986). *Ecological Theory and Integrated Pest Management Practice*. Wiley, New York

Krispyn, J. W. and Todd, J. W. (1982). The red imported fire ant as a predator of the southern green stinkbug on soyabean in Georgia. *J. Ga Entomol. Soc.*, **17**, 19–26

Kropczynska, D. and Tomczyk, A. (1986). Influence of spider mite (*Tetranychus urticae* Koch) infestation on the development and yield of bean. *Med. Fac. Landbouw. Rijksuniv. Gent*, **51**, 931–938

Ladd, T. L. and Rawlins, W. A. (1965). The effects of the feeding of the potato leafhopper on photosynthesis and respiration in the potato plant. *J. Econ. Entomol.*, **58**, 623–628

Lamb, K. P. (1953). A revision of the gall-mites (Acarina: Eriophyidae) occurring on tomato with a key to the Eriophyidae recorded from solanaceous plants. *Bull. Entomol. Res.*, **44**, 343–350

Lamp, W. O., Morris, M. J. and Armbrust, E. J. (1984). Suitability of common weed species as host plants for the potato leafhopper, *Empoasca fabae*. *Ent. Exp. Appl.*, **36**, 125–131

Lange, W. H. (1987). Insect pests of sugar beet. *Ann. Rev. Entomol.*, **32**, 341–360

Lange, W. H. and Bronsin, L. (1981). Insect pests of tomatoes. *Ann. Rev. Entomol.*, **26**, 345–371

Langenbruch, G. A., Kreig, A., Huger, A. M. and Schnetter, W. (1985). First field tests on the control of the larvae of the Colorado beetle (*Leptinotarsa decemlineata*) with *Bacillus thuringiensis* var. *tenebrionis*. *Med. Fac. Landbouw. Rijksuniv. Gent*, **50**, 441–449

Las, A. (1980). Male courtship persistence in the greenhouse whitefly, *Trialeurodes vaporariorum* Westwood (Homoptera: Aleyrodidae). *Behavior*, **72**, 107–126

Lashomb, J. H. and Ng, Y. S. (1984). Colonization by Colorado potato beetles, *Leptinotarsa decemlineata* (Say) (Coleoptera: Chrysomelidae), in rotated and non rotated potato fields. *Environ. Entomol.*, **13**, 1352–1356

Laska, P., Slovakova, J. and Bicik, V. (1980). Life cycle of *Trialeurodes vaporariorum* (Westw.) and its parasite, *Encarsia formosa* Gah., at constant temperature. *Sborni'k Praci' Priodovedecke Fak. Univ. Palacke'ho v Olomouci, Biol.*, **20**, 95–106

Ledieu, M. S. and Helyer, N. L. (1985). Observations on the economic importance of tomato leaf miner (*Liriomyza bryoniae*) (Agromyzidae). *Agr. Ecosyst. Environ.*, **13**, 103–109

Lee, Y. I., Kogan, M. and Larsen, J. R., Jr. (1986). Attachment of the potato leafhopper to soybean plant surfaces as affected by morphology of the pretarsus. *Ent. Exp. Appl.*, **42**, 101–107

Li, T. Y. and Maschwitz, U. (1983). Sexual pheromone in the greenhouse whitefly, *Trialeurodes vaporariorum* Westw. *Z. Angew. Entomol.*, **95**, 439–446

Lindquist, R. K. (1972). Evaluation of insect growth regulators for whitefly control on tomatoes. *Res. Sum., Ohio Res. Dev. Ctr.*, **82**, 31–35

Lye, Ben-Huai, Story, R. N. and Wright, V. L. (1988). Southern green stink bug (Hemiptera: Pentatomidae) damage to fresh market tomatoes. *J. Econ. Entomol.*, **81**, 189–194

McKinlay, R. G. (1985). Effect of undersowing potatoes with grass on potato aphid numbers. *Ann. Appl. Biol.*, **106**, 23–29

McKinlay, R. G. (1988). Insect pest control on potatoes. *Outlook on Agriculture*, **17**, 30–34

McKinlay, R. G. and Franklin, M. F. (1983). Potato aphid and leaf roll virus control: insecticide granules and sprays. *Proc. 10th Int. Cong. Pl. Prot.*, Vol. 3, p. 1204

McKinlay, R. G. and Kerr, D. S. (1985). Survey of the attitudes of seed potato growers in south-east Scotland to aphid control and their use of insecticides in 1977–81. *Res. Dev. Agric.*, **2**, 33–36

McKinney, K. B. (1938). Physical characteristics on the foliage of beans and tomatoes that tend to control some small insect pests. *J. Econ. Entomol.*, **31**, 630–631

McLain, D. K. (1981). Sperm precedence and prolonged copulation in the southern green stinkbug, *Nezara viridula*. *J. Ga Entomol. Soc.*, **16**, 70–71

McPherson, R. M. and Newsom, L. D. (1984). Trap crops for control of stinkbug in soybean. *J. Ga Entomol. Soc.*, **19**, 470–480

MacVean, C. M., Brewer, J. W. and Capinera, J. L. (1982). Field tests of anti-desiccants to extend the infection period of an entomogenous nematode, *Neoaplectana carpocapsae*, against the Colorado potato beetle. *J. Econ. Entomol.*, **75**, 97–101

Martin, N. A. and Workman, P. (1986). Control of greenhouse whitefly with oxamyl granules. *Proc. 39th New Zeal. Weed and Pest Control Conf.*, pp. 234–236

Martin, N. A., Workman, P., Burgess, E. P. and Wearing, C. H. (1984). Integrated pest control in greenhouse crops. *Proc. 37th New Zeal. Weed and Pest Control Conf.*, pp. 265–267

Mazzone, P. and Viggiani, G. (1985). New data on *Encarsia pergandiella* (Howard) (Hym.: Aphelinidae), an exotic parasite introduced into Italy against *Trialeurodes vaporariorum* (Westw.) (Hom.: Aleyrodidae). *Proc. Accad. Nazionale Italiano Entomol.*, 855–859

Mead, F. W. (1976). A predatory stinkbug, *Euthyrhynchus floridanus* (L.) (Hemiptera: Pentatomidae). *Entomol. Circ. Div. Plant Ind., Fla Dept. Agric. Consum. Serv.*, No. 174

Metcalf, R. L. (1986). The ecology of insecticides and the chemical control of insects. In *Ecological Theory and Integrated Pest Management Practice* (ed. M. Kogan). Wiley, New York, pp. 251–297

Mishra, P. N. (1986). Studies on bio-efficacy of some insecticides against the pest complex of tomato, *Lycopersicon esculentum* Mill. *Madras Agric. J.*, **71**, 673–676

Miyasaki, S. and Sherman, M. (1966). Toxicity of several insecticides to the southern green stinkbug, *Nezara viridula* L. *Proc. Hawaiian Entomol. Soc.*, **19**, 281–287

Mndolwa, D., Bishop, G., Corsini, D. and Pavek, J. (1984). Resistance of potato clones to the green peach aphid and potato leafroll virus. *Am. Pot. J.*, **61**, 713–722

Möller, F. W. (1971). [Hybridisation experiments within the complex of species around the green-striped potato aphid, *Macrosiphum euphorbiae* (Thomas).] *Beitr. Ent.*, **21**, 531–537

Mor, U. and Marani, A. (1984). Relationships between physiology of the cotton plant and development of the tobacco whitefly, *Bemisia tabaci*. *Phytoparasitica*, **12**, 141

Mote, U. N. (1979). Effects of few insecticides alone and in combination with agricultural spray oil on the control of whitefly, *Bemisia tabaci*, population and incidence of leaf curl virus. *Indian J. Plant Protect.*, **6**, 19–22

Musgrave, C. A., Poe, S. L. and Weems, H. V. (1975). The vegetable leaf miner, *Liriomyza sativae* Blanchard in Florida. *Entomol. Circ. Fla Dept. Agric. Consum. Serv.*, No. 162

Nabi, M. N. (1983). Field cage trials with thiotepa-sterilised males of the potato moth, *Phthorimaea operculella* (Zeller) (Lepidoptera: Gelechiidae). *Bull. Ent. Res.*, **73**, 405–409

Nabi, M. N. (1984). Resistance of potato lines to potato moth, *Phthorimaea operculella*. Tests of Agrochemicals and Cultivars No. 5 (*Ann. Appl. Biol.*, **104**, Supplement), pp. 122–123

Nedstam, B. (1985). Development time of *Liriomyza bryoniae* Kalt and two of its natural enemies, *Dacnusa sibirica* Telerga and *Cyrtogaster vulgaris* Walker at different constant temperatures. *Med. Fac. Landbouw. Rijksuniv. Gent*, **50**, 411–417

Noldus, L. P. J. J., Rumei, X. and van Lenteren, J. C. (1986a). The parasite–host relationship between *Encarsia formosa* and *Trialeurodes vaporariorum*. XIX. Feeding-site selection by the greenhouse whitefly. *J. Appl. Entomol.*, **101**, 492–507

Noldus, L. P. J. J., Xu, R. M. and van Lenteren, J. C. (1985). The parasite–host relationship between *Encarsia formosa* and *Trialeurodes vaporariorum*. XVII. Within-plant movement of adult greenhouse whiteflies. *Z. Angew. Entomol.*, **100**, 494–503

Noldus, L. P. J. J., Xu, R. M. and van Lenteren, J. C. (1986b). The parasite–host relationship between *Encarsia formosa* and *Trialeurodes vaporariorum*. XVIII. Between-plant movement of adult greenhouse whiteflies. *J. Appl. Entomol.*, **101**, 159–176

Oatman, E. R. and Kennedy, G. G. (1976a). Methomyl-induced outbreak of *Liriomyza sativae* on tomato. *J. Econ. Entomol.*, **69**, 667–668

Oatman, E. R. and Kennedy, G. G. (1976b). Effects of insecticides on populations of the vegetable leafminer and associated parasites on summer pole tomatoes. *J. Econ. Entomol.*, **73**, 61–66

Panzzi, A. R. and Slansky, F., Jr. (1985). Review of phytophagous pentatomids (Hemiptera: Pentatomidae) associated with soybean in the Americas. *Fla Entomol.*, **68**, 184–214

Papaioannou-Souliotis, P. (1979). Effects of the population of *Tetranychus urticae* (Koch) on bean plants (*Phaseolus vulgaris* L.). *Ann. Inst. Phytopathol. Benaki*, **12**, 18–43

Parrella, M. P. and Keil, C. B. (1984). Insect pest management: The lesson of *Liriomyza*. *Bull. Entomol. Soc. Am.*, **30**, 22–25

Parrella, M. P. and Robb, K. L. (1982). Leafminer attacking bedding plants in California. *Univ. Calif. Coop. Ext. Fall/Winter*, 2–4

Parrella, M. P., Robb, K. L. and Bethke, J. (1981). Oviposition and pupation of *Liriomyza trifolii* (Burgess). *Proc. Inst. Food Agric. Sci. Ind. Conf., Biol. Control of Liriomyza Leafminers -II*, Lake Buena Vista, Fla, pp. 50–55

Penman, D. R. and Chapman, R. B. (1983). Fenvalerate-induced distributional imbalances of two-spotted spider mite on bean plants. *Entomol. Exp. Appl.*, **33**, 71–78

Penman, D. R., Chapman, R. B. and Bowie, M. H. (1986). Direct toxicity and repellent activity of pyrethroids against *Tetranychus urticae* (Acari: Tetranychidae). *J. Econ. Entomol.*, **79**, 1183–1187

Penman, D. R., Chapman, R. B. and Jesson, K. E. (1981). Effects of fenvalerate and azinphos-methyl on two-spotted spider mite and phytoseid mites, *Tetranychus urticae*. *Entomol. Exp. Appl.*, **30**, 91–97

Perring, T. M. and Farrar, C. A. (1986). Historical perspective and current world status of the tomato russet mite (Acari: Eriophyidae). *Misc. Pub., Entomol. Soc. Am.*

Pickett, C. H. and Gilstrap, F. E. (1986). Inoculative releases of phytoseiids (Acari) for the biological control of spider mites (Acari: Tetranychidae) in corn. *Environ. Entomol.*, **15**, 790–794

Poe, S. L., Everett, P. H., Shuster, D. J. and Musgrave, C. A. (1978). Insecticidal effects on *Liriomyza sativae* larvae and their parasites on tomato. *J. Ga Entomol Soc.*, **13**, 322–327

Pollard, D. G. (1977). Aphid penetration of plant tissues. In *Aphids as Virus Vectors* (ed. K. F. Harris, and K. Maramorosch). Academic Press, New York

Port, C. M. and Port, G. R. (1986). The biology and behaviour of slugs in relation to crop damage and control. *Agric. Zool. Rev.*, **1**, 255–299

Powell, D. F. (1982). The eradication campaign against American serpentine leaf miner, *Liriomyza trifolii* at Efford Experimental Horticulture Station, Hampshire, England, U.K. *Plant Pathol.*, **30**, 195–204

Pozarowska, B. J. (1987). Studies on low temperature survival, reproduction and development in Scottish clones of *Myzus persicae* (Sulzer) and *Aulacorthum solani* (Kaltenbach) (Hemiptera, Aphididae) susceptible and resistant to organophosphates. *Bull. Ent. Res.*, **77**, 123–134

Radcliffe, E. B. (1982). Insect pests of potato. *Ann. Rev. Entomol.*, **27**, 173–204

Ragsdale, D. W., Larson, A. D. and Newsom, L. D. (1981). Quantitative assessment of the predators of *Nezara viridula* eggs and nymphs within a soybean agroecosystem using ELISA. *Environ. Entomol.*, **10**, 402–405

Raman, K. V. (1983). Sex pheromones aid in the control of potato tuber moth. *Proc. 10th Int. Cong. Pl. Prot.*, Vol. 1, p. 274

Raman, K. V. and Palacios, M. (1983). Approaches to integrated control of *Phthorimaea operculella*. In *Research for the Potato in Year 2000* (ed. W. J. Hooker). International Potato Center, Lima, Peru, pp. 155–156

Rataul, H. S. and Butter, N. S. (1975). Effects of different systemic granular insecticides on the population of whitefly, *Bemisia tabaci*, the vector of tomato leaf curl virus (TLCV). *J. Res. Punjab Agric. Univ.*, **12**, 382–385

Rice, R. E. and Strong, F. E. (1962). Bionomics of the tomato russet mite, *Vastes lycopersia* (Massee). *Ann. Entomol. Soc. Am.*, **55**, 431–435

Rivnay, E. (1962). *Field Crop Pests in the Near East*. Monographiae Biologicae (ed. W. W. Weisbach), vol. X. Junk, Den Haag

Robert, Y. (1987). Aphid vector monitoring in Europe. In *Current Topics in Vector Research*, Vol. 3 (ed. K. F. Harris). Springer-Verlag, New York, pp. 81–129

Rodriquez, J. G., Knavel, D. E. and Aina, O. J. (1972). Studies in the resistance of tomatoes to mites. *J. Econ. Entomol.*, **65**, 50–53

Rodriquez-Velez, J. (1974). Observations on the biology of the greenbug, *Nezara viridula* (L.) in the Valle del Fuerte, Sinaloa. *Folia Entomol. Mex.*, No. 28, 5–12

Roush, R. T. and McKenzie, J. A. (1987). Ecological genetics of insecticide and acaricide resistance. *Ann. Rev. Entomol.*, **32**, 361–380

Royalty, R. N. and Perring, T. M. (1987). Comparative toxicity of acaricides to *Aculops lycopersici* and *Homeopronematus anconai* (Acari: Eriophyidae, Tydeidae). *J. Econ. Entomol.*, **80**, 348–351

Ruberson, J. R., Tauber, M. J. and Tauber, G. A. (1986). Plant feeding by *Podisus maculiventris* (Heteroptera: Pentatomidae): effect on survival, development and preoviposition period. *Environ. Entomol.*, **15**, 894–897

Sasser, J. N. and Freckman, D. W. (1987). A world perspective on Nematology: the role of the society. In *Vistas on Nematology: A Commemoration of the Twenty-fifth Anniversary of the Society of Nematologists* (ed. J. A. Veech and D. W. Dickson). Society of Nematologists Inc., Hyattsville, Maryland, pp. 7–14

Schroder, R. F. W. and Athanus, M. M. (1985). Review of research on *Edovum puttleri* Grissell, egg parasite of the Colorado potato beetle. *Research Bulletin, Massachusetts Agricultural Experiment Station* No. 704, 29–32

Seinhorst, J. W. (1982). The relationship in field experiments between population density of *Globodera rostochiensis* before planting potatoes and yield of potato tubers. *Nematologica*, **28**, 277–284

Sekeroglu, E. and Ozgur, A. F. (1984). A new tomato pest in Cukurova, *Aculops lycopersici* (Massee) (Acarina: Eriophyidae). *Turkiye Bitki Koruma Dergisi*, **8**, 210–211

Sharaf, N. S. and Allawi, T. F. (1980). Studies on whiteflies on tomato in the Jordan Valley. 3. Laboratory and field experiments on the control of whitefly populations with organophosphorous insecticides and the incidence of the tomato yellow leaf-curl virus. *Z. Pflanzenkr. Pflanzenschutz*, **87**, 176–184

Sharaf, N. S. and Allawi, T. F. (1981). Control of *Bemisia tabaci*, a vector of tomato yellow leaf curl virus disease in Jordan. *Z. Pflanzenkr. Pflanzenschutz*, **88**, 123–131

Sharaf, N. S., Al-Musa, A. and Nazer, I. (1984). The impact of three irrigation methods on the whitefly (*Bemisia tabaci*) populations and the incidence of tomato yellow leaf curl disease in Jordan. *Dirasat*, **11**, 109–117

Shelton, A. M., Wyman, J. A. and Mayor, A. J. (1981). Effects of commonly used insecticides on the potato tuberworm and its associated parasites and predators in potatoes. *J. Econ. Entomol.*, **74**, 303–308

Shepard, M., Carner, G. R. and Turnipseed, S. C. (1977). Colonization and resurgence of insect pests of soybean in response to insecticides and field isolation. *Environ. Entomol.*, **6**, 501–506

Sikura, A. I. and Sikura, L. V. (1983). [The use of biopreparations.] *Zashchita Rastenii*, No. 5, 38–39

Silcox, C. A. and Ghidiu, G. M. (1986). Laboratory and field studies on the antifeedant effect of piperonyl butoxide against the Colorado potato beetle on eggplant. *J. Agric. Entomol.*, **3**, 135–142

Simchuk, P. A. (1985). [Parasites of the potato moth.] *Zashchita Rastenii*, No. 4, 32–33

Singh, Z. (1973). *Southern Green Stinkbug and its Relationship to Soybeans*. Metropolitan Book Co., Delhi

Smith, J. C. and Newsom, L. D. (1970). The biology of *Ambyseius fallacis* (Acarina: Phytoseiidae) at various temperature and photoperiod regimes. *Ann. Entomol. Soc. Am.*, **63**, 460–462

Smith, K. M. (1972). *A Textbook of Plant Virus Diseases*, 3rd edn. Longman, London

Snyder, J. and Carter, C. (1984). Leaf trichomes and resistance of *Lycopersicon hirsutum* and *L. esculentum* to spider mites. *J. Am. Soc. Hort. Sci.*, **109**, 837–843

Solovei, E. F. and Kol'tsov, P. D. (1976). The action of entomopathogenic fungi of the genus *Aschersonia* on the greenhouse whitefly. *Mikol. Fitopatol.*, **10**, 425–429

Sombatsiri, K. (1978). Ecological study on insect pests of chilli tomato and their control. In

Ann. Res. Rep., Kasetsart University, Thailand, pp. 40–41

Sorokin, N. S. (1981). [Ground beetles (Coleoptera, Carabidae)—natural enemies of the Colorado beetle *Leptinotarsa decemlineata* Say.] *Entomologicheskoe Obozrenie*, **60**, 282–289

Sorokin, N. S. (1982). [Irrigation and natural enemies.] *Zashchita Rasteniĭ*, No. 2, 30

Spaull, A. M., Trudgill, D. L. and Phillips, M. S. (1987). Behaviour of three new potato clones in integrated control systems for potato cyst nematodes *Globodera* species. *Proc., Crop Protection in Northern Britain 1987*. Scottish Crop Research Institute, Dundee, pp. 196–201

Spencer, K. A. (1965). Agromyzidae from the Azores and Madeira. *Biol. Mus. Munic. Funchal*, **19**, 104–110

Stacey, D. L., Wyatt, I. J. and Chambers, R. J. (1985). The effect of glasshouse red spider mite, *Tetranychus urticae*, damage on the yield of tomatoes. *J. Hort. Sci.*, **60**, 517–523

Stephenson, J. W. (1965). Damage to potatoes: Irrigation and slug attack. Slug parasites and predators. Rothamsted Experimental Station, *Report for 1964*, pp. 187–188

Stephenson, J. W. and Bardner, R. (1977). Slugs in agriculture. Rothamsted Experimental Station, *Report for 1976; Part 2*, pp. 169–187

Stoner, A. K. and Stringfellow, T. (1967). Resistance of tomato varieties to spider mites. *Proc. Am. Soc. Hort. Sci.*, **90**, 324–329

Storms, J. J. H. (1971). Some physiological effects of spider mite infestation on bean plants. *Neth. J. Plant. Pathol.*, **77**, 154–167

Suski, Z. W. and Badowska, T. (1975). Effect of host plant nutrition on the population of the two-spotted spider mite, *Tetranychus urticae*. *Ekol. Pol.*, **23**, 185–209

Tamada, T. and Harrison, B. D. (1981). Quantitative studies on the uptake and retention of potato leafroll virus by aphids in laboratory and field conditions. *Ann. Appl. Biol.*, **98**, 261–276

Taylor, C. E. (1986). Genetics and evolution of resistance to insecticides. *Biol. J. Linnean Soc.*, **27**, 103–112

Taylor, D. E. (1978). Entomological notes—tomato russet mite. *Rhodesia Agric. J.*, **43**, 244–245

Taylor, R. A. J. and Reling, D. (1986). The preferred wind direction of potato leaf hopper (*Empoasca fabae*) migrants and its relevance to the return migration of small insects. *J. Am. Ecol.*, **55**, 1103–1114

Temerak, S. A. (1983). Field management of the potato tuberworm, *Phthorimaea operculella* (Zeller), infesting certain potato varieties through manipulation of different levels of irrigation. *Bull. Soc. Ent. Egypte*, No. 62, 153–160

Thiery, D. and Visser, J. H. (1986). Masking of host plant odour in the olfactory orientation of the Colorado potato beetle. *Ent. Exp. Appl.*, **41**, 165–172

Thomas, D. C. (1947). Some observations on damage to potatoes by slugs. *Ann. Appl. Biol.*, **34**, 246–251

Todd, J. W. and Herzog, D. C. (1980). Sampling phytophagous Pentatomidae on soybean. In *Sampling Methods on Soybean Entomology* (ed. M. Kogan and D. C. Herzog). Springer-Verlag, New York, Heidelberg, Berlin, pp. 437–478

Trudgill, D. L. (1985). Potato cyst nematodes: a critical review of the current pathotyping scheme. *EPPO Bull.*, **15**, 273–279

Trudgill, D. L. (1986). Yield losses caused by potato cyst nematodes: a review of the current position in Britain and prospects for improvements. *Ann. Appl. Biol.*, **108**, 181–198

Trudgill, D. L., Mathias, P. L. and Tones, S. J. (1985). Effects of three rates of aldicarb and of different degrees of resistance and tolerance in potato cultivars on yields and post-harvest population densities of potato cyst nematodes *Globodera rostochiensis* and *G. pallida*. *Ann. Appl. Biol.*, **107**, 219–229

Trudgill, D. L., Phillips, M. S. and Alphey, T. J. W. (1987). Integrated control of potato cyst nematode. *Outlook on Agriculture*, **16**, 167–172

Trumble, J. T. (1985). Integrated pest management of *Liriomyza trifolii*: Influence of avermectin, cyromazine and methomyl on leafminer ecology in celery. *Agric. Ecosyst. Environ.*, **12**, 181–188

Tyron, E. H., Jr., Poe, S. L. and Cromroy, H. L. (1980). Dispersal of vegetable leaf miner, *Liriomyza sativae* onto a transplant production range. *Fla Entomol.*, **63**, 292–296

Tyron, H. (1917). Report of the entomologist and vegetable pathologist. *Queensland Ann. Rep. Dept. Agric. Stock.*, Brisbane, pp. 49–63 van de Veire, M. (1985). Control of Tetranychidae in crops—greenhouse ornamentals. In *Spider Mites. Their Biology, Natural Enemies and Control*, Vol. 1B (ed. W. Helle and M. W. Sabelis). Elsevier, Amsterdam, Oxford, New York, Tokyo, pp. 273–282

van de Veire, M., Hertveldt, L. and Aerts, J. (1974). The control of the greenhouse whitefly on gherkins with insect growth regulators and insecticides. *Med. Fac. Landbouw. Rijksuniv. Gent*, **39**, 1482–1489

Van Emden, H. F., Eastop, V. F., Hughes, R. D. and Way, M. J. (1969). The ecology of *Myzus persicae. Ann. Rev. Ent.*, **14**, 197–270

Veen, B. W. (1985). Photosynthesis and assimilate transport in potato with top-roll disorder caused by the aphid *Macrosiphum euphorbiae. Ann. Appl. Biol.*, **107**, 319–323

Veverka, K. and Oliberius, J. (1987). Side-effects of foliar application of urea + ammonium nitrate solution on the Colorado beetle. *J. Appl. Entomol.*, **103**, 119–124

Viggiani, G. and Mazzone, P. (1980). On the introduction into Italy of *Encarsia pergandiella* Howard, a parasite of *Trialeurodes vaporariorum* (Westw.). *Boll. Lab. Entomol. Agric. Filippo Silvestri', Portici*, **37**, 39–43

Waite, G. K. (1980). The basking behavior of *Nezara viridula* (L.) (Pentatomidae: Hemiptera) on soybeans and its implication in control. *J. Australian Entomol. Soc.*, **19**, 157–159

Wardlow, L. R. (1985). Pyrethroid resistance in glasshouse whitefly, *Trialeurodes vaporariorum* (Westw.). *Med. Fac. Landbouw. Rijksuniv. Gent*, **50**, 555–557

Warley, A. P. (1970). Some aspects of the biology, ecology and control of slugs in S.E. Scotland with particular reference to the potato crop. *PhD Thesis, University of Edinburgh*

Webb, R. E., Stoner, A. K. and Gentile, A. G. (1971). Resistance to leaf miners in *Lycopersicon* accessions. *J. Am. Soc. Hort. Sci.*, **96**, 65–67

Whitehead, A. G. (1977). Control of potato cyst nematode, *Globodera rostochiensis* Ro1, by picrolinic acid and potato trap crops. *Ann. Appl. Biol.*, **87**, 225–227

Whitehead, A. G. (1986). Chemical and integrated control of cyst nematodes. In *Cyst Nematodes* (ed. F. Lamberti and C. E. Taylor). Plenum Press, London, pp. 413–432

Wigglesworth, Sir Vincent B. (1965). *The Principles of Insect Physiology*. Methuen, London

Wilson, G. F. (1950). The broad mite. *J. Roy. Hort. Soc.*, **75**, 69–74

Wolfenbarger, D. A. and Wolfenbarger, D. O. (1966). Tomato yields and leaf miner infestations and a sequential sampling plan for determining need for control treatments. *J. Econ. Entomol.*, **59**, 279–283

Wysoki, M. (1985). Control of Tetranychidae—other outdoor crops. In *Spider Mites. Their Biology, Natural Enemies and Control*, Vol. 1B (ed. W. Helle and M. W. Sabelis). Elsevier, Amsterdam, Oxford, New York, Tokyo, pp. 375–381

Yang, Q. H. and Chen, C. X. (1982). A study on *Polyphagotarsonemus latus* Banks. *Kunchong Zishi*, **19**, 24–26

Zalom, F. G., Kitzmiller, J., Wilson, L. T. and Gutierrez, P. (1986a). Observation of tomato russet mite (Acari: Eriophyidae) damage symptoms in relation to tomato plant development. *J. Econ. Entomol.*, **79**, 940–942

Zalom, F. G., Wilson, L. T. and Hoffmann, M. P. (1986b). Impact of feeding by tomato fruitworm, *Heliothis zea* (Boddie) (Lepidoptera: Noctuidae), and beet armyworm, *Spodoptera exigua* (Hubner) (Lepidoptera: Noctuidae) on processing tomato fruit quality. *J. Econ. Entomol.*, **79**, 822–826

Zehnder, G. W. and Trumble, J. T. (1984). Spatial and diel activity of *Liriomyza* species (Diptera: Agromyzidae) in fresh market tomatoes. *Environ. Entomol.*, **13**, 1411–1416

CHAPTER 9

Pests of Umbelliferous Crops

P. R. Ellis and J. A. Hardman

9.1 INTRODUCTION

The Umbelliferae family of flowering plants is characterised by a distinctive inflorescence called an umbel, a word derived from *umbella*, Latin for 'a sunshade'. The typical umbel is a flat-topped flower cluster in which individual flower stalks grow to different lengths to raise all flowers to the same height. The Umbelliferae comprise about 300 genera with 2500–3000 species occurring in most parts of the world. Most species are found in temperate uplands and are relatively scarce in tropical zones. Apart from the umbel, members of this family also possess characteristic two-endospermic-seeded fruits and a distinctive chemistry which provides the basis for their unique flavour and fragrance. Most umbellifers are herbaceous annuals, biennials or perennials, but a few are woody shrubs. The majority of species form, initially, a rosette of leaves on a tap root and, later, a stem which elongates to bear flowers. Some species, however, form tubers, a few are rhizomatous and others are stoloniferous. Down through the ages, members of this family have been used by Chinese, Mexican Indian, Greek and Roman civilisations. A few species, such as carrots, celery and parsnips, are staple foods, while other species are grown as herbs and spices. A few umbellifers are important as flavouring and decorative components of foods and drinks. Several species are valuable sources of gums and medicines, while others are grown as ornamentals. More recently, in the search for additional food crops, considerable interest has been shown in arracacha, an Andean umbellifer used formerly by Inca Indians and now by peasant farmers. It is a root vegetable crop with great potential as a source of food. A list of the most important species grown in horticulture is provided in Table 9.1.

Umbelliferous crops have a few important pests and numerous occasional or minor pests. Certain pest species are oligophagous and attack only umbellifers, while others are polyphagous and attack a wide range of vegetable crops in different plant families. In this chapter only the more important pests of the Umbelliferae are described but lists of minor pests are provided. The section on cutworms is relevant to all vegetable crops.

Table 9.1   List of commoner culinary species of Umbelliferae

Species	Common name	Crop use
Aegopodium podagraria L.	Ground elder	Herb
Anethum graveolens L.	Dill	Spice
Angelica archangelica L.	Garden angelica	Herb
Anthriscus cerefolium (L.) Hoffm.	Garden chervil	Herb
Apium graveolens L. var. *dulce* DC.	Celery	Vegetable
Apium graveolens L. var. *rapaceum* (Miller) DC.	Celeriac	Vegetable
Arracacia xanthorrhiza Bancroft	Arracacha	Vegetable
Bunium bulbocastanum L.	Great earthnut	Vegetable
Chaerophyllum bulbosum L.	Turnip-rooted chervil	Vegetable
Conopodium majus (Gouan) Loret	Pignut	Vegetable
Carum carvi L.	Caraway	Spice
Coriandrum sativum L.	Coriander	Spice
Crithmum maritimum L.	Rock samphire	Vegetable
Cryptotaenia canadensis (L.) DC.	Canadian honewort	Herb
Cuminum cyminum L.	Cumin	Herb
Daucus carota L. ssp. *sativus* (Hoffm.) Arcangeli	Carrot	Vegetable
Eryngium campestre L.	Eryngo	Herb
Foeniculum vulgare Miller	Fennel	Herb
Foeniculum vulgare Mill. var. *azoricum* Thell.	(Finocchio)	Vegetable
Levisticum officinale Koch	Lovage	Herb
Ligusticum scoticum L.	Scots lovage	Herb
Myrrhis odorata Scop.	Sweet ciceley	Herb
Pastinaca sativa L.	Parsnip	Vegetable
Petroselinum crispum Nym.	Garden parsley	Herb
Petroselinum crispum Nym. var. *tuberosum* Crov.	Hamburg parsley	Vegetable
Pimpinella anisum L.	Anise	Spice
Sium sisarum L.	Skirret	Vegetable
Smyrnium olusatrum L.	Alexanders	Herb

9.2 MIRIDS

9.2.1 *Lygus rugulipennis* Popp.: Tarnished Plant Bug (Hemiptera: Miridae)

There are several species of mirid bugs, apart from *L. rugulipennis*, which attack umbelliferous crops, viz. *L. lineolaris* (Beauv.), American tarnished plant bug; *Lygocoris pabulinus* (L.), common green capsid; *Orthops campestris* (L.) and *O. kalmi* (L.), stack bugs; *Nysius ericae* (Schill.), false chinch bug; *Calocoris norvegicus* (Gmelin), potato capsid; and the southern European bug, *Graphosoma italicum* Müller. The biology of these insects is described in detail by Southwood and Leston (1959).

Geographical Distribution

Tarnished plant bugs and related mirids are widely distributed throughout the north temperate regions of the world and can be serious pests wherever vegetable crops are grown.

Description

Adult tarnished plant bugs are 5–6 mm long, oval-shaped in outline and brownish-green in colour with dark markings. They usually have a tarnished appearance: hence their common name (Figure 9.1). A small

Figure 9.1 Adult tarnished plant bug, *Lygus regulipennis*. © HRI

head projects in front of the body and bears a pair of long, jointed antennae. When disturbed, tarnished plant bugs, which are very active insects, run quickly across plants or fly away. Eggs are elongate and slightly curved. Each nymph is green in colour, with black spots on its thorax.

Life Cycle

Tarnished plant bugs overwinter as adults on evergreen foliage and in all types of leaf litter. In March and April they emerge from overwintering and begin feeding on umbelliferous plant tissues. During May females lay eggs which are inserted into the tissues of terminal shoots, buds or fruits. The eggs hatch after about 10 days. Nymphs develop through six instars, reaching maturity after about 1 month. Adults are produced by July. They produce, in turn, another generation during August and September. In warm regions further generations may be produced.

Plant Damage

Tarnished plant bug nymphs and adults have hollow piercing–sucking mouthparts (stylets) which are used to penetrate plant tissues and suck up cellular contents. While feeding, the bugs secrete a toxic substance from their salivary glands which kills cells surrounding feeding sites. Usually the first signs of damage are small brown spots on young leaves. Because the area surrounding each feeding site dies as a result of bug attack, affected plants cannot grow properly. When healthy tissues expand, holes replace former feeding sites, and leaves, buds and fruits all become malformed (Figure 9.2). Terminal shoots and flowers may be killed, resulting in loss of yield.

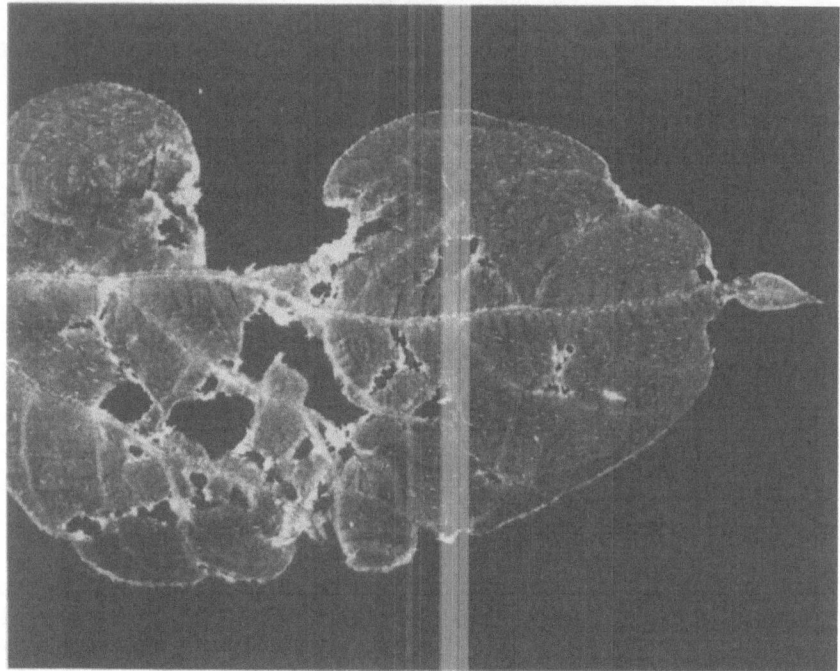

Figure 9.2 Leaf damage caused by the tarnished plant bug, *Lygus regulipennis*. © HRI

Tarnished plant bugs are polyphagous and attack a wide range of vegetable crops and weed species, including wild carrots. They puncture the tender stalks of celery plants, producing large, brown coloured, wilted spots and a blackening of joints, commonly known as 'black-joint'. This type of damage seriously reduces marketability. Tarnished plant bugs also cause damage to the flowers and developing fruits of many other umbel-liferous crops, reducing seed production.

Plant Resistance

Scott (1970) recorded differences in the susceptibility of carrot cultivars to feeding damage by *Lygus* Hahn species. Susceptibility was very low in CV Royal Chantenay, low in CV Nantes but high in CV Imperida.

Chemical Control

Chemical treatments with organophosphorus insecticides are aimed at nymphs and adults. Some aphicides applied as granules or sprays have been effective in killing plant bugs.

9.3. APHIDS

9.3.1 *Cavariella aegopodii* (Scopoli): Willow-carrot Aphid (Hemiptera: Aphididae)

Apart from *C. aegopodii*, certain other *Cavariella* Del Guercio species, as well as several species in other aphid genera, attack umbelliferous crops. All of these species and their host crops are listed in Table 9.2.

Geographical Distribution

The willow-carrot aphid is widespread throughout the temperate regions of the northern and southern hemispheres.

Description

The apterae (or wingless forms) of willow-carrot aphids are green or yellowish-green in colour and measure 1.0–2.6 mm long. Their greenish colour and small size make them inconspicuous on plants. The body of each aptera is elongate, oval-shaped and flattened dorsoventrally, and has an upper surface which is roughened with many small depressions. The alatae (or winged forms) are darker in colour than the apterae and range from 1.4 mm to 2.7 mm long. Because each alata has a black, dorsal abdominal patch, the alatae are more conspicuous on plants than are the apterae.

Life Cycle

The shiny, black eggs of willow-carrot aphids overwinter around the bud axils of willow trees, mainly *Salix fragilis* L. and *S. alba* L. Other *Salix* L. hosts may also be infested (Dunn, 1965). The eggs hatch in February or March and, if the overwintering host has not yet broken bud, the young

Table 9.2 Aphids of umbelliferous crops

Latin name and authority	Common name	Angelica	Anise	Caraway	Carrot	Celery	Chervil	Coriander	Dill	Fennel	Lovage	Parsley	Parsnip
Anuraphis subterranea (Walker)	Pear hogweed aphid												+
Aphis apigraveolens Essig						+					+		+
A. armoraciae Cowan[a]					+	+							+
A. citricola van der Groot	Spiraea aphid				+	+							+
A. decepta Hottes and Frison		+			+	+							+
A. fabae Scopoli	Black bean aphid				+	+							+
A. gossypii Glover[a]	Melon and cotton aphid				+	+							+
A. helianthi Monell[a]					+	+							
A. lambersi (Börner)	Permanent carrot aphid				+								
A. sambuci L.	Elder aphid					+							
Aulacorthum solani (Kaltenbach)	Glasshouse-potato aphid		+		+	+		+	+	+		+	+
Cavariella aegopodii (Scopoli)[a]	Willow-carrot aphid			+	+	+		+	+		+		+
C. archangelicae (Scopoli)		+											
C. konoi Takahashi		+											
C. pastinacae (L.)					+				+				
C. theobaldi (Gillette and Bragg)[a]	Parsnip aphid			+	+	+							+
Dysaphis angelicae (Koch)		+											
D. apiifolia (Theobald)	Hawthorn-parsley aphid				+	+			+	+			+
D. bonomii (HilleRisLambers)	Permanent parsnip aphid												+
D. crataegi (Kaltenbach)	Hawthorn-carrot aphid				+				+	+			
D. foeniculus (Theobald)										+			
D. ossiannilssoni Stroyan		+											
Hydaphis coriandri (Das)[a]			+	+				+		+		+	+
H. foeniculi (Passerini)[a]	Honeysuckle aphid	+	+	+	+	+			+	+		+	+
Macrosiphum euphorbiae (Thomas)	Potato aphid	+	+		+	+			+	+		+	+
Myzus ascalonicus (Doncaster)	Shallot aphid									+			

M. ornatus (Laing)　　Violet aphid
M. persicae (Sulzer)[a]　　Peach-potato aphid
Pemphigus passeki Börner
P. phenax Börner and Blunck
Rhopalosiphoninus latysiphon (Davidson)　　Carrot root aphid
Semiaphis dauci (F.)[a]　　Bulb and potato aphid
S. heraclei (Takahashi)

[a]Virus vector.
Main sources of information: Blackman and Eastop (1984); Hill (1987); Kennedy et al. (1962); and subject index of *Review of Applied Entomology*.

aphids feed through the bark of young shoots. These aphids and subse-quently hatching young aphids move after bud break to the newly emerged foliage and catkins, where they initiate colonies. In May winged forms migrate to carrots, the summer host, over a period of 5–6 weeks, reaching a peak in numbers in early June. During this spring flight, some winged aphids fly to other willows and initiate colonies which develop rapidly and produce adults in about 10 days. In late seasons the migration of aphids from willows may be delayed by 2 or 3 weeks. In Britain the numbers of progeny of spring migrants reach a peak during June, with alatae flying off to other umbelliferous crops (e.g. parsnips) and related wild flower species or else returning to willows. Populations decline following this emigration, and by July few aphids are to be found on carrots. Although many individuals migrate back to the winter host, others overwinter on umbel-liferous crops left in the ground or on hedgerow umbelliferous plants. In the following spring these colonies may produce alatae before the colonies on willows.

Plant Damage

Willow-carrot aphids attack all umbelliferous crops and herbs and many wild umbelliferous species (Dunn and Kirkley, 1966; Hardman *et al.*, 1985a). They not only damage crops directly through their feeding activities, but also are plant virus vectors: persistent carrot motley dwarf and carrot red leaf viruses; semi-persistent parsnip yellow fleck and *Anthriscus* yellows viruses; and several non-persistent viruses, including celery mosaic (Blackman and Eastop, 1984). Losses of carrot yield resulting from motley dwarf virus have been estimated to be as high as 40–60% in the most severe outbreaks in England (Watson *et al.*, 1964). During the 1940s carrot motley dwarf virus disease was so severe in Australia that the growing of carrots in some areas was abandoned (Stubbs, 1948). Willow trees are not a source of carrot motley dwarf virus. This virus is spread by willow-carrot aphids which overwinter on infected crops of carrots, parsley, etc., or on infected hedgerow umbelliferous plants.

Aphids feed by penetrating plant tissues with their hollow stylets and sucking up plant sap (see *Myzus persicae*, page 264). Excess sugars are excreted in the form of honeydew. Carrot plants attacked by willow-carrot aphids develop twisted and malformed foliage which often results in stunting and, in extreme cases, death of seedlings. The foliage commonly turns a reddish or yellowish colour and the honeydew gives infested plants a varnished appearance. On parsley direct feeding damage is usually less noticeable than on carrots, partly because the leaves are tightly curled, but also because the foliage is more dense. Carrot motley dwarf virus, however, is particularly damaging to parsley (Anon., 1979). Aphids' cast skins stick to honeydew and also become scattered around the plant base.

Monitoring and Prediction

Rabasse *et al.* (1976) experimented with traps of different designs in carrot fields and developed methods for monitoring movements of aphids into the crop.

Cultural Control

Although not practicable or environmentally acceptable, if willow trees, the overwintering hosts of the willow-carrot aphid, could be removed, the aphid problem would to a large extent be eliminated. Siting crops away from willows does reduce the chances of infestation. The main source of virus, and an important source of aphids, is overwintering umbelliferous crops. Any crops that can be disposed of should be cleared away. At present, overwintered umbelliferous crops are becoming more common in the UK because of the use of polythene covers in the production of early carrots and other crops. This novel agronomic practice may exacerbate aphid and virus problems. Because overwintered parsley often supports willow-carrot aphids, subsequent crops should be sited as far as possible from sources of infestation.

Certain umbelliferous species and crop cultivars possess some degree of resistance to the willow-carrot aphid and/or the complex of viruses transmitted by it. In Australia genes for resistance to carrot motley dwarf virus disease were bred into a carrot cultivar, Western Red, which has been used in the production of many new resistant cultivars. This virus is no longer a problem disease of carrots in Australia (Seddon, personal communication). Symptomless systemic infection by carrot motley dwarf virus has been reported in carrot CV. Gold Pak and celery CV. Spartan (Dixon, 1981). In tests of resistance to willow-carrot aphid infestation of representative cultivars of different carrot groups (Dunn, 1970), Autumn King and Berlicum were the most and the least susceptible cultivars to aphid attack, respectively. Although CV Autumn King was most suscepti-ble to aphid attack, it was tolerant of carrot motley dwarf virus infection. Wild umbelliferous plants differ greatly in their apparent relative attrac-tiveness to colonising willow-carrot aphids: hemlock was colonised little, but chervil attracted so many aphids that it was killed (Dunn and Kirkley, 1966). In more recent experiments two species of wild umbelliferous plants, false thorow-wax and Alexanders, failed to support apterous forms of *C. aegopodii* (Hardman *et al.*, 1985a).

Chemical Control

Several carbamate and organophosphorus insecticides, applied as medium- to high-volume sprays or as granules, are available to protect crops against willow-carrot aphid attack. Timing of treatments is critical and sprays need to be applied according to information on crop colonisation and/or trapping. The spring migration of aphids from willows to summer host crops can be controlled by two insecticidal treatments, the first applied 1 week after arrival of aphids on the summer host crop, and the second 14 days later. A third application may be required against severe infestations. Migrations between summer host crops—from previously infested crops to uninfested crops—can be more difficult to control. Ideally, sources of infestation need to be treated. Aphids can be prevented from migrating from their summer hosts to overwinter on willows and various umbelliferous crops by spray treatments of the summer host crops. It is particularly important to prevent parsley crops from becoming infested, because of the serious consequences of carrot motley dwarf virus infection for crop quality.

9.3.2 *Dysaphis apiifolia* (Theobold): Hawthorn-parsley Aphid or Rusty Banded Aphid (Hemiptera: Aphididae)

Geographical Distribution

D. apiifolia is distributed widely and has been recorded in Europe, the Middle East, north and south Africa, central Asia, Australia and North and South America.

Description

The apterae of the hawthorn-parsley aphid (1.4–2.6 mm long) are yellowish-grey to greenish-grey in colour and lightly covered in wax. They are found in dense colonies around the leaf bases of umbelliferous crops. The alatae (1.5–2.4 mm long) are dull green in colour, each with a black dorsal patch.

Life Cycle

The primary host of *D. apiifolia* is the hawthorn tree, *Crataegus monogyna* Jacq. The aphids produce deep-red-coloured, crumpled leaf galls on these trees in the spring. Alatae migrate from hawthorn to various secondary herbaceous hosts in late spring and produce colonies at or below soil level. The aphids on these secondary hosts are attended by ants which 'milk' the aphids of their honeydew. In return, the aphids are protected from their enemies by the ants, which build their nests around the aphids. *D. apiifolia* (*sensu lato*) breeds continuously by parthenogenesis (i.e. anholocycly) on secondary hosts. The subspecies *D. apiifolia petroselini* (Borner) has, by contrast, a holocyclic life cycle and alternates between primary (*Crataegus* L. sp.) and secondary (various umbelliferous) host plants and between sexual and asexual phases. Further details of the biology and life cycle of this aphid and related species are given by Stroyan (1963) and Blackman and Eastop (1984).

Plant Damage

The secondary, summer hosts of the hawthorn-parsley aphid include various wild umbelliferous species, and celery, coriander, fennel and parsley. The direct feeding activities of the aphids are not serious, but they are vectors of celery mosaic and celery crinkle leaf spot viruses, both viruses causing very important diseases of umbelliferous crops. The viruses, which appear to be a complex, are also transmitted by other aphid species, particularly *Aphis gossypii* Glover (see page 140) and *Myzus persicae* (Sulzer) (see page 263) (Dixon, 1981). The production of celeriac has been made extremely difficult by the virus complex in parts of the Netherlands (Maramorosch and Harris, 1981). In the UK the virus complex causes chlorosis and stunting of coriander, and a golden-yellow-coloured chlorosis and necrotic spotting of parsley (Dixon, 1981). The crinkle leaf strain of the virus complex causes a yellow coloured mottle to appear on celery leaves, with associated raised, blister-like areas and a general crinkling of the foliage.

Cultural Control

Control of the celery mosaic and celery crinkle leaf spot virus complex in California was achieved by enforcing a celery-free period to eliminate sources of infection. In the UK this approach is not considered likely to be effective, because of the widespread occurrence of wild umbelliferous species which provide a reservoir of aphids and virus complex (Dixon, 1981). Differences in varietal susceptibility of celery are caused by differences in the virulence of strains of the virus complex (Walkey and Cooper, 1971).

Chemical Control

Insecticides recommended for control of the willow-carrot aphid (see page 335) also reduce numbers of the hawthorn-parsley aphid.

9.4 PSYLLIDS

9.4.1 *Trioza apicalis* Forster: Carrot Psyllid (Hemiptera: Psyllidae)

Geographical Distribution

The carrot psyllid is principally a pest of umbelliferous crops in northern

Europe, particularly Scandinavia, although it is also found in Czecho-
slovakia and Russia.

Description

Carrot psyllids are similar in size to aphids and resemble small cicadas.
Adults are about 3 mm long, are green in colour and have two pairs of
membranous wings which extend beyond the tip of each adult's abdomen.
The forewings are tough and have prominent veins, while the hindwings
are very flimsy. The head is broad, with long, jointed antennae and
prominent, deep red eyes. The hindlegs are larger and more muscular than
the others and are used in leaping. The eggs, 0.3 mm long, are narrow and
elongated. The nymphs are yellow in colour and oval in outline, with a
flattened body bearing waxy filaments and long legs. The antennae of
nymphs are short and conical. A crest runs down the middle of each
nymph's back. There are five nymphal instars, and in later stages wing buds
are prominent.

Life Cycle

Carrot psyllids overwinter as adults in evergreen foliage and other dense
herbage and emerge during May. They are very active insects and move by
leaping or flying from plants, but they cannot sustain flight. After mating,
each female lays up to 750 eggs on the leaves of host plants. Eggs are laid
singly and are attached by one end so that they project vertically from leaf
surfaces. Nymphs hatch after 12–15 days and group themselves along leaf
veins. Adults and nymphs feed by inserting their stylets into leaf tissues.
Nymphs are developed fully after about 30 days. Towards the end of the
season, many adult psyllids congregate on the shoots of pine trees,
although some survive the winter on carrots left in the soil. Further details
of the biology and life cycle of this pest are provided by Balachowsky and
Mesnil (1936).

Plant Damage

Like aphids, carrot psyllids are phloem feeders. When feeding, psyllids
inject, along with their saliva, a toxin which disrupts the metabolism of host
plants. The damage to carrot plants resembles a systemic toxaemia and,
while the leaves on which the psyllids feed appear normal, young,
developing leaves assume a parsley-like appearance and become greatly
deformed. The toxin dramatically reduces plant growth, especially affecting the roots, and, in the severest cases, yields of carrots are practically nil
(Markulla *et al.*, 1976). The degree of injury was found to increase as the
duration of feeding increased. The damage caused by overwintered
females was found to be more severe than the damage caused by

overwintered males or nymphs and females produced during the season. The carotene content of the roots of attacked plants was found to be reduced by 30%. A secondary deleterious effect of psyllid feeding is the deposition of honeydew on the foliage of infested plants and the subsequent growth of sooty mould colonies which reduce crop yield. With carrot seedlings, one psyllid per root causes crop loss.

Control

Spray treatments with a wide range of carbamate and organophosphorus compounds control adult carrot psyllids. A number of parasitoids and predators also help to keep populations of psyllids in check.

9.5 HOPPERS

9.5.1 *Macrosteles fascifrons* (Stål.): Aster Leafhopper (Hemiptera: Cicadellidae)

Geographical Distribution

The aster leafhopper is distributed widely in Canada and the USA, and is an important pest of carrots, celery and many other crops—e.g. lettuces, onions, potatoes.

Description

The adult aster leafhopper is rather frog-like in appearance, with prominent eyes, a tapered body and transparent wings which, when not used for flying, project beyond the end of the abdomen (Figure 9.3). Resting insects

Figure 9.3   Adult aster leafhopper, *Macrosteles fascifrons*. © HRI

are inconspicuous, but when disturbed, they are highly active and leap into the air and fly to other resting places. They are greyish-green in colour and about 3 mm long. The eggs are laid in leaf tissues and are translucent when newly laid, becoming white in colour. They are smooth, elongated, slightly curved and tapered at the ends, measuring 0.8×0.2 mm (Hagel and Landis, 1967). The nymphs are small, wingless, greenish-coloured, very active insects which moult five times before becoming adults.

Life Cycle

Adult aster leafhoppers overwinter in warm regions but further north leafhoppers overwinter as eggs on autumn-sown cereal crops. In Canada the leafhoppers develop on cereal crops during May and then migrate to carrot and celery crops, where they insert their eggs singly into leaf tissues. The nymphs hatch, feed on leaf undersurfaces and become adults in about 2 weeks. During the summer up to five generations may develop on vegetable crops. Peak populations occur in late July and August in Ontario and the eastern USA. Further west, in Wisconsin and Minnesota, leaf-hoppers arrive on carrot and celery crops from the southern states before the hoppers on cereals have hatched from overwintering eggs (Chapman, 1985). Observation of this northerly migration has aided control programmes for this pest (see below).

Plant Damage

The direct damage done by aster leafhoppers—piercing plant tissues and feeding on sap (Figure 9.4)—is minor compared with the crop losses caused by leafhopper-transmitted aster yellows, a mycoplasma-like organism (MLO) pathogen. (Plant 'mycoplasmas' are generally referred to as 'mycoplasma-like organisms' or MLOs.) For all practical purposes, this disease is spread only by aster leafhoppers. Total losses of crops, such as carrots and celery as well as asters and lettuces, can result from infection, and up to 50% losses may occur in other crops, including potatoes. The symptoms of leafhopper feeding are mottled specks on the leaves of infested plants. The symptoms of aster yellows disease on carrots are excessive red- or yellow-coloured, frond-like shoots with hairy, bitter roots. On celery the stalks become deformed, twisted and unmarketable.

About 8 h feeding by leafhoppers on infected plants is required for individual leafhoppers to acquire the MLO in a sufficient amount to become an infective vector. It is transferred to healthy plants following a similar period of feeding. A long incubation period of 2–3 weeks is required after acquisition for the MLO to multiply inside the body of the leafhopper before it can infect a healthy plant. During this time transmis-sion to plants cannot occur. Continued multiplication of the MLO enables

Figure 9.4 Feeding punctures caused by the aster leafhopper, *Macrosteles fascifrons*. © HRI

each leafhopper to be infective for the whole of its life and to transmit the pathogen over great distances. Using information on varietal susceptibility and percentage infectivity of the leafhopper population as a whole, an aster yellows index has been devised to guide growers in the use of insecticidal treatments against aster leafhoppers.

Monitoring and Prediction

During spring the time of arrival of aster leafhoppers in the northern USA from the southern USA has been predicted accurately from surveys of leafhoppers' activity on the eastern side of their migration path (Chapman, 1985). Assessments of MLO infectivity of leafhoppers are also made. Cool, wet weather in the southern states coupled with strong winds exceeding leafhoppers' flight threshold prevent populations from migrating north and, as a result, few problems have been encountered in these seasons.

Cultural Control

Aster yellows infection is affected by density of crop plants: as the number of aster plants to a row increases, a corresponding increase occurs in the numbers of colonising leafhoppers until foliage completely covers the soil surface. Further increases in plant density do not then lead to increased numbers of leafhoppers and numbers of aster yellows infections also cease to increase. In consequence, growers are advised to sow crops at high density to decrease both leafhopper numbers and aster yellows infections.

Plant Resistance

When a range of carrot cultivars and breeding lines was tested for resistance to aster leafhoppers, some showed less than 10% of the aster yellows infections of the most susceptible lines (Chapman, 1985). Breeding programmes have produced some carrot lines which are practically immune to aster yellows. The relationship between the leafhopper and resistance to aster yellows infection in carrots parallels very closely the relationship between the carrot fly and resistance in carrots (see page 373). With both insects, the Nantes types of carrot have been found to be resistant and the Danvers cultivars have been found to be susceptible.

Chemical Control

As there is a long interval between colonisation of plants and transmission of aster yellows, insecticides can be used very effectively to prevent spread of the disease within crops. There is a shorter period in which insecticides may be used effectively against migrating infective leafhoppers. Pyrethroid, carbamate and organophosphorus compounds offer good protection of susceptible crops.

Integrated Control

Chapman (1985) provides details of an integrated control programme for aster leafhoppers in the USA. A study of their spring migration has enabled the time and severity of outbreaks to be predicted. At the predicted times vulnerable crops are examined twice weekly for the presence of leafhoppers, which are then checked for MLO infectivity. When the aster yellows index is exceeded, growers are advised to treat their crops with insecticides. By rigorously applying the integrated programme, the numbers of insecticidal treatments of susceptible carrots has been reduced from 13 on traditional spray schedules to 4. Only 2 treatments are now applied to resistant carrots.

9.6 BEETLES

9.6.1 *Listronotus oregonensis* (Le Conte): Carrot Weevil (Coleoptera: Curculionidae)

Geographical Distribution

The carrot weevil is distributed widely in north temperate regions of north America. It is an important pest of carrots, celery, parsley, parsnips and dill.

Description

The adult carrot weevil is dark brown in colour and about 6 mm long. The eggs are white in colour when first laid, later becoming darker. The legless larva, 7 mm long when fully grown and typically 'C'-shaped, has a chestnut-brown-coloured head and creamy white body. The larval body is distinctly segmented and dorsoventrally flattened. The pupa is white in colour and similar in length to the fully grown larva. It has vestigial legs and a white-coloured head with a red eye spot.

Life Cycle

Stevenson (1976) studied the seasonal appearance of carrot weevils in Ontario, Canada. The weevils overwinter in plant debris in the vicinity of host crops. As temperatures rise in the spring, the weevils emerge to mate after a few warm days. Although adult carrot weevils are capable of flying considerable distances, they rarely do so. They seem to prefer to remain at the soil surface near host plants. By mid- to late May, the adult females start laying eggs in small cavities hollowed out in petioles or tops of tap roots. Two to four eggs are laid in each cavity, which is then sealed with a black exudate. After 1–2 weeks the eggs hatch into larvae, which tunnel downwards into the tap root or crawl down the leaf petiole to the tap root before commencing to feed. Quite often several larvae invade a single root. In about 2 weeks the fully grown larvae leave the root and pupate in earthen cocoons in the soil. Adult weevils emerge in a further 2 weeks, to produce a second generation of larvae which attack plants during August. In Canada there is usually only one generation in a season, but further south a second generation is common and a partial third generation may also occur.

Plant Damage

Damage caused to host plants by adult carrot weevils is not serious. However, larval damage is much more serious. The larvae excavate characteristic tunnels in the upper third of carrot and parsnip roots. These tunnels form darkened areas (Figure 9.5). Parsley and other herb crops develop yellow leaves, drastically reducing marketability, as a result of larval injury to their roots. Damage may be so serious as to destroy young crops of celery. However, larger celery plants can tolerate a certain amount of root damage without their yields being affected.

Monitoring and Prediction

Perron (1971) evaluated methods for monitoring populations of carrot weevils in Canada: colour traps (comprising Plexiglas dishes with a range

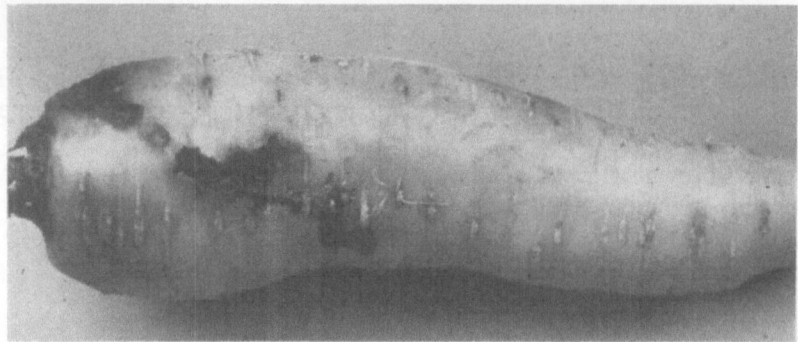

Figure 9.5 Carrot damage caused by the carrot weevil, *Listronotus oregonensis*.
© A. B. Stevenson

of coloured bases fixed to wooden stakes); light traps; plant traps (comprising carrots sunk to soil level in young crops); sweep nets; and root sampling. The use of 'plant traps' examined every 3–4 days for eggs has helped Canadian growers to determine the necessity and the timing of control treatments.

Cultural Control

Carrot weevil damage can be reduced by careful choice of sowing or planting dates. Stevenson (1976) has shown that late sowing of carrots in mid-July and of celery from July onwards reduced damage to a minimum. However, for most growers, late sowing would not be a practicable measure for preventing weevil damage. As weevil infestations tend to build up locally because they rarely fly, crop rotation in space and time can be effective.

Chemical Control

Susceptible crops can be protected from carrot weevil attack by applying sprays of organophosphorus compounds at 10–14 day intervals from seedling emergence.

9.7 MOTHS

A comprehensive list of lepidopterous species which attack umbelliferous crops is provided in Table 9.3. The more important pest species are discussed below.

9.7.1 *Hepialus humuli* (L.): Ghost Swift Moth or Ghost Moth (Lepidoptera: Hepialidae)

Geographical Distribution

The ghost swift moth is widespread in northern and central Europe, and its range extends as far as the Middle East. It occurs throughout most of the British Isles, but has not been recorded in Jersey or the Isles of Scilly.

Description

The female ghost swift moth has pale, ochreous-yellow-coloured forewings with oblique pinkish-brown bands. The wingspan is up to 57 mm wide. The hindwings are a light, ochreous-grey colour tinged with pink. The male moth has smaller wings, only 48 mm wing span, the forewing and hindwing being silvery-white-coloured, while the costa (leading edge of wing) and cilia (fringe of hairs on wing) are tinged with an ochre colour. In both sexes the head, thorax and abdomen are pale, ochreous-brown to yellow in colour and the antennae are very short and simple. The eggs are oval-shaped, 0.7×0.5 mm, and shining white in colour when first laid, but they become a shining black colour in a few hours after laying. The larva, with a shining-white or greyish body colour and a glossy, reddish-brown-coloured head, is highly characteristic. When fully fed, the larva is about 70 mm long, with a black-coloured dorsal line which ends on the sixth abdominal segment. The prothorax has a reddish-brown-coloured, shining plate. The pupa is elongate, about 24 mm long and dark brown in colour, with a series of spines and cutting plates on the abdominal segments.

Life Cycle

Adult ghost swift moths emerge in June and July and fly at dusk, the males having a hovering flight and emitting a scent of goats which is believed to attract females. After mating, the females lay eggs in flight, broadcasting them onto plants. Between 200 and 1600 eggs are laid per female moth, the eggs hatching in 15–20 days. The larvae usually live in subterranean tunnels from July to the following April or May, but some take 2 years to complete development. They feed on plant roots. Pupation takes place in a larval tunnel, and, just before emergence of the adult moth, the pupa actively propels itself to the surface of the soil, using its abdominal spines. There is only one generation each year.

Plant Damage

The larvae of ghost swift moths damage a wide range of plants from many families and are considered to be pests of grassland, lawns, several field

Table 9.3 Lepidoptera of umbelliferous crops

Latin name and authority	Common name	Angelica	Anise	Caraway	Carrot	Celery	Chervil	Coriander	Dill	Fennel	Lovage	Parsley	Parsnip
Hepialus humuli (L.)	Ghost moth				+	+							+
H. lupulinus (L.)	Common swift moth				+	+							+
Depressaria daucella (D. & S.)				+									
D. marcella Rebel				+									
D. pastinacella (Duponchel)			+		+								+
D. ultimella Stainton					+	+							
Agonopterix nervosa (Haworth)					+	+							
Epermenia aequidentella (Hofmann)					+	+							
E. chaerophyllella (Goeze)					+								
Argyrotaenia pulchellana (Haworth)	Grey-barred tortrix										+		
Epinotia thapsiana Zell.										+			
Cnephasia interjectana (Haworth)										+			
Aethes dilucidana (Stephen)	Flax tortrix			+	+								
A. francillana (F.)	Parsnip flower moth				+								+
A. williana (Brahm)	Carrot flower tortrix				+								
Pediasia caliginosella (Clemens)	Corn root webworm				+								
Sitochroa palealis (D. & S.)					+								
Margaritia sticticalis (L.)					+								
Ostrinia nubilalis (Hübner)	European corn borer					+			+	+			
Udea ferrugalis (Hübner)	Celery leaf tier				+	+			+				
Papilio machaon (L.)	Swallow-tail butterfly	+			+	+						+	
P. polyxene (F.)	Black swallowtail butterfly			+	+	+						+	+
P. zelicaon Lucas	Western parsley caterpillar			+	+	+			+			+	+
Agrotis segetum (D. & S.)	Turnip moth; common cutworm				+	+							+

	Dark swordgrass moth; black cutworm
A. ipsilon (Hufnagel)	Dark swordgrass moth; black cutworm
Euxoa nigricans (L.)	Garden dart moth
Noctua pronuba (L.)	Large yellow underwing
Xestia c-nigrum (L.)	Setaceous hebrew character
Hydraecia micacea (Esper)	Rosy rustic moth
Phlogophora meticulosa (L.)	Angle shades moth
Lacanobia oleracea (L.)	Bright-line brown-eye; tomato moth
Anagrapha falcifera (Kirby)	Celery semi-looper
Trichoplusia ni (Hübner)	Ni moth; cabbage looper
Autographa gamma (L.)	Silver Y moth
Spodoptera eridania (Cramer)	Southern armyworm
S. exigua (Hübner)	Beet armyworm
S. littoralis (Boisduval)	Cotton leaf worm
Platysentor sutor Guenee	Mediterranean climbing cutworm

crops and many garden and glasshouse plants. They feed on lateral roots and tunnel into bulbs, corms, rhizomes, tap roots and tubers. Vegetable crops attacked include lettuces, carrots, parsnips and potatoes. Further details of the bionomics of the ghost swift moth are provided by Edwards (1964).

Cultural Control

As with leatherjackets (see page 83) and wireworms (see page 33), also pests of grassland, any arable crop which follows grass is likely to be attacked by ghost swift moth larvae remaining in the soil. Therefore, land ploughed out of grass should not be cropped until the risk of crop damage has been minimised. For example, the sowing or planting of vegetables should be delayed until the larvae have completed their development. Regular cultivation of the soil will destroy some larvae and expose others to predators and adverse weather conditions. As with noctuid cutworms (see pages 352–364), land should be kept weed-free, as swift moths have numerous wild flower hosts and egg-laying females are attracted to weed cover.

Biological Control

Swift moths have numerous natural enemies, such as birds, mammals, parasitic and predatory insects and diseases, but none has yet been used in a biological control programme.

Chemical Control

Incorporation into the soil of organophosphorus or carbamate insecticides as emulsions or suspensions is recommended for the control of swift moth larvae.

9.7.2 *Hepialus lupulinus* L.: Garden Swift Moth (Lepidoptera: Hepialidae)

The garden swift moth has a very similar biology to the biology of the ghost swift moth but differs in the following respects.

Description

The adult garden swift moth is smaller than the adult ghost swift moth, the wingspan being 25–40 mm wide. The forewings are yellowish-brown in colour, often with whitish markings, although these markings are variable and may be lacking in the female moth. The hindwings are greyish-fuscous in colour. The head and body are pale-fuscous-coloured. The female moth is usually more drab and larger than the male moth. Female garden swift moths each lay between 50 and 200 eggs, between a quarter and an eighth of the number of eggs laid by ghost swift moths, and the larva and pupa are smaller than in the ghost swift moth. The 25-mm-long larva is a translucent white colour rather than the shining white colour of the ghost swift moth and the dark gut often shows through the cuticle. The pupa is about 15 mm long.

Life Cycle

Adult garden swift moths emerge in mid-May and June, a little earlier than adult ghost swift moths. Male garden swift moths do not hover like male ghost swift moths but instead actively seek out females. Garden swift moths frequently complete their development in a single year, as opposed to the 2 years needed for the complete development of ghost swift moths.

Plant Damage

Garden swift moth larvae are polyphagous and have been recorded attacking a wide range of vegetable crops, including artichokes, beans, carrots, celery, garlic, lettuces, parsnips, peas, potatoes, swedes and tomatoes.

Control

The measures recommended above (see page 348) to control the ghost swift moth are also recommended to control the garden swift moth.

9.7.3 *Epermenia chaerophyllella* (Goeze): Common Lancewing Moth (Lepidoptera: Epermeniidae)

Geographical Distribution

E. chaerophyllella is common in north and central Europe and parts of Asia. In Britain it ranges throughout England, Wales, Ireland and as far north as the Caledonian canal in Scotland.

Description

Meyrick (1895) described *E. chaerophyllella* and related species. The adult moth has a wingspan of 12–13 mm. The forewings are more or less narrowly elongate, pointed and brown-white in colour spotted with black. There are several black lines as well as two white spots on each forewing. The hindwings are greyish-coloured. The larva, which is about 10 mm long when fully grown, has a pale-brown-coloured head and a yellow-white or greenish-white body colour with a white dorsal line and brown spots on its abdomen. The larva (Figure 9.6) pupates in an open network cocoon.

Life Cycle

E. chaerophyllella overwinters as an adult. Moths emerge during April, to produce a generation of caterpillars that feeds in June. Moths appear again during July and August, to produce a second generation of caterpillars which feeds on plants in the autumn. These caterpillars produce the overwintering adult moths. Larvae live within silken webs woven around plant foliage and flower parts. Related species may spin threads and draw together leaflets and umbel parts (*E. insecurella* (Stainton)) or feed as leaf miners (*E. aequidentella* (Hofmann)).

Plant Damage

The caterpillars of *E. chaerophyllella* damage umbelliferous crops, including carrots and parsnips. They also attack several related wild species. Damage is done by larvae feeding on plant parts and distorting tissues by webbing (Figure 9.6). Seed production may be reduced by larval damage to flowers.

Figure 9.6 Larva of the common lancewing moth, *Epermenia chaerophyllella*, and damaged carrot foliage. © HRI

Control

There are no published recommendations for the control of *E. chaero-phyllella*.

9.7.4 *Depressaria pastinacella* (Duponchel) (Synonym: *D. heracliana* (De Geer)): Parsnip Moth or Parsnip Webworm (Lepidoptera: Oecophoridae)

Geographical Distribution

The parsnip moth is distributed throughout Europe and North America and is found in parts of Asia. It occurs widely in the British Isles and as far south as Madeira in Europe. It was first recorded in North America in Ontario during 1869 and is now common in southern Canada and the northern states of the USA east of the Mississippi river. The closely related species *D. daucella* (D. and S.), *D. daucivorella* Roy, *D. (Agonopterix) ferulae* Zell., *D. marcella* Rebel, *D. (Agonopterix) nervosa* Haw., *D. ultimella* Stainton and *D. veneficella* Zell. also infest umbelliferous seed crops and are particularly important pests in southern Europe.

Description

The adult parsnip moth has a wingspan of 27 mm. The forewings have a pale brown or greyish colour suffused with ochreous white and marked with a series of dark-brown, transverse flecks. The hindwings are whitish-coloured with pale brown veins and outer edges. The adult moth has a pale-brown-coloured head, thorax and abdomen. The egg is round and greenish-yellow in colour. The larva, which grows to a length of 15 mm, has a black-coloured head and a body colour which is bluish-grey above and yellowish-grey below the spiracles. The larval body has many short setae and dark-coloured spots. The pupa is shiny, reddish-brown in colour and covered with many setae. The pupal cremaster has eight hooked setae.

Life Cycle

Adult parsnip moths emerge from flower stems during late summer in North America or early autumn in Britain and seek overwintering sites such as under tree bark, leaf litter or in buildings. In the spring the moths resume activity and lay eggs in developing flower heads and on other parts of host plants. The larvae hatch and commence feeding on flowers and seeds, webbing together floral parts. When fully fed in July and August, the larvae either descend to the plant base and hollow out a pupation chamber in the stem or remain under the silken webs, where they pupate. There is only one generation each year.

Plant Damage

The parsnip moth is particularly damaging to parsnips, but it can also damage carrots, celery, fennel and parsley. The flower heads of many wild umbelliferous plants are also infested by parsnip moth larvae. The damage, which can seriously affect seed production, includes the larval eating of bracts, flowers, leaves and seeds, webbing of flower parts and mining of stems.

Control

Cultural methods include the removal and the destruction of infested flower heads as well as the destruction of wild umbelliferous plants in the neighbourhood of seed crops. Sprays with organophosphate insecticides and the bacterium *Bacillus thuringiensis* Berliner in June have provided control of parsnip moths in southern Europe. Flower heads should be examined regularly in the spring for the first signs of infestation.

9.7.5 *Agrotis segetum* **(Denis and Schiffermuller): Turnip Moth or Common Cutworm (Lepidoptera: Noctuidae)**

Geographical Distribution

The turnip moth is distributed widely in temperate regions of the Old World, occurring throughout Asia, Africa and Europe. It is not a pest in the Americas or in Australasia. In western Europe the turnip moth is a most serious and widespread pest of vegetable crops: it is the predominant cutworm in England and Wales (Sherlock, 1983) and in Denmark (Jorgensen, 1978).

Description

Adult turnip moths have a wingspan of about 40 mm. The forewings are pale, greyish-brown-coloured with dark-brown markings which include rings (stigmata) and lines (fascia) (Figure 9.7). The hindwings are white in colour on the male and, on the female, are suffused with brown, particularly towards the edges. The robust body and head of the adult moth are brown in colour and the collar region has a black band. Up to 1000 ribbed eggs, each with a reticulate pattern, are laid by every female moth in small irregular masses. The eggs are globular, about 0.5 mm in diameter and milky-white in colour, later turning cream-coloured with reddish-yellow markings and an orange band. Fully grown turnip moth larvae are about 40 mm long. The larval body is plump and rather greasy in appearance (Figure 9.8). It is greyish-brown in colour, sometimes tinged

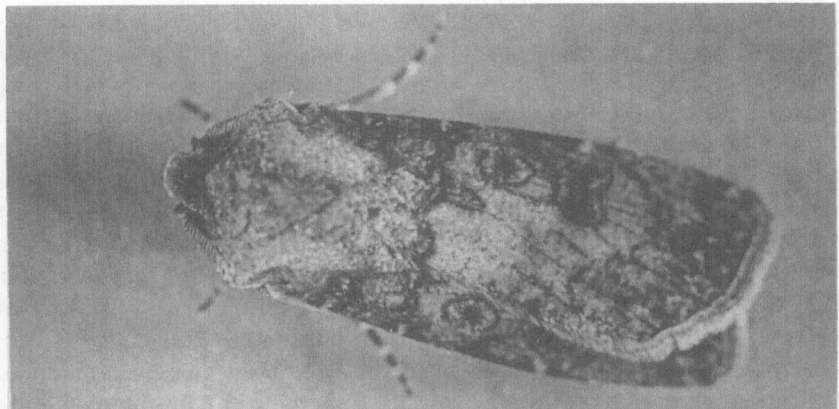

Figure 9.7 Adult turnip moth, *Agrotis segetum*. Note the characteristic stigmata and fascia on the wings. © HRI

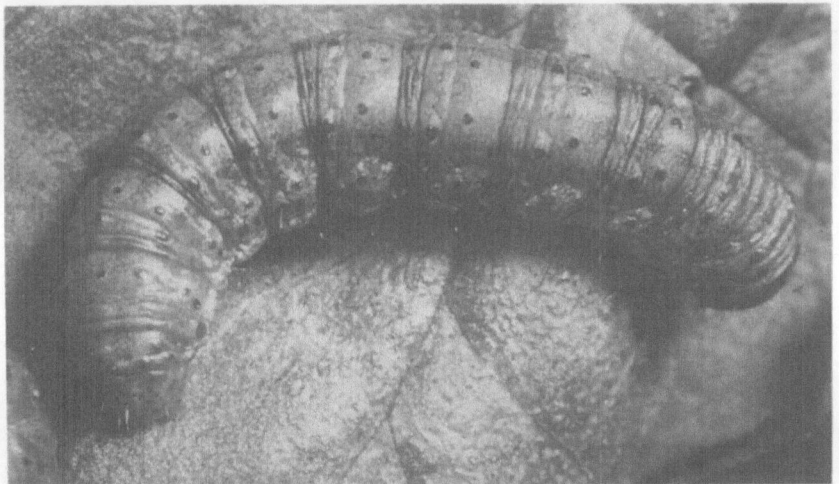

Figure 9.8 Caterpillar of the turnip moth, *Agrotis segetum*. © HRI

with green, and has a dark-coloured dorsal line and a lighter-coloured spiracular line. The thoracic legs of the larvae are yellowish-brown in colour and the abdominal prolegs are pale, greyish-brown in colour. The pupa, about 15 mm long, has a reddish-brown colour with dark brown spiracles. The pupal cremaster has two short, divergent prongs.

Life Cycle

Adult turnip moths emerge from overwintering larvae or pupae during May and June and fly at night. Females lay their eggs on plants, soil or plant debris. The eggs hatch in 10–28 days and the first two larval instars feed on plant foliage. In the later instars the larvae descend to the soil and feed at night as typical cutworms, remaining by day in soil crevices or under stones or plant debris. Feeding occurs at or below soil level. These cutworm caterpillars either develop quickly, to produce a second generation in the same year, or develop slowly throughout autumn and winter, to become fully developed during the following spring. About 80% of the food consumed by all larval instars is eaten by the final instar (Hill, 1987). In Britain there is a partial second generation in September or October, while in southern Europe there are two generations. Up to five generations occur in warmer climatic zones. The larvae pupate in the soil in earthen cells. The time spent in this stage varies considerably, depending on the time of year, but it may be as short as 10 days or as long as several months. In warm conditions the life cycle of the turnip moth can be completed in 6 weeks. Further details of the insect's biology and life cycle are provided by Heath and Emmett (1979) and Carter (1984).

Plant Damage

The turnip moth is the most economically important cutworm in Europe. Its umbelliferous host plants are listed in Table 9.3. Most damage to crops in northern Europe occurs in late summer and early autumn. Seedlings and young plants are severed from their tap roots and die (Figure 9.9).

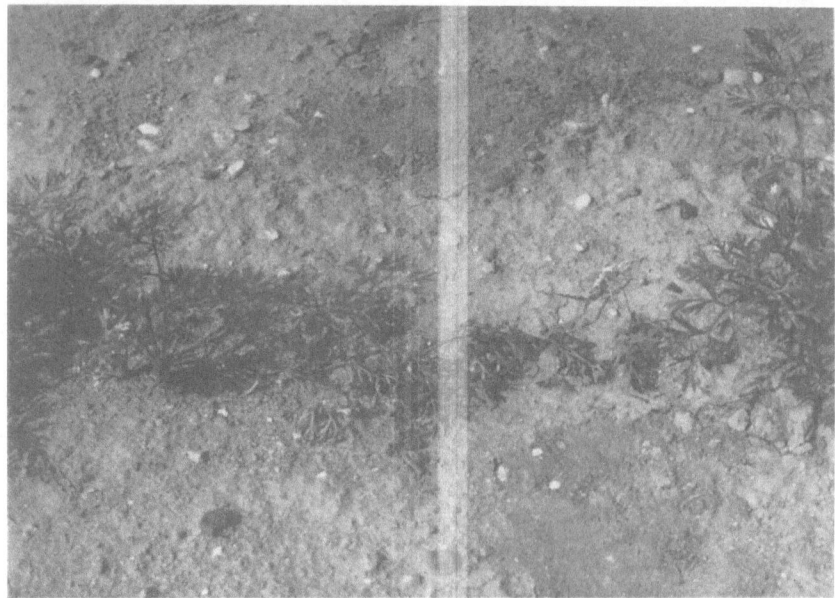

Figure 9.9 Damage to seedlings (wilting) by the turnip moth caterpillar, *Agrotis segetum*. © HRI

Cutworms create cavities in stems, rhizomes, tubers and roots of larger plants (Figure 9.10). This damage is similar to slug damage (see page 291) but it is not accompanied by the slime characteristic of slug activity. Row crops such as red beets, celery, leeks, onions and carrots, as well as other crops with slender tap roots (for example, lettuces), are particularly vulnerable to cutworms moving along rows and cutting plants off one after another. Damage to root crops such as potatoes, swedes and turnips may not be evident until harvest, when the marketability of the produce may be reduced severely. Cutworm damage is most severe in light, sandy soils, where the larvae can burrow easily. In Britain cutworm outbreaks occur sporadically. Records of their importance as vegetable pests show that high infestations damage up to 50% of red beet crops and up to 20% of carrot, leek and lettuce crops. The proportion of roots damaged within these crops may exceed 10% for red beet, carrot, leek and lettuce (Bowden *et al.*, 1983).

Monitoring and Prediction of Cutworm Activity

Turnip moth attacks are forecast in Europe in order that growers can more accurately time insecticidal sprays or, in some years, avoid the use of chemicals. Early monitoring of moth activity was based on light-trap catches, but the development of baited pheromone traps has improved the efficiency of forecasting (Esbjerg, 1983; Emmett, 1984). Mathematical models have been developed to forecast cutworm attacks (Mikkelsen and Esbjerg, 1981; Bowden *et al.*, 1983; Esbjerg, 1983). Inputs for these models include: effects of temperature on the rate of insect development; behavioural characteristics of the larvae on plants; and the susceptibility of

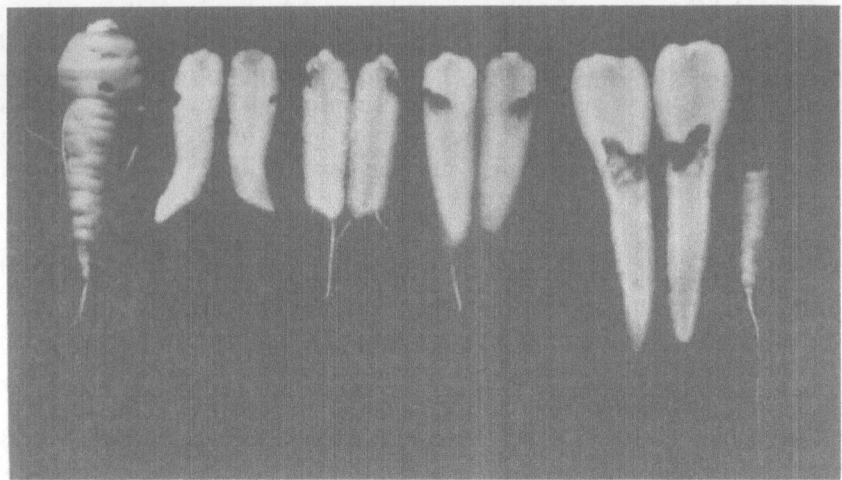

Figure 9.10 Damage to carrot roots caused by larvae of the turnip moth, *Agrotis segetum*. © HRI

young larvae to rain (Zethner and Esbjerg, 1978). A cumulative survival index of caterpillars up to the third instar of larval development may also be used to estimate the degree of risk to crops in any given year. The models suggest that weather conditions alone are responsible for major cutworm outbreaks and that, for epidemics, two consecutive years of favourable weather are required (Emmett and Rhodes, 1987).

Cultural Control

Weedy land harbours most cutworms, as the adult moths appear to prefer sites providing dense plant cover for egg laying. Crops immediately following dense weed cover are therefore much more likely to be seriously damaged by cutworms than are crops planted in weed-free soil. Although it may be possible to hand-pick caterpillars from the soil near infested plants in gardens and allotments, this technique is not practicable on a field scale. Ploughing injures some larvae and brings many others to the soil surface, where they are vulnerable to desiccation and predators. As rain reduces larval survival, irrigation of susceptible vegetable crops provides a useful method of managing infestations, and, in some countries, fields are flooded to drown cutworms and other soil-inhabiting pests.

Biological Control

A granulosis virus has been used effectively against cutworms of the turnip moth and other cutworms of noctuid moths in experiments in Denmark, Pakistan, Russia and Sweden (Zethner, 1980). In field trials in Denmark, the numbers of cutworms and damage caused by them were reduced by 80% on red beet, carrot and potato crops by use of aqueous suspensions of the virus applied as sprays.

Several different parasitoids and predators of the turnip moth have been recorded. In Hungary, larvae of *A. segetum* were infested by several dipterous parasitoids (including *Peleteria nigricornis* (Meigen) (Tachinidae)), which may be important in suppressing cutworm outbreaks (Homonnay and Csehi, 1967). However, no predators or parasitoids are currently used in the biological management of turnip moth cutworm numbers.

Chemical Control

In the past, cutworms were controlled with poison baits containing Paris green mixed with moistened bran or beet pulp and spread over infested areas or placed under covers to retain moisture (Jones and Jones, 1984). HCH was substituted for Paris green but baits are now rarely used. High-volume sprays of DDT have also been effective, but treatments aimed specifically at young larvae before they move underground are now based on high-volume sprays of carbamate, organophosphorus or pyrethroid insecticides.

Integrated Control

In many temperate countries spray warnings based on pheromone trap catches and meteorological data are issued by newsletter, pest intelligence reports, radio or television. If relatively few larvae are expected to survive, only the most susceptible crops require treatment.

9.7.6 *Agrotis ipsilon* (Hufnagel): Dark Swordgrass Moth or Black or Greasy Cutworm (Lepidoptera: Noctuidae)

Geographical Distribution

The dark swordgrass moth is almost world-wide, but absent from some tropical countries. It occurs throughout Europe as far north as 65 degrees in Norway. It is believed to migrate annually between North Africa and Europe. The dark swordgrass moth is particularly common and serious in the USA. Most infestations in Canada result from moths which migrate from the USA in the spring.

Description

Adult dark swordgrass moths have a wingspan of 45 mm. The forewings are dark purple-brown to pale brown in colour, suffused with dark purplish-brown, and have a distinctive pattern made up of brown lines and black streaks and rings. The hindwings are white in colour to translucent, with brown lines and veins. The head and thorax are purplish-brown-coloured and the collar is reddish-brown with a black band. The abdomen is greyish-brown in colour (Carter, 1984). The eggs are about 0.6 mm in diameter and pale yellow in colour at first, but later they become brownish-grey. They are conical, strongly ribbed and reticulated. Up to 1800 eggs are laid by each female either singly or in small groups around the base of host plants. The eggs hatch in 3–13 days. The larva is grey- or purple-brown in colour with dark-coloured lines along the spiracles and back. The head is dark-brown-coloured with a yellow-brown frons. The larvae become a rich purple-brown colour dorsally and a yellow or green colour ventrally in the last two instars and have a dark-coloured, greasy appearance. Fully grown larvae are about 35 mm long. The pupa is a reddish-brown colour with dark brown spiracles. The pupal cremaster has two spines.

Life Cycle

Adult dark swordgrass moths fly at night and migrate considerable distances. They emerge from overwintering larvae or pupae in late May or early June and the females, after mating, lay their eggs on grass leaves, weeds or soil surface. In Britain most eggs are laid between July and September. The eggs hatch in 5–7 days and the young larvae feed on plant foliage before descending to the soil and exhibiting typical cutworm behaviour of hiding in soil crevices, etc., during the day and feeding during the night. Larval development takes 28–34 days under warm conditions. The larvae may overwinter or pupate and produce another generation. Fully grown larvae pupate as deep as 12 cm in the soil, each earthen cell having a very slight cocoon. Pupation lasts 10–30 days. The numbers of generations vary in temperate regions: 1–2 in Britain; 2 in Canada; 3–4 in Bulgaria; and 4 in Israel, Russia and the USA. Each generation lasts at least 32 days.

Plant Damage

The dark swordgrass moth is polyphagous on herbaceous plants. Seedlings as well as roots and tubers are damaged, and crops which are attacked include red beet, crucifer, lettuce, potato and sugar beet, as well as many cereal and row crops and even herbaceous plants and trees. However, in Britain it does not cause significant damage to vegetable crops.

Control

The measures recommended above (see page 356) for the control of the turnip moth are also recommended for the control of the dark swordgrass moth. In Egypt two preparations of *Bacillus thuringiensis* Berliner have been effective against dark swordgrass moth larvae. A nematode, *Hexamermis arvalis* Poinar and Gyrisco (Enoplida: Mermithidae), parasitises larvae in the USA and the larvae also support the hymenopterous parasitoid *Apanteles ruficrus* Haliday (Braconidae) in Australia.

9.7.7 *Agrotis exclamationis* (L.): Heart and Dart Moth (Lepidoptera: Noctuidae)

Geographical Distribution

The heart and dart moth is distributed throughout the Palaearctic region from Great Britain to Japan.

Description

Adult heart and dart moths have a wingspan of about 40 mm. The forewings vary in colour from pale brown through shades of greyish-brown to mahogany or dark fuscous. Each forewing is suffused with a darker brown colour and has a whitish base as well as a distinctive pattern of dark-brown lines and rings. The hindwings are whitish-coloured with brown veins. The head and body are brown in colour and the collar has a dark brown band. The egg is globular, strongly ribbed and finely reticulated, with a flat base, and about 0.5 mm in diameter. At first the eggs are white in colour, each marked with a purple ring and several spots, but before hatching they become grey or dull pink. The larva is brownish-coloured with brown, dorsal pear-shaped marks, bordered on each side with a dark-edged, yellow stripe. The spiracles are large, black in colour and ringed with white. The head is a pale, yellowish-brown colour with dark markings. Fully fed larvae may be 38 mm long. Heart and dart moth larvae can be distinguished from turnip moth larvae by the much larger size of the spiracles and also by the characteristic pear-shaped, dorsal markings. Fully grown larvae pupate in earthen cocoons. The pupa is a brown colour, smooth, stout and shining, with a sharp anal point and large spiracles. The pupal cremaster has two spines.

Life Cycle

Adult heart and dart moths emerge from pupae produced from overwintering larvae during June and July in Britain and may live for several months, feeding on nectar and flying mainly at night. They lay eggs in small groups on leaves, stems and soil. Hatching takes 10–14 days. The larvae are typical cutworms, feeding initially on low herbage and later hiding below the soil surface during the day and feeding on plants during the night. They pupate in earthen cocoons 7–10 cm below the soil surface in late winter. A partial second generation occurs in Britain with moths emerging in September or October.

Plant Damage

Heart and dart moth larvae feed on a wide range of plants, including many weeds, garden flowers, vegetable crops and seedling trees. Seedlings are severed and holes are excavated in potato tubers and roots of swedes, turnips and carrots. However, the heart and dart moth rarely causes severe damage.

Control

The measures recommended above (see page 356) for control of the turnip moth are also recommended for control of the heart and dart moth.

9.7.8 *Euxoa nigricans* (**L.**)**: Garden Dart Moth (Lepidoptera: Noctuidae)**

Geographical Distribution

The garden dart moth is distributed throughout Europe and Eurasia, ranging from Finland in the north to Portugal in the south-west and from the Ural Mountains in the north-east to Iran in the south-east.

Description

Adult garden dart moths have a wingspan of 32–40 mm. The forewings are reddish-brown in colour with darker brown markings, while the hindwings are ochreous white with brown spots. The wings of females are darker-coloured than the wings of males. The eggs, about 0.5 mm in diameter, are globular, flat-based, lightly reticulated and initially shining white in colour, later turning yellowish. The larva has a dorsal pale or dark ochreous brown colour with greenish sides. The larval body has greenish-grey-coloured lines edged with black and a double whitish line low down on each side. The head of the larva is reddish-brown in colour with blackish-brown markings. Fully grown larvae are about 40 mm long. The pupa is a shining reddish- or yellowish-brown colour with dark spiracles. The pupal cremaster has tapered, divergent spines.

Life Cycle

Garden dart moths emerge from pupae produced from overwintering larvae in July or August and fly at night. Eggs are laid on host plants and do not hatch until the following spring. The larvae feed from March to June and, when fully grown, pupate in silken cocoons which have soil particles incorporated along with the filaments.

Plant Damage

The garden dart moth is polyphagous and damages many important crops, particularly clover and members of the Umbelliferae. Because the larvae are active in the spring in temperate regions, most damage is done to seedlings, which are cut off and left on the soil surface. A single caterpillar can destroy many seedlings, and populations as small as 12000/ha can cause serious damage. At one time the garden dart moth was a very important pest of sugar beet, carrot and onion crops, particularly on light soils and where crops were backward or late-sown. It is now a less important pest.

Control

The measures recommended above (see page 356) for control of the turnip moth are also recommended for control of the garden dart moth.

9.7.9 *Euxoa tritici* (L.): White-line Dart Moth (Lepidoptera: Noctuidae)

Geographical Distribution

The white-line dart moth has a Eurasiatic distribution ranging from south Norway, eastern Sweden and Finland in the north to north Africa in the south and eastwards to Siberia.

Description

The adult white-line dart moth has a wingspan of 28–40 mm. The forewings vary in colour from pale brown through reddish brown, grey-brown to dark chocolate or blackish brown. Each forewing is marked with lines, rings and, sometimes, wedge-shaped patterning. The hindwings of the male moth are whitish-coloured with brown veins. By comparison, the female moth's hindwings are greyer. The eggs are globular, flat-based, finely reticulated and shining white in colour. They are laid singly or in small groups and each female moth may produce up to 1500. The larva is a greyish-brown colour, pale below the spiracular line. It has a dark-edged, pale, dorsal line and two dusky lines on each side. The head of the larva is rounded and yellowish-brown in colour, with dark-brown markings. Fully grown larvae are 40 mm long. The pupa is glossy, smooth and a yellow-brown colour, with dark, reddish-brown spiracles. The pupal cremaster has two short spines. Pupation occurs below the soil surface.

Life Cycle

Adult white-line dart moths fly in July and August, mainly at night. Eggs are laid in August and September. The white-line dart moth may overwinter in the egg stage in southern Europe, but elsewhere it overwinters in the larval stage. The larvae feed from March to June.

Plant Damage

The larvae of the white-line dart moth feed as typical cutworms and may occur with other noctuid cutworms, including *Agrotis vestigialis* (Hufn). and *Euxoa nigricans* (L.). They attack a wide range of plants, including cereals, tomatoes, vegetables, vines and young trees. Like *E. nigricans*, *E. tritici* can be particularly damaging to root crops on sandy soils. On wild host plants it is chiefly found on coastal sandhills, but it is also found on heaths and moors as well as in gardens.

Control

The measures recommended above (see page 356) for control of the turnip moth are also recommended for control of the white-line dart moth.

9.7.10 *Noctua pronuba* (L.): Large Yellow Underwing Moth (Lepidoptera: Noctuidae)

Geographical Distribution

The large yellow underwing moth is widespread in the Palaearctic region and is found as far north as Iceland and Finland. It also occurs in north Africa and was reported recently from Nova Scotia, Canada.

Description

The adult large yellow underwing moth has a wingspan of 50–60 mm. This species of moth is sexually dimorphic, and, although the sexes may be similar in size and shape, they differ in other respects. For instance, dark-coloured specimens are males, while pale ones are females. The forewings range from yellowish brown to dark brown in colour and may have spots and ring markings. The hindwings are an orange-yellow colour, each with a black border. The head and thorax are a similar colour to the forewings, while the abdomen is brown. The eggs are about 0.75 mm in diameter, hemispherical, ribbed and reticulated above the middle and smooth below. Initially, they are yellowish white in colour, but later become reddish grey. The larva is robust and may grow to a length of 50 mm. It varies in colour from yellow-brown to green and is marked with black above the spiracles. The underbody of the larva is a pale colour. The lines along the dorsal surface and both sides of the upper part of the larval body are narrow and pale-coloured. The head of the larva is a light-brown colour and has reddish-brown markings. The pupa is smooth, glossy and reddish brown in colour, with dark-brown spiracles. The pupal cremaster has two spines and two pairs of minute bristles.

Life Cycle

Large yellow underwing moths have a prolonged emergence in Britain, beginning in mid-June and peaking in August. They are active during the night, feeding on nectar, hiding during the day in leaf litter. When disturbed, they fly off wildly, displaying the bright-orange patches on their hindwings. Female moths each lay up to 1000 eggs in flat masses of 100–150 on the undersides of plant leaves, most eggs being laid in July and August. After 10–28 days, the eggs hatch and larvae commence feeding on above-ground plant parts, with later larval instars feeding on below-ground plant parts as typical cutworms. Fully grown larvae remain active during the winter months in Britain and pupation occurs in oval cells in the soil during late spring.

Plant Damage

The large yellow underwing moth is polyphagous, feeding on most agricultural crops and many wild and garden herbaceous plants. Larvae attack the roots and foliage of plants. Early brassicas and lettuces may be attacked by the larvae, which can considerably damage leaves and stems.

Control

The measures recommended above (see page 356) to control the turnip moth are also recommended to control the large yellow underwing moth.

9.7.11 *Peridroma saucia* **(Hübner): Pearly Underwing Moth or Variegated Cutworm (Lepidoptera: Noctuidae)**

Geographical Distribution

The pearly underwing moth has been reported to be a serious pest in at least ten countries across four continents. It is virtually cosmopolitan, migrating annually to most temperate areas from warmer regions. However, the moth is only a moderately common immigrant to the Netherlands and Denmark and is hardly known in Norway and Sweden, although in some years it reaches the Faroes and Iceland. In Britain it is most common in the south-west (Heath and Emmett, 1979). The moth is widespread in the temperate zones of North America.

Description

The wingspan of adult pearly underwing moths is 45–56 mm. The forewings are reddish brown in colour, suffused with whitish scales, and have dark-brown to black markings. The fascia and stigmata are usually indistinct. The hindwings are translucent, are white in colour, are suffused with brown at the margins, particularly in the female, and have brown veins. The head and thorax are a reddish-brown colour, the thorax having a silvery grey longitudinal crest. Each egg is small, flattened and button-shaped, with a large number of prominent ribs radiating from a raised centre on its upper surface. The eggs are delicately reticulated, light yellow in colour and glossy in appearance. Fully grown larvae reach a length of 42 mm. Each larva is plump and tapers slightly towards a small head, which is pale brown in colour with black markings. The body is a reddish-grey colour or brown dorsally and paler along the sides. An obvious, undulating stripe, dark-brown to black in colour above and pale beneath, is visible above the larval spiracles and there are nine pairs of lateral white-coloured spots. Pupation takes place in an oval-shaped, earthen cocoon. The pupa is a reddish-brown colour and the pupal cremaster has two spines.

Life Cycle

Small numbers of adult pearly underwing moths migrate from the European continent to Britain each year in the spring, with larger numbers arriving later. The moths may be present between May and November but rarely overwinter. Each female lays between 300 and 2000 eggs in batches on host plants. The eggs hatch in 10–14 days into larvae which feed on plant foliage in the early instars before assuming the more typical cutworm feeding habit in later instars. The first brood of larvae in June and July gives rise to a second generation from September to November in Britain, while in the USA there may be three or four generations a year.

Plant Damage

The pearly underwing moth is a severe pest in many temperate countries, particularly North America, where large populations occur sporadically. Before the Second World War the moth was reported to be responsible for $2.5 million worth of damage to crops in the USA each year (Metcalf and Flint, 1939). It is the most important cutworm species attacking vegetables in the USA and is as important on glasshouse-grown vegetables as on field-grown vegetables. The vegetable-host range of the pearly underwing moth has been reviewed by Ring *et al.* (1976). The most severe damage occurs on potatoes and tomatoes, but up to 200 larvae have been collected around single cabbage plants, an indication of the high populations which can occur in some years.

Control

The measures recommended above (see page 356) to control the turnip moth are also recommended to control the pearly underwing moth.

9.8 FLIES

A comprehensive list of the pest flies of umbelliferous crops is provided in Table 9.4. The more important pest species are described below.

9.8.1 *Euleia heraclei* **(L.) (Synonym:** *Philophylla heraclei* **(L.)): Celery Fly or Celery Leaf Miner (Diptera: Tephritidae)**

Geographical Distribution
The celery fly is distributed widely in Europe, Canada and the USA.

Description
Adult celery flies are about 5 mm long, with a wingspan of about 10 mm (Figure 9.11). The head of the adult fly is large and has two eyes, each with

Figure 9.11   Adult celery fly, *Euleia heraclei*. © HRI

a deep metallic-green hue. The fly's body is a tawny-brown colour and bears the distinctive wings, which are mottled in various shades of brown separated by hyaline areas and are iridescent. There are two well-defined forms of the adult fly, one having a light reddish-brown-coloured body and wing markings and the other being very dark brownish black in colour. The egg, which measures 0.5 × 0.2 mm, is white in colour, smooth, elongate and oval in shape, and has a mushroom-shaped micropyle at the narrow

Table 9.4 Diptera of umbelliferous crops

Latin name and authority	Common name	Angelica	Anise	Caraway	Carrot	Celery	Chervil	Coriander	Dill	Fennel	Lovage	Parsley	Parsnip
Dasineura angelicae Rübsaamen	Parsnip flower midge	+											
Kiefferia pericarpicola (Bremi)	Gall midge			+	+							+	+
Lasioptera carophila Loewi	Celery fly			+	+						+	+	+
Euleia heraclei (L.)	Carrot fly	+	+	+	+	+	+	+	+	+	+	+	+
Psila rosae (F.)					+	+					+	+	+
Phytopsila carota I. H. & K.					+								
Liriomyza pusilla (Meigen)	Vegetable leaf miner					+							
L. sativae Blanchard	American serpentine leaf miner					+							
L. trifolii (Burgess)						+							
Melanagromyza apii Hering						+							
M. splendida Frick					+	+							
Napomyza carotae Spencer	Carrot miner				+	+							
N. lateralis (Fallén)	Calendula fly				+								
Phytomyza ferulae Hering													
Pegohylemyia fugax (Meigen)					+								

end. The eggs are usually laid on the undersurface of leaves, but may be inserted in host-plant leaf tissues. The larvae develop through three instars and, when fully mature, measure about 8 mm in length. They have a translucent white or pale-green colour and can be observed inside leaf tissues in which they are feeding. The gut is visible through the translucent larval body as a black line. The final stage of the life cycle is the pupa, which develops inside the puparium, a case formed from the wrinkled skin of the fully grown larva. The puparium is oval-shaped, pale yellow or light brown in colour and 3–5 mm long.

Life Cycle

Adult celery flies emerge in late April, May or June, depending on environmental conditions. After mating, the female locates a host plant and lays eggs, inserting them with its ovipositor one at a time into the leaf tissues. Up to 100 eggs may be laid by each female. The eggs hatch in 1–2 weeks and the larvae mine inside the leaf tissues, quite often communally. After 14–19 days the fully grown larvae pupate either within the leaf tissues or else in the soil, having, in the case of soil pupation, cut emergence holes in the leaves to allow them to drop to the soil surface. Second-generation flies emerge 3–4 weeks later from July onwards, their progeny giving rise either to a third generation in the autumn or to pupae which overwinter in leaf debris or in the soil.

Plant Damage

The celery fly attacks celery, parsley and parsnip crops as well as various wild Umbelliferae, including angelica and hogweed. The insertion of the adults' ovipositor into leaf tissues during egg laying causes small brown necrotic puncture marks. The larvae cause most damage by excavating the tissues between the upper and the lower leaf epidermis. Larval damage appears as large blistered areas which are cream, pale green or straw in colour and contrast markedly with the green colour of adjacent, undamaged, chlorophyll-rich tissues (Figure 9.12). The damaged areas shrivel and become brown in colour and affected leaves may crumple and die. Larvae can be present in leaves as late as December. In very severe attacks affected plants are reduced in vigour: celery plants fail to produce normal stems and parsnip roots are reduced in size. Leaf mines disfigure parsley foliage, reducing marketability.

Cultural Control

Good crop hygiene and husbandry can reduce populations of celery flies. Infested plant material should be collected and burned to destroy larvae and pupae. Damaged foliage is observed readily and, in gardens and allotments, the larvae mining inside plant tissues may be eliminated by removing and destroying damaged leaves.

Figure 9.12 Leaf mines (blisters) caused by larvae of the celery fly, *Euleia heraclei*. © HRI

Chemical Control

Insecticidal treatments are recommended when there is more than one mine per five leaves per plant on celery and parsnip and when the mining is likely to reduce the marketability of parsley crops. Plants should be examined regularly during May and June and damage levels monitored. Several organophosphorus compounds are available for use in high-volume sprays when damage exceeds the thresholds quoted above.

9.8.2 *Psila rosae* (F.): Carrot Fly or Carrot Rust Fly (Diptera: Psilidae)

Geographical Distribution

The carrot fly is distributed widely in northern temperate regions of America (between 40 degrees north in northern USA and 50 degrees north in Canada), Europe (between 68 degrees north in Norway and 36 degrees north in Spain) and, probably, Asia. In Britain the carrot fly is practically ubiquitous in lowland areas. The insect is also found in temperate regions of New Zealand.

Description

Adult carrot flies are small (6–8 mm long) two-winged flies with a shining black body colour, a reddish-brown head, and yellow legs as well as iridescent wings (Figure 9.13). The sexes can be separated by the shape of

Figure 9.13 Adult carrot fly, *Psila rosae*. © HRI

the abdomen: in the female it has an elongated pear shape because of the ovipositor, while in the male it is more cylindrical and rounded at the tip. Adult *P. nigricornis* Meigen, the chrysanthemum stool miner, may be differentiated from adult *P. rosae* by the colour of the third antennal segment, which is black in *P. nigricornis* and partly yellow in *P. rosae*. However, these two closely related species may be distinguished positively only by the size and the structure of the male genitalia, details of which are described and illustrated by Collin (1944). The elongate eggs are 0.15 mm in diameter, 0.6–0.7 mm long and white in colour. They are sculptured with a reticulate pattern and have pronounced longitudinal ribbing. At one end of each egg there is a micropylar cap bearing a circular plug with eight sockets around its rim. The legless, headless larva is creamy white in colour and develops through three instars. Fully grown larvae are 8–10 mm long. The larval body is tapered towards the anterior end and possesses a pair of sclerotised oral hooks used to rasp plant tissues. The fully developed larva pupates within the last larval cuticle, which forms a puparium about 5 mm long and 1.5 mm in diameter. This puparium changes colour from yellow to brown as it hardens.

Life Cycle

Adult carrot flies emerge in the spring, the exact time of emergence varying from late April in southern England and late May in Canada and the northern United States to mid-June in Scotland. Mating is believed to take place in hedgerows or other sheltered areas, and females then move into host crops to lay their eggs between 1 and 4 days after emergence. The eggs are inserted with the female's ovipositor singly or as small groups of 2–3 eggs into the soil around the stems of host plants—as many as 150 may be laid in total by a single female but about 40 is more usual. After 7–14 days the eggs hatch and the young, tiny larvae move down into the soil to feed on the lateral roots of the host plants. Older larvae tend to burrow into the tap roots of carrot plants, although these larvae are also able, like the younger larvae, to feed at the soil–root interface on more fibrous-rooted umbellifers. The time taken to complete larval development varies considerably, depending on soil temperature and food availability, ranging from 6 weeks in the summer to 3 months during the winter. The pupal stage also varies considerably in duration from 25 days in the summer to several months in the winter. Pupae may aestivate or enter diapause, depending on environmental conditions.

The number of carrot fly generations varies each year according to climate and latitude: in northern USSR there is one complete generation and a partial second, while in Spain there may be three complete generations. Further information on the biology of the carrot fly is given by Dufault and Coaker (1987), and a comprehensive bibliography has been published by Hardman *et al.* (1985b).

Plant Damage

Damage results from the feeding activities of the carrot fly larvae only—adult flies feed on the nectar of wild flowers. The young larvae feed at first on the fine lateral roots of host plants and probably also browse root hairs. As the larvae develop, their oral hooks are used increasingly to rasp the plants' root cortex, resulting in channels and cavities. Injuries to large and swollen tap roots are collectively termed 'mine' Figure 9.14) and several different types can be recognised (Ellis *et al.*, 1978). Rust-coloured exudates are released from damaged root tissues and dry at the edges of the mines. The tip of each tap root appears to be a preferred feeding area probably for two reasons: it is more nutritious to the larvae than other regions of the root because the metabolic activity at the tip is greater; and there is little secondary thickening of tissues.

The first symptom of carrot fly attack on young plants is wilting. The foliage may then collapse, the leaves showing the typical yellow and red colour symptoms associated with water stress. Attacked plants that are not killed often produce distorted, forked and, therefore, unmarketable roots.

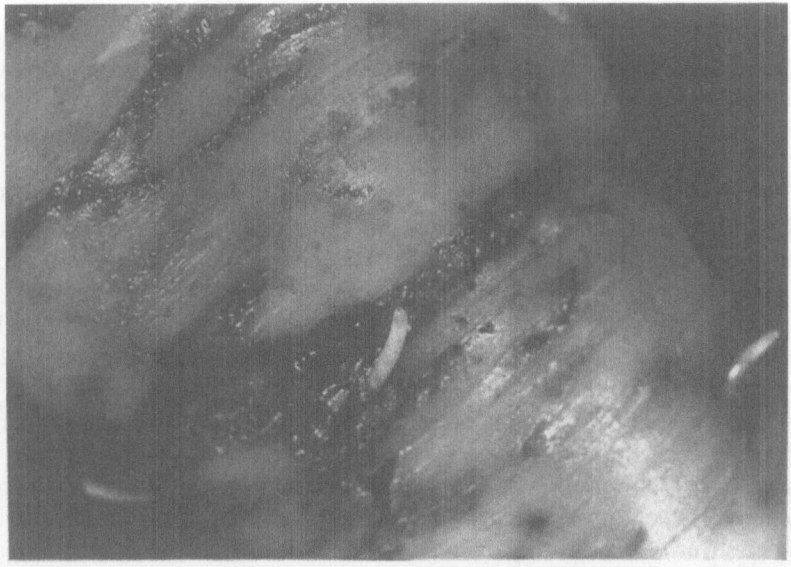

Figure 9.14 Larvae of the carrot fly, *Psila rosae*, in mines at the root surface.
© HRI

Large roots may survive carrot fly attack but the mining and associated root infections render the roots unmarketable. Damage to parsnips is very similar to carrot damage but it is often concentrated in the upper regions of long roots. Celery plants are damaged seriously, the maggots boring into the roots, crowns and leaf stalk bases. Umbelliferous herbs such as parsley (*Petroselinum crispum* (Miller) A. W. Hill) and wild hosts which have fibrous roots such as fools parsley (*Aethusa cynapium* L.) may be killed by carrot fly attack, or if these plants are not killed, they are at least likely to be reduced in vigour. Carrot fly is the most serious pest of carrots and certain other umbelliferous crops in many countries in temperate regions. The pest is so devastating in Britain that only very few crops are not treated with insecticide.

Monitoring and Prediction

Adult carrot fly activity has been monitored for over 30 years in the UK by sampling hedgerows and ditches alongside previously infested fields with sweep nets and counting numbers of flies caught. More recently, yellow sticky or water traps have been used, and, in several countries, growers are now supplied with trapping information about the activity of carrot flies by means of newsletter, radio, telephone or facsimile machine, or even by a system of signals such as flags, so as to time insecticidal treatments and, in some cases, to help in deciding whether treatment is needed at all.

Cultural Control

Carrot flies are weak fliers and, under most circumstances, probably migrate only short distances, preferring apparently to move from hedgerows into crops and then returning to the shelter of trees and herbs. Whenever possible, succeeding and neighbouring crops should be separated spatially from each other and maincrops should not be drilled adjacent to early crops. At least 1 km separation is likely to be needed between crops to break the sequence of host crops in the life cycle of a local population of carrot flies.

As carrot flies aggregate and rest in sheltered areas, reductions of shelter may reduce overall crop damage but, in many areas, this action would be considered environmentally unacceptable.

Crop hygiene can reduce carrot fly populations. For example, severely infested crops should not be ploughed in or permitted to remain unharvested, as either of these events will only increase the numbers of larvae completing their development. These crops should be lifted and destroyed.

The severity of carrot fly damage can be reduced by sowing host crops to avoid peak egg-laying activity. This method of cultural control has become more of an established practice with carrot growers in the UK since the advent of a wide range of quick-maturing cultivars. To avoid the worst of late season damage, carrots should be harvested by about mid-October in Europe and transferred to purpose-built stores. Early harvesting and storage of roots are practised in some countries (for example, Denmark), but in other countries, because the expense of storage facilities has been prohibitive, crops often overwinter in the field only to sustain further larval damage, especially during mild winters. The intercropping of carrots with other crops such as onions has been suggested as a method of reducing damage but it has produced conflicting results. Maximum reductions in carrot fly damage were obtained when young onions were intercropped with carrots at the time of the first-generation carrot fly attack.

Polyethylene barriers have been used successfully to prevent cabbage root flies laying their eggs in vegetable brassica crops and, as a consequence, there is now considerable interest in barriers for carrot fly control. In gardens, 80-cm-high plastic barriers erected around carrot beds or erected along the rows of plants have reduced damage. However, this method is obviously impracticable on a field scale. Plastic film or plastic mesh covers, used to produce early carrots, have also prevented carrot fly damage.

Biological Control

The carrot fly has many natural enemies, including predatory insects, spiders and vertebrates, as well as parasitic insects, bacteria, fungi and viruses. Although there have been several attempts to mass-rear and release biological control agents, none has proved economic.

Plant Resistance

Resistance to carrot fly damage has been identified in commercially acceptable carrot cultivars and in wild umbelliferous species (de Ponti and Freriks, 1980; Hardman and Ellis, 1982; Ellis *et al.*, 1984). The resistance of the carrot cultivars is manifested mainly as antibiosis: when the most and the least resistant cultivars are compared, numbers of larvae able to develop and crop damage are reduced by up to 50%. Generally, the Nantes types of carrot have been found to be resistant and the Danvers cultivars have been found to be susceptible. The degree of resistance has been improved slightly by carrot breeding. Greater degrees of resistance to carrot fly attack have been discovered in wild umbelliferous species and, as a result, certain *Daucus* L. species which are compatible with the cultivated carrot, *Daucus carota* ssp. *carota* L., have considerable potential in breeding programmes. Progeny developed from crosses between North African *Daucus capillifolius* Gilli and cultivated carrot varieties have been shown to be agronomically promising and moderately resistant to carrot fly attack.

Chemical Control

Growers rely on carbamate and organophosphorus insecticides to protect carrot, celery, parsley, parsnip and various herb crops from attack by the various generations of the carrot fly. First-generation attack is prevented by seed treatments or granules incorporated into the soil at drilling by in-furrow, bow-wave or vertical-band techniques. None of these approaches is sufficiently persistent to protect crops from the attack of later generations of the carrot fly and, therefore, supplementary mid-season treatments are recommended.

Integrated Control

The benefits of combining partially resistant carrot cultivars with reduced doses of insecticides have been demonstrated, only one-third of the commercially recommended dose of an organophosphorus insecticide being required to protect the partially resistant variety Sytan from carrot fly attack. Similarly, the benefits of combining partial resistance with careful choice of sowing and harvesting times have been shown to be considerable.

9.8.3 *Napomyza carotae* Spencer: Carrot Miner (Diptera: Agromyzidae)

Geographical Distribution

The carrot miner causes considerable damage to carrot crops in Germany, Switzerland and the Netherlands (Hassan, 1971).

Description

The most detailed accounts of the carrot miner are given by Wiesmann (1961) and Spencer (1966). Adults' wingspan varies from 2.5 mm in the male to 3 mm in the female. The most noticeable diagnostic feature of the adult carrot miner is a fringe of white pubescence on the third segment of each antenna. The front of the head and either side of the thorax are yellow in colour, while the rest of the head and body are greyish-black. The legs are black in colour with bright yellow knees. The eggs are 0.4 mm long. The larva usually has a milky-white appearance, but appears to be coloured with patches of red-orange as a result of carrot particles within its gut. Fully grown larvae are 5.5 mm long. The pupa is white in colour, slender and up to 5.5 mm long.

Life History

Using their ovipositors, female carrot miners make holes in the leaves of host plants. After 4–5 days feeding on the sap exuding from the leaf punctures, they are able to lay their eggs. The short stem to which leaf bases are attached appears to be the preferred site for egg laying but often the leaf is used, with the average number of eggs laid per female carrot miner being about 175. Eggs are laid singly and take 5 days to hatch. The young larvae mine near the upper surface of each leaf, with the mines initially running towards the main vein and later down the petiole to the tap root. Larval development takes 53–68 days at 23–24°C. When eggs have been laid in a mature plant, larval development can be completed in a leaf stem only: larvae will not mine down to the tap root. Pupation occurs in a host-plant stem or root and lasts 7–8 days. Adults of the first generation appear in early May in central Europe and emergence continues until early June. Most eggs are laid in mid- to late May, with pupae subsequently being found in late July. Adults of the second generation emerge from early August through to mid-September. A partial third generation of flies may be on the wing in early October.

Plant Damage

The carrot miner occurs on carrots and has been found to develop successfully on *Anthriscus sylvestris* (L.) Hoffm. (cow parsley), *Apium graveolens* L. (celeriac), *Pastinaca sativa* L. (parsnip), *Heracleum* L. spp. (cow parsnip) and *Petroselinum crispum* (Miller) A. W. Hill (garden parsley). It is also believed to occur in the wild on *Daucus carota* L. (carrot) and on other Umbelliferae.

On carrots larvae of the carrot miner mine the midribs of leaves, moving down into the petioles and stems and eventually entering the tap roots. The larvae tend to damage the upper part of each root. The damage is not deep, but superficial. As roots expand, the mines gape open to form galleries. The mining damage deforms roots and facilitates the entry of bacterial rots. In store, damaged carrots deteriorate rapidly. Up to 10 larvae may be found in a single root. Early carrot crops suffer most from miner attack.

Control

The carrot miner can be controlled with contact and systemic organophosphorus insecticides. Hassan (1971) devised a method, based on the appearance of the characteristic feeding marks of the adult females on the leaves of the carrot crop, to forecast damage and to determine the time for treatment of crops. Good control was achieved by applying a single spray 2–3 weeks after the appearance of the feeding marks.

9.8.4 *Liriomyza trifolii* (Burgess) (Synonyms: *Oscinis trifolii* Burgess, *Liriomyza alliovora* Frick): American Serpentine Leafminer (Diptera: Agromyzidae)

Several leafminers attack umbelliferous crops (Table 9.4). The American serpentine leafminer is polyphagous and attacks many important vegetable crops, including carrots, celery and tomatoes (see page 304).

9.9 NEMATODES

9.9.1 *Heterodera carotae* Jones: Carrot Cyst Nematode (Tylenchida: Heteroderidae)

Geographical Distribution

The carrot cyst nematode occurs locally in temperate regions as an important pest of the carrot crop. It is widely distributed in Europe and probably in north Asia, and has recently been discovered in the USA.

Description

H. carotae is a typical cyst nematode in which the female's swollen body hardens to form a tough protective cyst around the egg mass. As it matures, the cyst changes in colour from white to brown. *H. carotae* has small cysts compared with certain other species (<0.5 mm; compare potato cyst nematode—see page 282) and the surface structures of the cyst help to distinguish this species from closely related types. The presence of copulatory spicules with bidentate tips on the male also helps to distinguish this species from other *Heterodera* Schmidt species. The carrot cyst nematode is highly host-specific and has been recorded only from carrot and certain other *Daucus* L. species.

Life Cycle

H. carotae larvae hatch from eggs in response to root diffusates released from their host plants—carrots and other related species. The larvae move towards actively growing host plants and invade the roots. Males are more active than females and, when mature, move in search of females, which are sedentary and partly embedded in root tissues. Males take about 20 days to develop and are short-lived, remaining active for only 9–10 days. Shortly after the females burst through the root cortex, the egg sacs develop. After mating, the female dies and her body, together with the egg sac, forms the protective enclosure or cyst for the eggs.

Plant Damage

Up to 60% of fields in western France are infested with populations of the carrot cyst nematode, reaching levels of 230 cysts/100 g soil (approximately 100 larvae/g). Populations of >50 larvae/g soil can destroy carrot crops (Bossis, 1988).

Control

Several different methods have been tried in France to reduce carrot cyst nematode problems. The pest is highly host-specific and the crop has a low tolerance level: therefore, the idea of using 'trap' crops to induce mass hatching of larvae is not practicable. A 5 year rotation (or 10 years where infestations are severe) is recommended in carrot-growing districts. Fumigants are expensive and difficult to apply, but can be effective, although recolonisation of soils occurs rapidly when carrots are grown and populations may ultimately exceed initial levels. Research is also being done on biological control agents and tests for screening for host resistance to the pest (Bossis, 1988).

REFERENCES

Anon. (1979). *Willow-carrot Aphid*. Ministry of Agriculture, Fisheries and Food Advisory Leaflet No. 603. HMSO, London, 6 pp.

Balachowsky, A. and Mesnil, L. (1936). *Les Insectes Nuisibles aux Plantes Cultivées. Leurs Moeurs, leur Destruction*. Busson, Paris

Blackman, R. L. and Eastop, V. F. (1984). *Aphids on the World's Crops: An Identification and Information Guide*. Wiley, Chichester

Bossis, M. (1988). Control of *Heterodera carotae* Jones (1950): Effectiveness of available methods and prospects. In *Progress on Pest Management in Field Vegetables* (ed. R. Cavalloro and T. S. C. Peleren). Balkema, Rotterdam, pp. 187–193

Bowden, J., Cochrane, J., Emmett, B. J., Minall, T. E. and Sherlock, P. L. (1983). A survey of cutworm attacks in England and Wales, and a descriptive population model for *Agrotis segetum* (Lepidoptera: Noctuidae). *Annals of Applied Biology*, 102, 29–47

Carter, D. J. (1984). *Pest Lepidoptera of Europe with Special Reference to the British Isles.* Junk, Dordrecht

Chapman, R. K. (1985). Fight back against aster yellows. *American Vegetable Grower*, 33, 6–9

Collin, J. E. (1944). The British species of Psilidae (Diptera). *Entomol. Month. Mag.*, 80, 214–224

Dixon, G. R. (1981). *Vegetable Crop Diseases.* Macmillan, London

Dufault, C. P. and Coaker, T. H. (1987). Biology and control of the carrot fly, *Psila rosae* (F.). 3. *Agricultural Zoology Reviews*, 2, 97–134

Dunn, J. A. (1965). Studies on the aphid, *Cavariella aegopodii* Scop. 1. On willow and carrot. *Annals of Applied Biology*, 56, 429–438

Dunn, J. A. (1970). The susceptibility of varieties of carrot to attack by the aphid, *Cavariella aegopodii* (Scop.). *Annals of Applied Biology*, 66, 301–312

Dunn, J. A. and Kirkley, J. (1966). Studies on the aphid, *Cavariella aegopodii* Scop. *Annals of Applied Biology*, 58, 213–217

Edwards, C. A. (1964). The bionomics of swift moths. 1. The ghost swift moth, *Hepialus humuli* (L.). *Bulletin of Entomological Research*, 55, 147–160

Ellis, P. R., Hardman, J. A., Dowker, B. D. and Horobin, J. F. (1984). Progress in the studies of resistance to carrot fly (*Psila rosae*) in carrots. In *Breeding for Resistance to Insects and Mites.* IOBC-WPRS Bulletin V11/4, pp. 47–48

Ellis, P. R., Wheatley, G. A. and Hardman, J. A. (1978). Preliminary studies of resistance in carrots to carrot fly attack. *Annals of Applied Biology*, 88, 159–170

Emmett, B. J. (1984). Pheromones in UK farm pest control. In *Statistical and Mathematical Methods in Population Dynamics and Pest Control* (ed. R. Cavalloro). Proceedings of a meeting of the EC Experts Group, Parma, 26–28 October, 1983, pp. 45–47

Emmett, B. J. and Rhodes, J. (1987). *An Economic Evaluation of the ADAS Cutworm Prediction Service.* ADAS report

Esbjerg, P. (1983). Fangst af agerugler (*Agrotis segetum*) og nedborsmaling som baggrund for knopormevarsling. *Tidsskrift for Planteavl*, 87, 371–377

Hagel, G. T. and Landis, B. J. (1967). Biology of the aster leafhopper, *Macrosteles fascifrons* (Homoptera: Cicadellidae), in Eastern Washington, and some overwintering sources of aster yellows. *Annals of the Entomological Society of America*, 60, 591–595

Hardman, J. A. and Ellis, P. R. (1982). An investigation of the host range of the carrot fly. *Annals of Applied Biology*, 100, 1–9

Hardman, J. A., Ellis, P. R., Kempton, D. P. H., Reseigh, L. C. and Saw, P. L. (1985a). Host ranges of carrot fly and willow-carrot aphid. *Report of the National Vegetable Research Station for 1984*, p. 41

Hardman, J. A., Ellis, P. R. and Stanley, E. A. (1985b). *Bibliography of the Carrot Fly, Psila rosae (F.).* Vegetable Research Trust, Wellesbourne

Hassan, S. A. (1971). Forecasting of damage and determining the appropriate time to control the carrot miner fly *Napomyza carotae* (Diptera: Agromyzidae). *Zeitschrift für Angewandte Entomologie*, 68, 68–73

Heath, J. and Emmett, A. M. (1979). *The Moths and Butterflies of Great Britain and Ireland.* Volume 9: *Sphingidae—Noctuidae, Noctuinae and Hadeninae.* Curwen Books, London

Hill, D. S. (1987). *Agricultural Insect Pests of Temperate Regions and Their Control.* Cambridge University Press, Cambridge

Homonnay, F. and Csehi, E. (1967). The damage caused by *Agrotis segetum* and the role of parasites in suppressing an outbreak of it. *The XVIIth Scientific Conference on Plant Protection* (ed. L. Szalay-Marzso). Vol. 1, p. 351

Jones, F. G. and Jones, M. G. (1984). *Pests of Field Crops.* 3rd edn, Edward Arnold, London

Jorgensen, A. S. (1978). The species of cutworms (*Agrotis* spp.) found in Danish agricultural crops. *Zeitschrift für Angewandte Entomologie*, 87, 76–81

Kennedy, J. S., Day, M. F. and Eastop, V. F. (1962). *A Conspectus of Aphids as Vectors of Plant Viruses.* Commonwealth Institute of Entomology, London

Maramorosch, K. and Harris, K. F. (1981). *Plant Diseases and Vectors: Ecology and Epidemiology*. Academic Press, New York

Markulla, M., Laurema, S. and Tiittanen, K. (1976). Systemic damage caused by *Trioza apicalis* on carrot. In *The Host Plant in Relation to Insect Behaviour and Reproduction* (ed. T. Jermy). Plenum Press, New York, pp. 153–155

Metcalf, C. L. and Flint, W. P. (1939). *Destructive and Useful Insects, Their Habits and Control*. McGraw-Hill, New York

Meyrick, E. (1895). *A Handbook of British Lepidoptera*. Macmillan, London

Mikkelsen, S. A. and Esbjerg, P. (1981). The influence of climatic factors on cutworm (*Agrotis segetum*) attack level, investigated by means of linear regression models. *Tidsskrift for Planteavl*, **85**, 291–301

Perron, J. P. (1971). Insect pests of carrots in organic soils of south-western Quebec with special reference to the carrot weevil, *Listronotus oregonensis* (Coleoptera: Curculionidae). *Canadian Entomologist*, **103**, 1441–1448

Ponti, de, O. M. B. and Freriks, J. C. (1980). Breeding of carrot (*Daucus carota*) for resistance to carrot fly (*Psila rosae*). In *Integrated Control of Insect Pests in The Netherlands* (ed. A. K. Minks and P. Gruys). Pudoc, Wageningen, pp. 169–172

Rabasse, J. M., Brunel, E., Delecolle, R. and Rouze Jouan, J. (1976). Influence de la dimension de pieges a eau colores en jaune sur les captures d'aphides dans une culture de carotte. *Annales de Zoologie—Ecologie Animale*, **8**, 39–52

Ring, R. W., Johnson, B. A. and Arnold, F. J. (1976). Host range of the variegated cutworm on vegetables: A bibliography. *Bulletin of the Entomological Society of America*, **22**, 409–415

Scott, D. R. (1970). *Lygus* bugs feeding on developing carrot seed: plant resistance to that feeding. *Journal of Economic Entomology*, **63**, 959–961

Sherlock, P. L. (1983). The natural incidence of disease in the cutworm *Agrotis segetum* in England and Wales. *Annals of Applied Biology*, **102**, 49–56

Southwood, T. R. E. and Leston, D. (1959). *Land and Water Bugs of the British Isles*. Warne, London

Spencer, K. A. (1966). A clarification of the genus *Napomyza* Westwood (Diptera: Agromyzidae). *Proceedings of the Royal Entomological Society of London, Series B, Taxonomy*, **35**, 29–40

Stevenson, A. B. (1976). Seasonal history of the carrot weevil, *Listronotus oregonensis* (Coleoptera: Curculionidae) in the Holland Marsh, Ontario. *Proceedings of the Entomological Society of Ontario*, **107**, 71–78

Stroyan, H. L. G. (1963). *The British Species of Dysaphis Borner (Sappaphis Auctt. nec Mats.)*. Part II. *The Subgenus Dysaphis Sensu Stricto*. HMSO, London

Stubbs, L. L. (1948). A new virus disease of carrots: its transmission, host range and control. *Australian Journal of Scientific Research*, **1**, 303–332

Walkey, D. G. A. and Cooper, V. C. (1971). Virus disease of celery. *Report of the National Vegetable Research Station for 1970*, pp. 111–112

Watson, M., Serjeant, E. P. and Lennon, E. A. (1964). Carrot motley dwarf and parsnip mottle viruses. *Annals of Applied Biology*, **54**, 153–166

Wiesmann, R. (1961). *Phytomyza lateralis* Fall., ein werig beachteter mohren – und Karottenschadling. *Mitteilungen aus der Entomologischen Gesellschaft Basel*, **11**, 39–62

Zethner, O. (1980). Control of *Agrotis segetum* (Lep: Noctuidae) in root crops by granulosis virus. *Entomophaga*, **25**, 27–35

Zethner, O. and Esbjerg, P. (1978). Cutworm attack in relation to rainfall and temperature during 70 years. *Proceedings of the Nordic Symposium on Climatic Changes and Related Problems—Danish Meteorological Institute, Copenhagen, April 1978. Climatological Papers*, Vol. 4, pp. 103–108

Glossary

aestivation	: state of dormancy during summer or dry season
a.i.	: active ingredient
ala(e)	: wing-like projection or structure
alata(e)	: in aphids' life cycles, winged form; usually winged, parthenogenetic female
alate	: winged
anholocyclic	: in aphids' life cycles, parthenogenesis occurs throughout the year without the occurrence of any sexual reproduction
anterior	: nearer head end
antibiosis	: association between organisms which is antagonistic, with one producing chemical compounds (antibiotics) harmful to the other(s)
antixenosis	: plant properties resulting in negative reactions or total avoidance by insects
anus	: opening (usually posterior) of alimentary tract to exterior
aphicide	: a pesticide which specifically kills aphids
aptera(e)	: in aphids' life cycles, wingless form; usually wingless, parthenogenetic female
apterous	: wingless
arcuate	: arch- or bow-shaped
arrhenotoky	: parthenogenetic reproduction which results in the appearance of males only
arthropod	: member of the animal phylum Arthropoda, jointed-limbed organisms possessing a hard exoskeleton (e.g. insects, crustaceans, etc.)
autoecious	: in aphids' life cycles, an annual alternation does not occur between primary and secondary host plants
bursa	: pouch or sac-like cavity, usually genital in animals
cadaver	: a dead body
cauda(e)	: tail or tail-like appendage
chitin	: a polysaccharide of N-acetylglucosamine units; it is the chief constituent of arthropod exoskeletons
chorion	: in insects, the superficial outer coat of the egg, which is non-cellular and has been secreted by the ovary around the ovum

corpora allata : endocrine glands associated with insects' brains which secrete juvenile hormone

cremaster : cluster of hooks at posterior end of butterfly or moth pupa

crepuscular : describes animals which fly before sunrise or at twilight

cuticle : outer skin (in animals) or wall (in plants)

cysts : in nematodes, bladder- or sac-like structures which form after the death of the adult females and protect the enclosed, fertilised eggs

cytokinesis : synonymous with cell division

diapause : in insects' life cycles, a hormonally controlled period of arrested growth and development

dimorphic : having two different forms

dorsal : in or on upper surface of an animal

dorsum : upper surface of an animal

eclosion : hatching from egg (or pupal case)

edaphic : describes any physical, chemical or biotic factor of the soil which influences plant growth

elytra(ron) : the anterior, leathery or chitinous wings of beetles which protect the underlying, membranous, hindwings used in flight; at rest, the elytra meet in a straight line down the middle of the dorsum of the insect

entomogenous : growing and developing in or on an insect

epidermis : outer layer(s) of skin

exoskeleton : skeleton present on outside of an organism (e.g. arthropods)

fascia : transverse band or broad line

femur(mora) : usually stoutest segment of (insect) leg, attached to body through articulation with coxa and trochanter

frass : solid excrement of insect larvae

fundatrix(ices) : in aphids' life cycles, a parthenogenetic female form which develops from a fertilised egg

gubernaculum : in mammals, cord supporting testes in scrotal sac

genotype : genetic constitution of an organism

glabrous : smooth, hairless

gravid : describes female organism containing fertilised eggs

groundkeepers : potato tubers which are missed during harvesting and are left in the soil, to grow subsequently into new plants

gynandromorph : animal having male and female features within the one body

gynopara(e) : in aphids' life cycles, a parthenogenetic female form which produces sexual female forms

hemelytron(tra)	:	forewing of heteropteran bug, having a distal membranous section
heteroecious	:	in aphids' life cycles, an annual alternation occurs between primary and secondary host plants
holocyclic	:	in aphids' life cycles, parthenogenesis is interrupted by an annual phase of sexual reproduction, with the appearance of male and female sexual forms and the production of eggs
hyperplasia	:	excessive multiplication of cells
hypertrophy	:	excessive increase in size of cells
instar	:	a larval stage in insect development
juvenile hormone	:	during insect development, juvenile hormone maintains larval characteristics during each moult until, eventually, adult metamorphosis occurs
karyotype	:	appearance, arrangement and number of chromosomes in a cell
mandibles	:	in insects, one of a pair of mouthparts (other one is maxillae) used for crushing food
meristem	:	groups of actively dividing cells in plants, e.g. root and shoot tips
mesonotum	:	dorsal part of posterior segment of the insect thorax (metathorax)
metamorphosis	:	change in form of an organism from larva to adult
multivoltine	:	in insects' life cycle, many generations a year (or season)
Nearctic	:	zoogeographical region including Greenland, north America and northern Mexico
nematicide	:	a pesticide which specifically kills nematodes
New World	:	America
notum	:	dorsal portion of insect segment
Old World	:	world before discovery of America
oogenesis	:	formation, development and maturation of the ovum
organism	:	any living animal or plant
ovate	:	egg-shaped, with broader end at base
ovipara(e)	:	in aphids' life cycles, a sexual female form which mates with sexual male form and lays fertilised eggs
oviposition	:	process of egg laying
ovum	:	female gamete (e.g. egg, egg cell)
paedogenesis	:	reproduction occurring while animal is still sexually immature: in insects, as larvae
Palaearctic Region	:	zoogeographical region including Europe, North Africa, Western Asia, Siberia, Northern China and Japan

parasitoid	: commonly, a hymenopterous (or wasp) parasite of insects which eventually kills its host
parthenogenesis	: reproduction by development of individuals from eggs which have not been fertilised (e.g. aphids' life cycles)
pathogen	: disease-causing organism
pericycle	: outer, cellular layer of stele, between endodermis and conducting tissue
perineal	: of perineum (part of animal's body surface bounded by scrotum or vulva to front, anus to back, and surface of thigh to side)
phenotype	: visible characteristics of an organism; the observable result of the interaction between the genotype of an organism and its environment
phasmids	: pair of posterior sense organs in nematodes, possibly chemosensory
photoperiod	: period of daylight in a 24 h cycle of day and night
polyandrous	: female has several male mating partners at one time
polygynous	: male has several female mating partners at one time
polymorphism	: usually, occurrence of different forms of individuals within a species
polyphagous	: consumes many different kinds of food
posterior	: nearer tail end
pubescent	: covered with soft hair or down
rhizosphere	: volume of soil immediately surrounding and influenced by plant roots
sclerites	: chitinous plates separated by thinner membranes in arthropod exoskeleton
scutellum	: posterior part of insect notum
seta(e)	: in insects, a slender, hair-like extension to the epidermis
sexuales	: in aphids' life cycles, typically appear in the last generation of the life cycle and consist of the true female and the male. The female is called an ovipara because she lays eggs instead of bearing live young
sexupara(e)	: in aphids' life cycles, either apterous or alate viviparae that typically develop in the last generation of the summer viviparae
siphuncle(culi)	: honeydew tube of an aphid
spicule(s)	: small, pointed process, composed of calcium or silica, found in invertebrates
spiracle	: in arthropods, the exterior opening of the tracheal respiratory system
stadium(ia)	: interval between insect larval moults
stele	: primary, conducting tissues in stems and roots of vascular plants

stem mother or fundatrix	: in aphids' life cycles, the form hatching in the spring from the overwintering egg and reproducing parthenogenetically (without mating with a male) and viviparously (young are born alive)
stigma(mata)	: in Lepidoptera, specialised patch of usually dark scales on forewings
striation	: a linear mark, ridge or furrow
summer viviparae	: in aphids' life cycles, consist of apterous (wingless) and alate (winged) summer viviparous females that develop from the young produced by the stem mothers and by succeeding generations
syncitium	: a multinucleate mass of protoplasm which is not divided into separate cells
tergite	: a dorsal sclerite
tolerance	: plant responses resulting in the ability of the plant to withstand pest infestation and support populations that would severely damage susceptible plants
univoltine	: in insects' life cycles, one generation a year (or season)
vasiform	: vessel-shaped
vector	: any organism that transmits a pathogen
ventral	: in or on lower surface of an animal
vermiform	: worm-shaped
viviparous	: producing young alive rather than laying eggs
vulva(e)	: in nematodes, opening of ovary to exterior

Index

Index